Environmental Technology Applications in the Retrofitting of Residential Buildings

Environmental Technology Applications in the Retrofitting of Residential Buildings

Editors

Mark B. Luther
Igor Martek
Mehdi Amirkhani
Gerhard Zucker

MDPI • Basel • Beijing • Wuhan • Barcelona • Belgrade • Manchester • Tokyo • Cluj • Tianjin

Editors
Mark B. Luther
Deakin University
Australia

Igor Martek
Deakin University
Australia

Mehdi Amirkhani
University of South Australia
Australia

Gerhard Zucker
Austrian Institute of
Technology
Austria

Editorial Office
MDPI
St. Alban-Anlage 66
4052 Basel, Switzerland

This is a reprint of articles from the Special Issue published online in the open access journal *Energies* (ISSN 1996-1073) (available at: https://www.mdpi.com/journal/energies/special_issues/energy_environmental_buildings).

For citation purposes, cite each article independently as indicated on the article page online and as indicated below:

LastName, A.A.; LastName, B.B.; LastName, C.C. Article Title. *Journal Name* **Year**, *Volume Number*, Page Range.

ISBN 978-3-0365-5269-9 (Hbk)
ISBN 978-3-0365-5270-5 (PDF)

Contents

About the Editors

Mark B. Luther

Honorary Professor Dr. Luther is the former Consortium Director of MABEL (the Mobile Architecture and Built Environment Laboratory at Deakin University, Geelong Australia). He is a registered architect in Michigan (USA) and Victoria (Australia). He has lectured in the School of Architecture and Built Environment at Deakin University, in graduate and undergraduate courses on lighting, acoustics and building services, as well as a course on Sustainable Future concepts. Through the MABEL program, Dr. Luther has been involved in measuring the Indoor Environmental Quality (IEQ) of over 35 buildings throughout Australia (Darwin to Hobart). Projects have included schools, offices, airports, sports centres, hospitals, and houses. MABEL's projects have partnered with the Australian Greenhouse Office (now DEWHA); Sustainability Victoria; the Department of Human Services; the CSIRO; the Building Commission (Victoria); and the commercial industry. Presently, Dr. Luther is the Director of Environmental Energy Services, a company which designs and prototypes projects such as the Environmental Service Pod and hydronic ceiling conditioning panels, among other things. Dr. Luther has produced over 100 papers that have been published in international journals and presented at conferences. He has been invited as a keynote speaker to Europe and the USA, as well as throughout Australia.

Igor Martek

Dr. Igor Martek is currently an academic at Deakin University, Australia. He earned his PhD in 'Enterprise Strategies in International Construction' from the University of Melbourne. He also has an MBA from the Australian Graduate School of Management, University of NSW, and an MA in International Relations from the Australian National University, Canberra. His first degree is a Bachelor of Architecture (Honours) from the University of Melbourne. Igor has worked extensively in industry evaluating, generating, managing and turning around large capital projects in various locations around the world. He has worked in Europe, including Eastern Europe, the Maghreb, Levant, China, Korea, and Singapore. In Kuwait, he worked for the Ministry of Finance, assessing and developing projects financed by the Kuwait Fund. He worked in Japan for over ten years as Managing Director of Far-East operations for a British consultancy, advising major global enterprises on strategy and competitiveness. Igor is a visiting scholar to countries as far apart as China, Poland and Indonesia. He has authored over 150 peer-reviewed articles and his latest book, published by Routledge, is titled 'International Construction Management.'

Mehdi Amirkhani

Dr. Mehdi Amirkhani is an academic at the University of South Australia (UniSA). Before joining UniSA, he worked as a Research Fellow and Educator at the University of Melbourne, Deakin University, Victoria University, Queensland University of Technology (QUT), Griffith University, and some universities in Iran. Mehdi has significant experience teaching environmental technology, including units centred around lighting, energy estimation, thermal comfort, acoustics (speech privacy and speech intelligibility) and computational simulation programs. In particular, his expertise includes strategies for applying environmental technologies in sustainable design and building retrofitting that facilitates occupants' health and well-being. Dr. Mehdi Amirkhani is an active researcher in building environmental technologies, building systems and services, building retrofitting, sustainable design, net-zero energy buildings and indoor environmental quality. His

PhD thesis established an innovative approach to lighting design in office buildings that optimised the visual comfort (electric and daylight) and energy consumption performance of commercial façades. Mehdi is a Nationwide House Energy Rating Scheme (NatHERS) Accredited Assessor and has more than twelve years of practical experience as an architect in Australia and overseas. He is also the Founder and Director of Abian Design Studio. Mehdi is an editor and reviewer of several scientific journals and international conferences. He is a board member of the Zero Energy Mass Custom Home (ZEMCH) Network; a Co-director of ZEMCH Network, Australia; an Associate Fellow of the UK Higher Education Academy (AFHEA); and a member of the Architectural Science Association (ASA).

Gerhard Zucker

Gerhard Zucker is a Senior Researcher at the Austrian Institute of Technology (AIT) in the field of smart buildings. He leads the research field of digital building technologies and works on machine learning, building automation and controls, and data analytics to optimize the energy performance of buildings. The most recent developments focus on the integration of building information modelling (BIM) as a method to address the performance goals of buildings. This includes data modelling in the open BIM standard Industry Foundation Classes (IFC) and the definition of standardizable data in IFC to close the gaps in the digital value change of the planning, construction and operation of buildings. The results of this research are further developed in closed cooperation with the Austrian standardization ASI—Austrian Standards International. Zucker finished his diploma in 1998 at the Vienna University of Technology and his Dr. techn. (PhD), viva-voce exam with excellence at the Vienna University of Technology in 2006. He is Associate Editor with the IEEE Transactions on Industrial Informatics (IEEE TII), has authored more than 20 scientific journals, and has contributed to three books in the field of building automation and artificial intelligence.

energies MDPI

Editorial

Special Issue "Environmental Technology Applications in the Retrofitting of Residential Buildings"

Mark B. Luther [1], Igor Martek [1,*], Mehdi Amirkhani [2] and Gerhard Zucker [3]

1 School of Architecture and Built Environment, Deakin University, Geelong, VIC 3220, Australia
2 UniSA Online, Science Technology Engineering and Mathematics (STEM), University of South Australia, Adelaide, SA 5000, Australia
3 Center for Energy, Austrian Institute of Technology, Giefinggasse 2, AT-1210 Vienna, Austria
* Correspondence: igor.martek@deakin.edu.au

1. Introduction

The architectural and construction professions are in a state of major transition. The imperative that the industry must embrace is 'sustainability', and with that imperative is the requirement to develop applications that make buildings energy-efficient, eco-friendly, resilient, inclusive, and overall 'greener'.

For much of history, building was understood as merely a craft with no overt underlying philosophy. This changed when, in 1414, a manuscript written some 2000 years ago by the Roman engineer Marcus Vitruvius was 'rediscovered' in the library of Saint Gall Monastery, Switzerland. Vitruvius' dictum '*firmitas, utilitas, venustas*' demands that all buildings fulfill three requirements. They must be structurally sound, serve the purpose for which they were built, and inspire the human soul. Works such as Vitruvius' led to the Renaissance, and with it, the rise of science and the broader debate about humankind's place in and responsibility for the world in which we all live.

As simple as Vitruvius' message may appear to be, it is fair to say that all architectural theorizing that has followed in the six centuries since Vitruvius was brought into the spotlight have revolved around defining and redefining his three terms. Environmental awareness first came into prominence in the 1970s, and this soon filtered into the world of construction. *The Limits of Growth*, published in 1972, popularized the terms 'sustainable development,' and 'sustainable life-style.' In 1987, the UN's Brundtland Commission released its report, *Our Common Future*, in which it defined sustainability as '*the development that meets the needs of the present without compromising the needs of future generations*' [1]. Today, it is hard to imagine architects or builders daring to operate in opposition to that sentiment. Vitruvius' requirements that buildings exhibit 'strength, utility and beauty' are now bound be expectations that they be environmentally friendly, non-polluting, carbon-neutral, and even recyclable. That is, only sustainable buildings are beautiful and fitting to the moral human.

The question then shifts as to how best to create an environmentally friendly built environment. Much attention in this regard has been levied on new buildings. Indeed, the race to showcase flagship corporate headquarter edifices with five-star sustainability ratings is very much the new game in town. While such activity positively brands those enterprises that invest in such 'virtue signaling,' the overall impact of such icons on cities and on the environment in general remains minimal. The truth is that cities, and the bulk of the materials, energy, and resources taken up by them, lie in its residences and houses. And while we should be constructing new houses with the best sustainability practices available, the truth also remains that most of the city is already built—80% of which are established homes [2].

The greatest hope that we have in effecting a transition to an environmentally friendly, sustainable built environment, therefore, lies in the mission of refurbishing and retrofitting

Citation: Luther, M.B.; Martek, I.; Amirkhani, M.; Zucker, G. Special Issue "Environmental Technology Applications in the Retrofitting of Residential Buildings". *Energies* **2022**, *15*, 5956. https://doi.org/10.3390/en15165956

Received: 11 August 2022
Accepted: 12 August 2022
Published: 17 August 2022

existing residential buildings. How ought this be done? The collection of papers in this Special Issue offers valuable insights into a sustainable way forward.

2. Special Issue Content

This collection showcases 11 studies that investigate applications of environmental technology tailored to improving the sustainable performance of existing residential buildings.

The first paper, "Mapping research trends in residential construction retrofitting", Ref. [3] is a scientometric literature review that lays out the state-of-play regarding research in this field. Only as recently as 2011 were more than 20 studies published annually, revealing the novice nature of this important research domain. Of these, the majority relate to energy consumption and efficiency, with scant mention of related relevant themes, such as air quality, insulation, energy policy, or occupant behavior. Thus, residential retrofitting remains a topic ripe for further investigation.

The second paper, "Radiant conditioning retrofitting for residential buildings", Ref. [4] begins with the premise that conditioning system retrofitting affords the greatest opportunities for winning significant comfort and energy savings outcomes. The technology is already out there in the marketplace but has yet to be implemented widely in homes. In fact, lightweight radiant systems can be as much as 40% more energy-efficient than conventional systems, while delivering response times of under 15 min.

The third paper, "Guidance on implementing renewable energy systems in Australian homes", Ref. [5] is an empirical case study that establishes guidelines for determining the size of photo-voltaic (PV) arrays to be installed in homes, along with an assessment of the viability of accompanying battery installations. PV size is shown to depend on four factors: daytime house load, night-time load, PV solar rating, and tariff rates charged.

The fourth paper, "Applying solar PV to heat pump and storage technologies in Australian homes", Ref. [5] looks at achieving a net zero energy goal for houses. It concludes that for a modest-sized house in Melbourne, Australia, it is possible to meet electrical, heating, and cooling loads with a heat pump that draws as little as 1 kW. While conventional systems tend to be designed to service maximum loads, net zero can in fact be achieved by judicious utilization of existing off-the-shelf PV arrays, heat pumps, and batteries.

The fifth paper, "Automatically creating HVAC control strategies based on building information modeling (BIM)", Ref. [6] presents a method for optimizing building heating using BIM data. While BIM data are created during the design and planning phases of a building, they can also be re-accessed in commissioning. Heat provisioning and distribution can be programmed to prioritize non-fossil fuel energy sources over fossil fuels, through intelligent BIM centered controls.

The sixth paper, "IFC BIM model enrichment with space function information using graph neural networks", Ref. [7] aims to improve room function classification precision. This study compliments the preceding paper in that enhancing information on rooms, such as kitchen, lounge and bedrooms, are a precursor to enhanced BIM modeling performance. A three-step method is proposed and then tested on residential building IFC models, showing an improvement in accuracy.

The seventh paper, "Development and evaluation of occupancy-aware HVAC control for residential building energy efficiency and occupant comfort", Ref. [8] recognizes that residential HVAC systems tend to be manually controlled and are therefore prone to be wasteful in heating or cooling houses beyond what is needed. Smart devices can be used to measure occupancy patterns and develop occupancy predictions, which can be linked to HVAC controls. Empirical tests on six homes in Colorado, USA, found that savings from the use of predictive algorithms that optimize thermal comfort against energy costs can be as much as 13%.

The eighth paper, "Efficiency of energy consumption between reinforced concrete structure and cross-laminated timber-based hybrid structure in east Asian cities", Ref. [9] proposes a novel approach to multi-story residential refurbishment. When reinforced concrete (RC) structures come to the end of their useful life, they can be torn down or renovated. While wooden structures are known to have superior energy saving properties as compared with RC structures, this study establishes that hybrid refurbished structures in which RC beams and columns are retained, while floors and walls are replaced by new timber laminates, results in lighter, cheaper buildings, with energy performance superior to the replaced fully RC structures, yet only marginally less efficient than full wooden buildings.

The ninth paper, "Impact of enclosure boundary patterns and lift-up design on optimization of summer pedestrian wind environment in high-density residential districts", Ref. [10] tackles the problem of low wind velocities at street level in dense, high-rise residential environments. Summer heat can be alleviated by light cooling winds. However, the presence of dense high-rise buildings stifle street-level air movement, raising temperatures and increasing pedestrian discomfort. Using simulations, this study reveals that astute enclosure design of buildings can improve wind velocity by 70%, with accompanying comfort levels reaching 200%.

The tenth paper, "Camera-driven probabilistic algorithm for multi-elevator systems", Ref. [11] offers a new means for optimizing the allocation and dispatch of elevators in high-rise complexes. Information obtained through surveillance cameras, feed into an artificial intelligence algorithm, was shown though simulations to reduce lift passenger travel times by up to 40%.

The eleventh paper, "Bushfire: retrofitting rural and urban fringe structures", Ref. [12] reviews extant research on bushfire behavior and its impact on residential structures. As a consequence of global warming, Australia and other parts of the world increasingly suffer from the devastation of bushfires. This final paper documents the main cause of structural damage to be attributable to fire embers, while the main cause of human fatalities arises from smoke inhalation. In view of these findings, houses at risk ought to be appropriately sealed to keep embers and smoke at bay, and practical remedies for retrofitting are offered.

3. Closing Remarks

Buildings consume 50% of the world's energy, while the urban environment generates 80% of the world's greenhouse emissions. Added to this, 80% of the world's buildings are people's homes. Existing residential buildings, therefore, represent the real coal face in responding to climate change. However, as this Special Issue reveals, research into this area has only just begun. The papers presented here add to the knowledge in finding needed effective responses. They also reveal much work is yet to be done.

It is hoped that the reader will take away a better understanding of the innovations that have been taking place that impact residential retrofitting. It is also hoped that the reader will see more clearly the many directions and opportunities laid bare inviting future research in this field.

Funding: This research received no external funding.

Conflicts of Interest: The authors declare no conflict of interest.

References

1. Martek, I.; Hosseini, M.R.; Shrestha, A.; Zavadskas, E.K.; Seaton, S. The sustainability narrative in contemporary architecture: Falling short of building a sustainable future. *Sustainability* **2018**, *10*, 981. [CrossRef]
2. Tam, V.W.; Zeng, S. Sustainable performance indicators for Australian residential buildings. *J. Leg. Aff. Disput. Resolut. Eng. Constr.* **2013**, *5*, 168–179.
3. Amirkhani, M.; Martek, I.; Luther, M.B. Mapping Research Trends in Residential Construction Retrofitting: A Scientometric Literature Review. *Energies* **2021**, *14*, 6106. [CrossRef]
4. Do, H.Q.; Luther, M.B.; Amirkhani, M.; Wang, Z.; Martek, I. Radiant Conditioning Retrofitting for Residential Buildings. *Energies* **2022**, *15*, 449. [CrossRef]
5. Simko, T.; Luther, M.B.; Li, H.X.; Horan, P. Applying Solar PV to Heat Pump and Storage Technologies in Australian Houses. *Energies* **2021**, *14*, 5480. [CrossRef]

6. Sporr, A.; Zucker, G.; Hofmann, R. Automatically Creating HVAC Control Strategies Based on Building Information Modeling (BIM): Heat Provisioning and Distribution. *Energies* **2020**, *13*, 4403. [CrossRef]
7. Buruzs, A.; Šipetić, M.; Blank-Landeshammer, B.; Zucker, G. IFC BIM Model Enrichment with Space Function Information Using Graph Neural Networks. *Energies* **2022**, *15*, 2937. [CrossRef]
8. Turley, C.; Jacoby, M.; Pavlak, G.; Henze, G. Development and Evaluation of Occupancy-Aware HVAC Control for Residential Building Energy Efficiency and Occupant Comfort. *Energies* **2020**, *13*, 5396. [CrossRef]
9. Tsai, M.-T.; Lin, W.-T. Efficiency of Energy Consumption between Reinforced Concrete Structure and Cross-Laminated Timber Based Hybrid Structure in East Asian Cities. *Energies* **2022**, *15*, 165. [CrossRef]
10. Jiang, Z.; Gao, W. Impact of Enclosure Boundary Patterns and Lift-Up Design on Optimization of Summer Pedestrian Wind Environment in High-Density Residential Districts. *Energies* **2021**, *14*, 3199. [CrossRef]
11. Bapin, Y.; Alimanov, K.; Zarikas, V. Camera-Driven Probabilistic Algorithm for Multi-Elevator Systems. *Energies* **2020**, *13*, 6161. [CrossRef]
12. Costin, G.P. Bushfire: Retrofitting Rural and Urban Fringe Structures—Implications of Current Engineering Data. *Energies* **2021**, *14*, 3526. [CrossRef]

energies

MDPI

Review

Mapping Research Trends in Residential Construction Retrofitting: A Scientometric Literature Review

Mehdi Amirkhani [1,2,*], **Igor Martek** [1] **and Mark B. Luther** [1]

[1] School of Architecture and Built Environment, Deakin University, Geelong, VIC 3220, Australia; igor.martek@deakin.edu.au (I.M.); mark.luther@deakin.edu.au (M.B.L.)
[2] Abian Design Studio, Melbourne, VIC 3031, Australia
* Correspondence: m.amirkhani@deakin.edu.au or mehdi@abiandesign.com.au

Abstract: The world is looking to reduce carbon emissions, prevent global warming, and become more energy sustainable. Despite the various strategies for mitigating climate change, the fact remains that 80% of greenhouse gas emissions are attributable to activities associated with the built environment, and this is where a concentrated focus is needed. Moreover, most buildings are residential, not commercial or industrial. In essence, ways must be found to reduce energy consumption and CO_2 emissions from existing houses and apartments globally if sustainability is to be realised. The recognised way to achieve this is through the retrofitting of existing residential buildings. Studies in this area have increased in recent times, but the extent of the work remains unmapped and undescribed. If further progress is to be made in this field, researchers' knowledge domain so far must be documented. This literature review delivers that goal. A scientometric evaluation of research on residential retrofitting is here presented. VOSviewer, Gephi, and CiteSpace are the software packages used. Findings identify retrofitting as an emerging theme, taking off only as recently as 2017. The breadth of research is very limited, primarily concerned with calibrating trade-offs between energy costs and thermal comfort. Emerging and new opportunities to expand retrofitting research are identified. Finally, while several journals accommodate publications on this topic, analysis reveals Energy and Buildings to be the significant citation source.

Keywords: residential construction; retrofitting; sustainable construction; literature review; research trends; scientometric review

Citation: Amirkhani, M.; Martek, I.; Luther, M.B. Mapping Research Trends in Residential Construction Retrofitting: A Scientometric Literature Review. *Energies* **2021**, *14*, 6106. https://doi.org/10.3390/en14196106

Academic Editor: F. Pacheco Torgal

Received: 23 August 2021
Accepted: 22 September 2021
Published: 25 September 2021

1. Introduction

The earth is warming, and this is generally attributed to human activity. Greenhouse gases are identified as the principal culprit, with CO_2 being the primary greenhouse gas. CO_2 is emitted as a by-product from the burning of fossil fuels. For the most part, fossil fuels are associated with vehicle transportation, car, train and aeroplane use [1]. However, the built environment is the most significant energy consumer, both in terms of initial building and ongoing maintenance. The manufacture of building materials—such as bricks and timber—consumes as much as 10% of the total energy produced [2]. Additionally, buildings last a long time and require vast amounts of energy to operate and maintain. Keeping the lights on and air-conditioners running in buildings consume over 50% of the world's energy output. More dramatic still, 80% of all greenhouse gas emissions are directly attributable to activities associated with the urban environment. If that were not enough, buildings deplete resources at rapacious levels. A total of 15% of global water resources go into building construction, such as in making concrete. Buildings consume 25% of world timber stocks, 30% of all other natural resources, and 40% of human-made materials. Moreover, 50% of all the waste dumped into landfills come from building demolition. The impact of buildings on the built environment is nothing short of devastating [3].

Of course, none of this has gone unnoticed. Significant initiatives to fight 'global warming' and roll back greenhouse gas emissions have been pursued worldwide. The

Paris Climate Accord is perhaps the best known, with almost all major developed and many developing countries 'promising' to cut back on fossil-based energy use. Australia, for example, has committed to reducing its CO_2 output to 28% of its 2005 levels by 2030. [4]. This is a considerable reduction, and if it is to have any chance of being realised, it clearly must target the built environment. To that end, governments around the world have sponsored building rating tools. The idea is to assess aspects of a building that impact energy use, consolidate these assessments, and deliver a rating. A higher '5-star' rating, in this scheme, signifies that a building is a better energy performer than, say, a '3 star' rated building. The primary rating tools are the Building Research Establishment Assessment Method (BREAM), established in 1990 in the UK and regarded as the first. Leadership in Energy and Environmental Design (LEED) emerged in 1998 as the US answer to BREAM. Green Star appeared in 2003 as a BREAM adaption for Australian conditions [5–7]. There are, in fact, as many as 600 energy rating instruments worldwide [8,9]. Each rating tool assesses different aspects of a building's energy performance and weighs them differently. The variations are arguably necessary to reflect the different building and environmental conditions, but it begs the question as to whether these tools deliver a reliable or objective measure given that different tools deliver different findings. There are further problems. One that stands out is that the ratings are typically predictive, rating a building off-plan rather than measuring actual performance. Indeed, when actual performance is measured, results have been shown to differ from predictions. Despite the marketing claims that they have penetrated the built environment market, these tools have had limited impact. Generally, they are used by prominent building owners who wish to showcase their social conscience. For example, in Australia, only 0.5% of the Australian building industry has sought a rating from Green Star [10]. We also know that in as many as 50% of these cases, the motivation for seeking an energy rating was indeed to highlight ownership of a 'trophy sustainability icon' [11].

Nevertheless, there is a bigger problem again. The problem is that most of the built environment is not made up of offices, factories or public facilities, but rather residences—houses and apartments. In many countries, dwellings make up at least 80% of the built environment [12]. Moreover, the obvious fact remains that most building stock is not recent, but stock that already exists. The bulk of buildings were built before the 1980s before any energy efficiency guidelines or regulations were introduced. [5] This all points to a pivotal yet neglected priority in construction sustainability research. The clear stand-out candidate demanding attention in this regard is residential retrofitting. By far, the biggest category of buildings to be found strewn across the urban landscape are houses—places where people live. Although building new buildings more sustainably is undoubtedly laudable, the existing building stock also needs attention. Indeed, finding ways to refurbish existing residential buildings sustainably may be the single most crucial quest before the construction community today. Added to this, we have to date no comprehensive assessment of the state of research regarding the retrofitting of residential builds. This paper addresses that need.

2. Literature Review

Building retrofitting is the key to reducing a buildings energy consumption while at the same time improving indoor environmental quality and minimising harmful emissions. It refers to the post-occupancy modification of thermal elements (walls, roofs and floors), services (heating, ventilation, air-conditioning, lighting, and water services), as well as fittings (windows and doors). Various terms are used in reference to building retrofitting, including *refurbishment*, *renovation*, *retro-commission*, and *tune-up*. Building refurbishments generally consider aesthetics and tenant amenities. However, they can also involve upgrades to the building systems in ways that potentially enhance building performance. Although *renovation* is similar to *refurbishment*, with the terms sometimes used interchangeably, renovation is more closely associated with buildings. *Retro-commissioning* is the process of regularly testing and adjusting building systems on a three to five-year cycle to ensure appropriate functional performance. *Tune-up* is a generic term that encompasses existing buildings systems or aspects of retro-commissioning and retrofit.

Research suggests that using appropriate environmental technologies in buildings can reduce energy consumption between 35% and 80% [13]. For example, Hughes and Shonder [14] report retrofitting 4003 homes at Fort Polk, Louisiana, by switching natural gas into a ground source heat pump for water and space heating, cutting energy usage by one third. A subsequent study demonstrated a 35% reduction in house energy usage by simply using modest insulation upgrades and air sealing [15]. A comprehensive retrofit of two apartment buildings in Basel, Switzerland, slashed heating requirements (hot water, ambient heating, and ventilation) by 100%, making the buildings effectively zero thermal energy houses [16]. Francisco, Palmiter [17] argue that sealing ductwork alone can reduce houses' annual heating and air-conditioning energy usage by 15% to 20%. Voss [18] investigated the impact of retrofitting 14 residential building projects by adding solar design concepts, including solar collectors to preheat domestic hot water (DHW) while also supplementing indoor heating demand, enclosing balconies with openable glass elements, and adding solar wall-heating with transparent insulation. The projects returned energy savings of up to 70%. A study in North China revealed that retrofitting a multi-story residential building could diminish energy usage by about 50% while still providing an acceptable indoor thermal environment [19]. Though energy reduction of as much as 70–85% is possible in residential buildings through retrofitting, the investment can be very high [20]. Variable user behaviour along with building management can also significantly impact the energy savings of buildings [21]. Studies have shown that counterproductive user interventions in design conditions can nullify predicted energy savings by up to 75% [22,23].

It is estimated that building energy usage will increase over 2005 levels by as much as 150% by 2050 and rise further to 220% by 2095 [24]. Globally, China ranks second in building energy consumption and first in regard to residential buildings [25]. Residential buildings in the European Union (EU) are responsible for more than 25% of the total energy usage in the region [26]. Due to the high energy conservation potential, investing in building retrofitting has become a major global priority. Studies in the EU predict that retrofitting residential buildings would reduce energy usage by 78% [27]. In Russia, 60% of apartment buildings need extensive retrofitting [28]. In Melbourne, Australia, the number of buildings that have undergone retrofitting work has grown over the last few years, with the upgrading of lighting and HVAC systems being among the most common retrofit activities [29]. The replacement of a broken service proves to be the most significant motivator for retrofitting. According to Shah [30], retrofitting can be undertaken at several levels, including light, medium, extensive, or comprehensive retrofitting, as well as through demolition. Zou, Stewart [31] also divide retrofitting into three levels: *existing building commissioning*, *standard retrofitting*, and *deep retrofitting*. *Existing building commissioning* can be achieved with low risk and capital outlay by enhancing building operation and maintenance procedures. The first level of retrofitting can deliver energy reduction of up to 25%. *Standard retrofitting* replaces existing services to enhanced energy performance, improving performance by up to 45%. *Deep retrofitting* is the integration of whole building improvements, leading to energy savings of over 45%. The US Green Building Council (USGBC) defines major or deep retrofitting as renovating HVAC elements, major envelope alterations, and significant internal rehabilitation [30]. Similarly, the Building Research Establishment (BRE) defines a major retrofitting as those results in the provision, extension or modification of thermal elements or building services [30]. The building's performance condition determines the level of retrofitting.

Surveys conducted in European countries as diverse as Bulgaria, Poland, and Portugal reveal that homeowners are primarily motivated to invest in environmental retrofitting technologies that improve energy usage and comfort [32]. Specifically, the most common information sought by homeowners engaged in retrofitting were expected benefits in terms of reduced heating bills and associated technical information needed to avoid mistakes.

Retrofitting buildings can involve substantial costs and a complex decision-making process requiring various stakeholders, including landlords, tenants, property managers, developers, architects, energy consultants, and local councils. However, these stakeholders are generally privy to different knowledge and technical information and respond dissimilarly in the retrofitting process. For example, most consultants use more complicated tools than regular architects would use. It is relevant to note that architects approach design holistically based on their experiential knowledge and are not primarily focused on improving buildings' energy performance. Overall, barriers such as the complexity of retrofits and financing, combined with the complications of determining the efficacy of potential energy-saving strategies, may prevent retrofitting. In summary, while buildings can be expected to perform better due to retrofitting, results are heavily contingent on the approaches taken and the means by which they are applied. The best cost-benefit solutions are not widely understood or documented. Although cognizant of the potential windfalls, homeowners, landlords, and investors remain uncertain as to how to proceed. At the same time, industry practitioners tend to peddle pre-packaged solutions, while a wider appreciation as to the best, unequivocally accepted retrofitting approaches remains undescribed.

Thus, while building practitioners and stakeholders acknowledge the need to use environmental technologies in retrofitting residential buildings, there is still a lack of knowledge of the efficiency of these systems and their applications to improve residential buildings' energy and comfort performance. This research explores the extent of these deficiencies. It adopts structured, quantitative methods that generate an objective, comprehensive portrait of the existing state of research knowledge in environmental technology applications in retrofitting residential buildings. This paper describes the state of play as documented by research academics regarding the environmental performance of residential buildings. The various approaches that have been undertaken are explored, including renewable energy systems, ventilation strategies, innovative conditioning systems, construction retrofits, new sensor technologies, optimised control strategies and more. This review identifies systems with high energy-saving potential along with those exhibiting high vulnerabilities. This study's outcomes may be used to inform future research directions in retrofitting residential buildings and aid funding efforts by policymakers and practitioners.

3. Method

This review-based study adopts a holistic approach to assess research outputs in environmental technology applications in retrofitting residential buildings, published in quality academic journals. The overall workflow of this study is shown in Figure 1 and contains five stages.

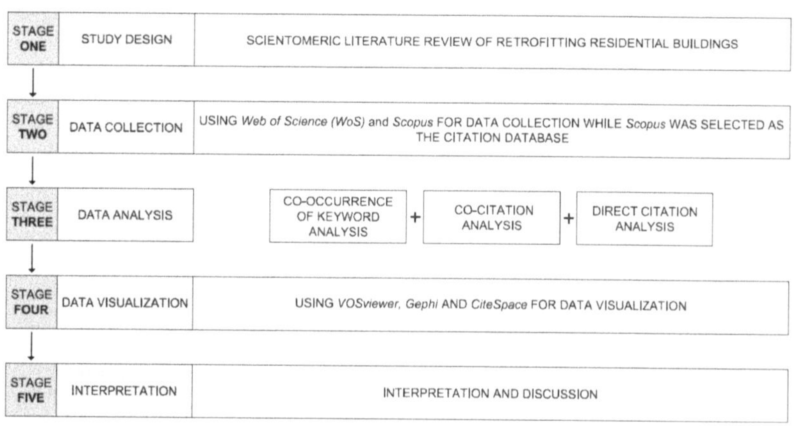

Figure 1. The research flowchart reviewing the literature on the retrofitting of residential buildings.

3.1. Data Collection

Only articles published in peer-reviewed journals were considered, as these have been peer-assessed for reliability and quality. Several databases are available for science mapping, including Web of Science (WoS), Scopus, ScienceDirect, SpringerLink, ProQuest, PubMed, Google Scholar, Dimensions Microsoft Academic, Patent Derwent innovation index, BOOK Citation index, and others. Nonetheless, WoS and Scopus are the most widely used databases in science mapping due to the accepted view that they contain higher data quality. WoS has more than 21,000 peer-reviewed journals that have been indexed in the database since 1900, with almost 1.9 billion cited references from over 171 million records [33]. In 2004, Elsevier launched Scopus, which covers more than 23,500 peer-reviewed journals [34]. It has over 75 million records going back as far as 1788 across different subject areas, such as health, life, physical and social science [34]. Although most databases include authors, affiliations, abstracts, and keywords, WoS and Scopus contain further information, particularly citation data. However, Li, Goerlandt [35] argues that while both WoS and Scopus are popular within the scientific community, Scopus has surpassed WoS over the last five years. As such, this research uses the Scopus database to retrieve and download articles. The search terms were "energy/retrofitting/residential buildings", "retrofit/residential", and "retrofitting/residential". The findings were limited to peer-reviewed journal articles. The research scope includes title, abstract and keywords. The record content covers citation information, bibliographical information, abstract, keywords, and funding details. Of some 1500 articles extracted, these were reviewed manually for relevance with the title and abstracts being read for content. As a result, the overall pool of retrieved articles was reduced to 511, being deemed relevant to the study at hand. These were downloaded and the entire article set subjected to fine-grained analysis.

3.2. Literature Review Metrics

Scholars use variable quantitative and qualitative literature reviewing methods to understand and organise recent findings. Among these, science mapping is the approach most used. It relies on mathematical statistics and visualisation techniques to evaluate a research domain's structural aspects and research policies [36]. Science mapping illustrates the correlations between disciplines, fields, and individual publications using a spatial approach [37]. The scope of science mapping research can be research, a scientific discipline, or topic areas considering a specific research question [38]. Science mapping also has a proven history in picturing systematic patterns in large bibliographical units and comprehensive literature review bodies [36,39]. The overall workflow in a science mapping analysis incorporates several steps, including data retrieval, preprocessing, network extraction, normalisation, mapping, analysis, visualisation, as well as interpreting and generating conclusions from the results [40]. It employs three metrics, including bibliometrics, scientometrics, and informetrics [39]. However, these metrics do not have clear boundaries and are generally used interchangeably.

According to Pritchard [41], bibliometrics is "the application of mathematical and statistical methods to books and other media of communication". Bibliometrics provides objective analysis and is generally used for quantitative research evaluation of academic outputs, especially in big datasets [42,43]. It can introduce a transparent, systematic and reproducible review process according to the statistical measurement of scientific activity, science, and scientists [44]. Bibliometrics focuses predominantly on the production, distribution, dissemination, and usage of data delivered in any document (journal, book, conference, patent, or website) [45].

Nalimov and Mulchenko [46] first introduced scientometrics and defined it as "applying quantitative research methods on the development of science as an information process". In other words, scientometrics is used to study science as an economic activity or discipline [47]. It is part of the sociology of science and has application to science policy-making. Although scientometrics is indistinguishable from bibliometrics in that it deals with quantitative analysis of publications or other forms of communication [48], it also measures scientific productivity and utility [44]. Figure 2 illustrates the different methods used in the scientometric analysis.

Figure 2. Methods applied in the scientometric analysis [49].

Informetrics, which come from the German term "informetrie", was first introduced by Nacke [50] and was widely accepted by the early 1990s [47]. Informetrics focuses on the discovery of mathematical models that describe the properties of information [51]. It is the study of quantitative aspects of data in any social group, not just among scientists, and in any form, not just bibliographies and records [48]. It investigates the quantitative aspects of spoken or informal communication, records, and information [48]. As such, it can be used to analyse information that lies outside the boundaries of both bibliometrics and scientometrics metrics.

The current study focuses on journal articles where the co-occurrence of keywords and document co-citation analysis was conducted, along with measurements of direct citation assessment of outlets and co-authorship. These techniques are widely accepted and recommended in construction research [52,53]. Moreover, they are also regarded as the core components of scientometric analysis, helping scholars comprehend the current research status [54]. Word co-occurrence analysis (also called co-word analysis) helps map the strength of correlation between research outlets in textual data [55]. It is based on the number of co-occurrences of two words appearing in the same outlet, abstract or keyword units. Co-word analysis can expose and visualise the interaction between research topics [56]. Here, we applied the co-word analysis concept to measure the co-occurrence of keywords specified by authors; the author keywords. Co-citation analysis is defined as two publications that are cited together in one research outlet [57]. If two publications have been regularly co-cited, they are highly correlated, and there is thus a robust semantic similarity between them. Due to co-citation analysis characteristics, many scholars use the approach to investigate the core concerns of a field [58]. The direct citation analysis (also known as a cross citation or inter-citation) illustrates the direct citing correlations between research outlets without creating connections based on a third-party paper [59]. In scientometric analysis, two authors have collaborated if they co-author papers. Co-authorship measurement has been divided into three levels, including micro-level (author analysis), meso level (institutes or cities analysis), and macro-level (country/regional analysis). Micro-level measurement clusters paper authors by common paper, which is used to understand who highly productive authors are in a scientific domain of inquiry and to determine teams of collaborating researchers. Likewise, meso and macro-level measurements are used to determine active contributors in a scientific domain of inquiry and how they are connected at institution, city, or country levels.

3.3. Data Visualisation

The visualisation technique is the most effective method for extracting data from an enormous amount of complex information and displaying them more clearly. It uses computer tools to illustrate the internal structure of information in a way that helps users develop a qualitative understanding of the data and better conduct subsequent outlier detection, pattern recognition, and much more. The visualisation technique also illustrates the correlation between data objects, the research field's development process, and the development trend. Several tools can be used for data visualisation, including VOSviewer and CiteSpace.

VOSviewer is an application used for constructing and visualising bibliometric networks at an aggregated level [60]. VOSviewer also provides a clustering function, which assigns keywords to clusters based on their co-occurrence [60]. CiteSpace is a Java application for evaluating and visualising a co-citation network [61]. This application's primary utility is to facilitate the investigation of emerging trends in a field of knowledge [62]. According to Chen [61], CiteSpace is suited for co-citation analysis. It automatically clusters collected documents and visually displays them. In the current study, both VOSviewer and CiteSpace were selected due to their suitability for extensive mapping networks and text mining capabilities.

3.4. Technical Details of Data Analysis

The technical details of the data analysis methodology used are as follows:

- Co-occurrence of keyword analysis: Author keywords were used to present a reproducible visualisation of the keywords using VOSviewer. A total of 1507 keywords were extracted from the dataset while fractional counting was deployed. The default minimum value of 5 was set for keyword occurrence, and a total of 51 keywords, which were connected to each other, were included in the network. The network created by VOSviewer was subsequently submitted to Gephi for further analysis.
- Co-citation analysis: This assessment was undertaken in VOSviewer. The unit of measurement was set to *cite references*, and the counting method was set to *fractional counting*. The minimum number of citations of cited papers was set to 3. The clustering function was used in the CiteSpace application, and the labels for clusters were generated using the log-likelihood ratio (LLR) algorithm with the indexing term.
- Direct citation analysis: All references were submitted to VOSviewer for direct citation analysis while the measurement unit was *sourced*. The minimum value for both the number of documents of a source and the number of citations of a source was set to 10.
- Co-authorship analysis: This measurement was undertaken in VOSviewer while the unit of analysis was set to *authors* and the counting method set to *fractional counting*. VOSviewer was used for co-authorship analysis while the unit of measurement being *organisations*, and the counting method was set to *fractional counting*. The minimum number of documents and citations of an organisation was set to 2 and 10, respectively. Additionally, co-authorship assessment was conducted with the unit of analysis being *countries* and the counting method set to *fractional counting*. The minimum number of documents of a country and the minimum number of citations were set to 10 and 15, respectively.

4. Results and Discussion

A total of 511 collected articles were assessed across five parameters: published journal, co-authorship, co-occurring keywords, article citations, and regions.

4.1. Research on Retrofitting Residential Buildings

Research activity in the field of residential retrofitting was evaluated based on the number of publications. As it happens, Vliet [63] conducted the first study on the retrofitting of residential buildings in 1979, focusing on retrofitting apartment buildings with solar, heating, cooling and hot water. The research level and the future development trend in a particular field can be understood by plotting the quantity of literature over time and conducting multivariate measurements. Figure 3 illustrates the number of published works on the retrofitting of residential buildings between 1979 and 2021. It is apparent from this line graph that while only a few research works have been published each year to 2010 (a total of 63 research works), there has been a sudden increase in the number of published studies since that time, especially after 2017. These results suggest that there has been a recent, meaningful increase in research interest in retrofitting residential buildings and that this trend can be expected to continue.

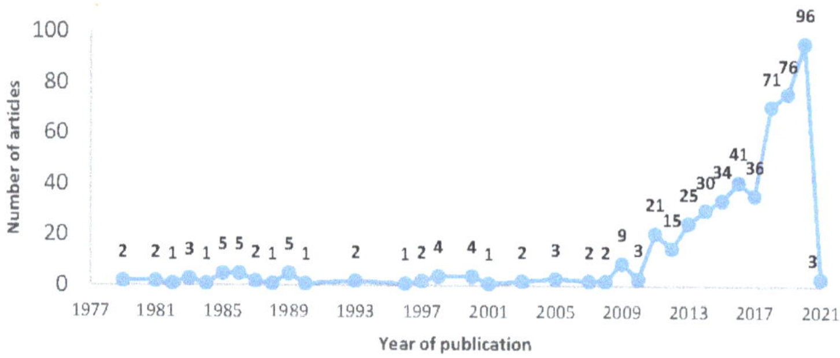

Figure 3. Yearly trend in the number of papers published on the retrofitting of residential buildings.

4.2. Co-Occurrence of Keyword Analysis

A co-occurrence analysis was conducted using author keywords in VOSviewer. A network of correlated keywords provides a clear understanding of scientific knowledge production in terms of relationships, patterns, and intellectual organisation of the topics covered [64]. The link strength between keywords is based on the number of documents in which both keywords occur together. Using Gephi, the following terms were merged for better analysis:

- Building retrofitting: "building retrofit", "building retrofitting", "retrofit", "retrofitting", "renovation", "refurbishment", and "optimisation".
- Energy retrofitting: "building energy retrofitting", "energy retrofitting", "energy retrofit", and "energy retrofits".
- "Residential building", "residential buildings", "residential", "residential sector", "residential building stock", "housing", "housing stock", "social housing", "buildings", and "building stock".
- "Energy-saving", "energy savings" and "energy conservation".
- "Energy efficiency measures" and "energy efficiency".
- "Energy", "energy consumption", and "energy performance".

A network comprised of 28 nodes and 145 links was generated, as shown in Figure 4, demonstrating the main areas of research identified in the field of residential retrofitting. The size of nodes illustrates its importance based on its repetition on different documents. The thickness of lines demonstrates the correlation between nodes based on the number of documents in which both occur.

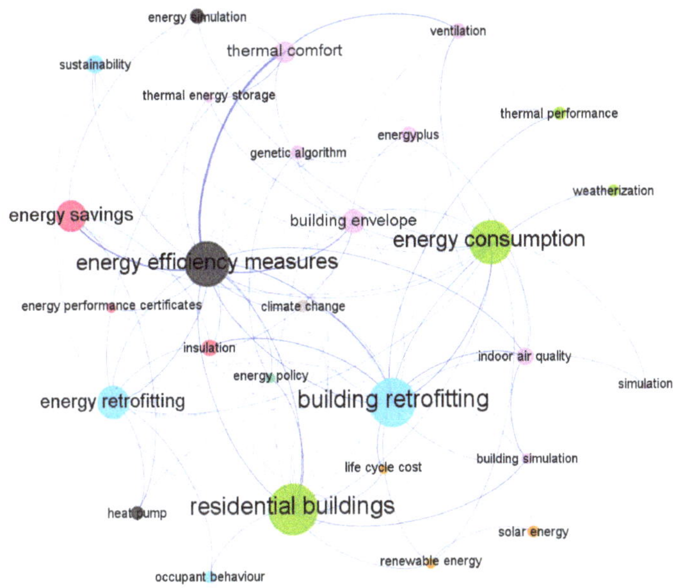

Figure 4. Main research areas in retrofitting residential buildings.

Ties between nodes can be assessed in a variety of ways when analysing network structures. Nonetheless, some measures, such as degree centrality (DC), weighted degree centrality (WDC), and relative importance, are the most widely accepted large-scale network metrics used in extracting information from a network. DC is based on a graph-theoretical network assessment approach [65]. It uses a directive and simple description of centrality and prioritises nodes in the network by counting the number of connections linked to a node. However, DC does not distinguish between the quantity and the quality of the links. Opsahl, Agneessens [66] improved DC and proposed WDC, which introduces the concept of strength, stating that the node would be more significant if it gains greater strength in the network. In other words, WDC is the sum of weighted values of the links (edges) connecting the node and its neighbours. Table 1 demonstrates the outcomes of the analysis of the network. The correlation of the research areas, as illustrated in Figure 4, and the ranking of the leading research areas, as shown in Table 1, reveal some interesting findings, reflecting gaps in the literature in the field of residential building retrofitting.

The centralising themes are 'residential buildings' and 'building retrofitting'. These themes are heavily investigated in energy, energy efficiency measures, energy consumption, and energy retrofitting. The emphasis here is primarily on 'building envelope', with the dual aims of improving 'thermal comfort' and 'energy savings'. However, residential retrofitting involves more than striking a balance between energy costs and comfort by upgrading building envelopes. Consequently, the analysis indicates that the subject of residential retrofitting is highly under-researched while at the same time revealing those areas within the domain that have attracted fledgling studies. Research themes seeded and warranting further attention include building and energy simulation, insulation, ventilation, and indoor air quality. Areas will almost no attention include energy performance certification, weatherisation, occupant behaviour, heat pumps, and perhaps surprisingly, solar energy.

Table 1. The central focus of retrofitting residential buildings.

Author Keyword	Degree Centrality	Weighted Degree Centrality	Relative Importance
residential buildings	31	19.83	1
building retrofitting	29	24.08	2
thermal comfort	12	13.67	3
sustainability	9	4.67	4
ventilation	8	5.33	5
energy efficiency measures	27	32.24	6
weatherisation	5	2.67	7
energy consumption	26	18.25	9
thermal performance	6	3.67	10
solar energy	5	2.00	11
indoor air quality	9	7.99	12
thermal energy storage	4	4.00	13
energy savings	18	10.08	14
energy retrofitting	19	12.83	15
occupant behaviour	5	2.75	16
genetic algorithm	7	4.99	17
energyplus	8	4.99	18
insulation	8	6.99	19
energy simulation	7	5.00	20
simulation	2	2.00	21
heat pump	6	3.67	22
energy performance certificates	4	1.99	23
climate change	6	4.5	24
energy policy	4	4.00	25
renewable energy	4	4.00	26
building simulation	5	5.00	27
life cycle cost	3	3.00	28
building envelope	13	10.08	29

4.3. Co-Citation Analysis

Co-citation analysis is a scientometric approach for exploring and organising core subjects and the knowledge structure in a discipline [57]. It can aggregate articles with related research content into clusters. We conducted co-citation and cluster analysis using CiteSpace to identify core factors in the retrofitting of residential buildings. The 'lookback year' for references was set to 10 years. The maximum number of links per node and the link retaining factor was set to 10 and 3, respectively. CiteScape identified 15,415 references, being 100% of the original references, as valid references. Figure 5 illustrates the co-citation analysis clusters, which consist of 711 nodes and 2307 connections. #0 is the largest cluster with the maximum number of studies. CiteSpace automatically assigns labels to the identified clusters based on their structure rather than their content [62].

The modularity score (the symbol is *Q*) measures the extent to which a network can be divided into independent components (e.g., clusters) [67]. Its score ranges from 0 to 1. A high modularity score indicates that the network components are isolated and have clear boundaries between the clusters. The modularity score (the symbol is *Q*) of the current study clusters is 0.3442, showing that the boundaries between the clusters are not very clear [62].

The silhouette metric is useful in approximating the hesitation involved in identifying a cluster's nature [68]. The silhouette score ranges from −1 to 1, with a higher score indicating a better separation from other clusters. In this study, the average silhouette value of all clusters is 0.9265, which indicates that the clusters are highly homogeneous, i.e., articles appear to be written with little relation to other studies. Work in this area appears fragmented, limited in scope, and sometimes repetitive, tending not to be undertaken as development or extension of previous work.

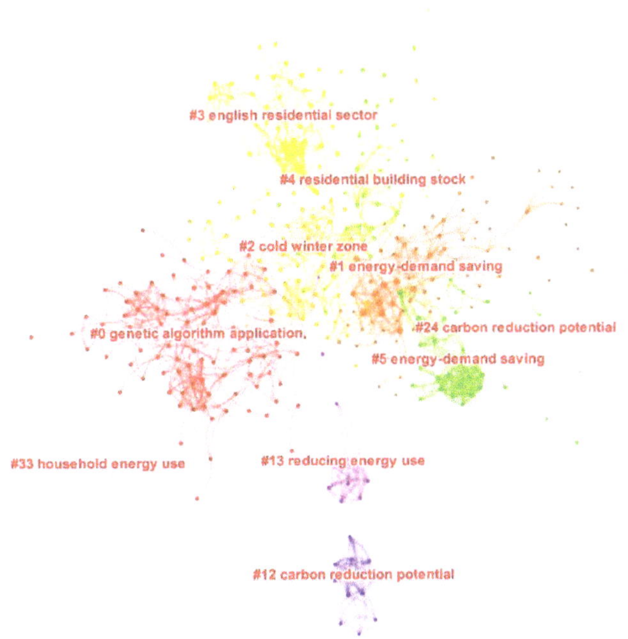

Figure 5. The cluster of co-citation analysis.

In this paper, the co-citation analysis of studies contains 107 clusters of different sizes, which together present the landscape of knowledge development in the field of retrofitting residential buildings. Detailed information of the first six clusters, which contain 337 nodes accounting for about 47% of the total nodes, is provided in Table 2.

Table 2. Main clusters and their content focus.

Clusters	Size	Silhouette	Mean Year	The Main Focus of the Cluster
0	94	0.91	2015	genetic algorithm application, data-driven approach
1	85	0.903	2010	energy-demand saving, genetic algorithm application
2	61	0.943	2014	cold winter zone, hot summer, low carbon heating
3	51	0.962	2012	English residential sector, energy efficiency investment
4	46	0.858	2011	residential building stock, cost-benefit analysis

Cluster 0 has 94 cited references, with an average publication year of 2015. Generic algorithms are based on the Darwinian "natural selection" concept to generate optimal solutions [69]. This cluster's first two highly cited references are Noris, Adamkiewicz [70] and Asadi, Silva [71]. Noris, Adamkiewicz [70] retrofitted 16 apartments for low-income residents in three buildings, using different approaches to minimising energy usage while enhancing IEQ. Their work's outcomes indicate general improvements in comfort conditions, carbon dioxide concentrations, bathroom humidity, acetaldehyde, particle matter mass (2.5 μm), and voltage organic compounds. Asadi, Silva [71] proposed a multi-objective optimisation model using artificial neural networks and generic algorithms to evaluate options in a building retrofitting project. Their work starts with the single optimisation of objective functions, focusing on a school building's performance and characteristics as a case study, including retrofit cost, energy usage, and thermal discomfort hours. After that, the proposed model was used to investigate the interaction between these conflicting objectives and evaluate their trade-offs.

Cluster 1 has 85 cited references, published in 2010, on average, with a research focus on energy-demand saving. One of the most cited studies in this cluster is Ma, Cooper [72]. Their research proposes a systemic method to properly select and identify the best retrofit possibilities for existing buildings. They discussed the building retrofit issues that impact retrofit decision-making. Additionally, major retrofitting activities, including energy auditing, quantification of energy benefits, building performance evaluation, economic analysis, risk measurement, measurement, and energy savings verification, are briefly reviewed in their work.

Cluster 2 has 61 cited references, with an average publication year of 2014, which probes aspects of the cold winter zone. This cluster's first two highly cited references are Kavgic, Mavrogianni [73] and Ballarini, Corgnati [74]. Kavgic, Mavrogianni [73] critically evaluated the existing bottom-up building physics-based residential energy models and identified the next generation of coupled energy-health bottom-up building stock models. Bottom-up models are built from information on a hierarchy of disaggregated elements combined according to assessments that influence energy consumption. Nic and Mark [75] argue that these models are also accepted as an approach to identify the most cost-effective alternatives to achieve given carbon decline targets using the best available technologies and processes. Ballarini, Corgnati [74] proposed a method for the identification of 18 reference buildings among different categories drawn from European residential building stock from which they developed a harmonized structure focusing on potential energy savings and diminished carbon dioxide emissions.

On average, cluster 3 has 61 cited references, published in 2012, with the English residential sector as the research subject. This cluster's first two highly cited references are Nair, Gustavsson [76] and Dowson, Poole [77]. Nair, Gustavsson [76] evaluated the factors that influence the selection of energy efficiency measures to reduce energy consumption using data collected from 3000 detached house owners in Sweden. They argue that personal attributes, including education, income, age, and contextual factors, such as thermal discomfort, house age, perceived energy cost, and past investment, affect homeowners' preferences for a specific energy efficiency measure. Dowson, Poole [77] investigated the existing UK housing stock's thermal performance along with the challenges of retrofitting these to reduce energy usage and carbon dioxide emissions. They found that there is a strong relationship between the age of dwellings and their thermal performance. They also argue that the high capital cost of some retrofitting strategies, such as new generation of double-glazed windows, and wall insulation, as well as their long financial payback period are the major barriers.

Cluster 4 has 46 cited references, with an average publication year of 2011. It probes residential building stock and cost-benefit analysis. This cluster's first two highly cited references are Sunikka-Blank and Galvin [78] and Mata, Sasic Kalagasidis [79]. Sunikka-Blank and Galvin [78] compared the energy performance ratings of 3400 German homes for space and water heating, using existing data on actual measured usage. They argued that there is a gap between energy performance and the measured energy consumption. They also concluded that building users generally behave more economically in homes that experience greater thermal discomfort issues after retrofitting. Mata, Sasic Kalagasidis [79] evaluated the energy consumption and carbon dioxide emissions of 1400 Swedish residential buildings. They used bottom-up modeling technology to evaluate 12 energy-saving measures (ESMs). They argued that the application of selected ESMs could diminish energy demand by 53%, and the level of carbon dioxide by 63%. Some measures, such as heat recovery, decreased the indoor temperature, while upgrading the U-values of the building envelope and the windows returned the maximum impact.

4.4. Direct Citation Analysis

Generally, three methods are used to explore the citation correlation between sources: direct citation, bibliographic coupling, and co-citation analysis [60]. The direct citation method is used to explore correlations between two publications where one publication cites the other. The bibliographic coupling method is used to identify correlations between publications that cite the same publication. The co-citation method is used to explore correlations between publications that are cited by the same publication. Waltman and Eck [80] argue that co-citation and bibliographic coupling correlations are indirect correlations, and therefore, provide less accurate data on the relatedness of publications than direct citation correlations. Additionally, Shibata, Kajikawa [81] claim that direct citation correlation is the best and quickest method to identify emerging research domains to compare co-citation and bibliographic coupling correlations. Nonetheless, direct citation analysis has limitations. For example, some publications might not have any direct citation correlations with other items, and therefore, cannot be assigned to a source.

As can be seen from Figure 6, the publication 'Energy and Buildings' is the premier source of reference material used by subsequent publications on residential retrofitting. 'Journal of Cleaner Production, 'Energy Policy', 'Sustainable Cities and Society', Applied Energy', and 'Energy' are essential additional sources of research citation. The size of nodes represents the citation of publications to compare with others. The bigger the node size, the more citation received. The thickness of lines is based on the number of citations between the two nudes.

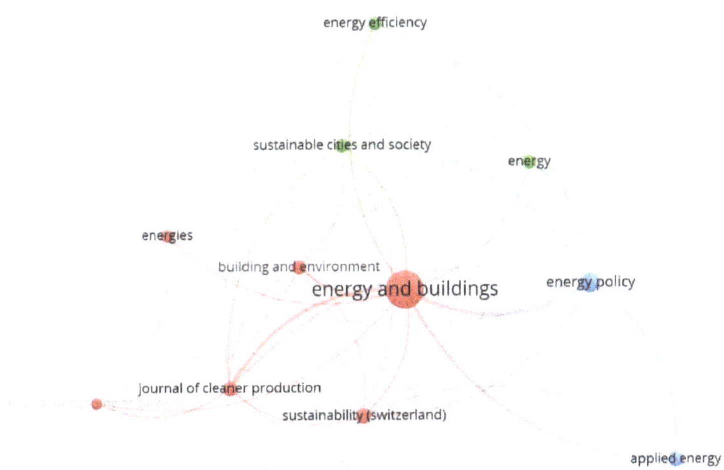

Figure 6. Direct citation analysis linkages between publications.

5. Findings

Sustainability, climate change, and in particular the impact of human-induced global warming, is seen by many as the crisis of our times. There is a global collective call for nations to reduce greenhouse gas emissions, and the further challenge faced is one of how this can possibly be achieved. The main culprit is CO_2 emissions, and the most direct route to CO_2 reductions is the transition from fossil fuels (gas and petroleum) to so-called renewable energies (solar, wind, biofuels, wave, etc.). Less appreciated outside of the building community is that it is the built environment that consumes the vast bulk of total world energy usage. Moreover, building activity itself is the main contributor to other features of the world's poor sustainability outcomes: high waste, resource degradation, ground and air pollution, environmental damage, while the building industry itself harbours notoriously poor social equity practices; particularly in developing nations.

Although this is generally appreciated, the obvious fact is that if poor sustainability outcomes along with high energy usage and high CO_2 emissions are to be addressed, attention must not only focus on the built environment per se, not only on the residential sector, which constitutes the bulk of construction activity but on the retrofitting of the current residential building stock. To be clear, running and maintaining the world's existing residential buildings is by far the greatest source of energy consumption, waste production and greenhouse gas generation.

Interestingly, the first finding of this study is that research on building retrofitting hardly touches on this core fact. That is the holistic potential of building retrofitting as a strategic means to address 'global warming' goes largely undiscussed. This ought to be explored in greater depth.

What we do see, however, is a new and growing interest in the theme of retrofitting. Prior to 2011, publications on retrofitting averaged only a couple per year. However, over the last decade, publications have skyrocketed. Indeed, last year (2020), publications reached a current peak of 96 papers. Why this recent fast-growing interest? A qualitative analysis of the papers reveals that interest in retrofitting is essentially directed by the perceived need to find ways to offset rising energy costs, primarily in water heating and ambient residential cooling and heating. Consequently, the bulk of papers deal with evaluations of alternative systems that reduce energy costs while maintaining similar service delivery outcomes. It also appears that these elements are the focus because homeowners and occupiers drive the retrofit agenda in their effort to curb energy costs. Their awareness of rising energy costs is constrained by what is obvious to them: water heating and HVAC systems.

This is an important finding as much more can be done to improve the thermal and energy performance of residential buildings than simply replacing water heaters and aircon systems. This study reveals some of these other directions; but the scarcity of breadth of alternate topics, combined with the scant numbers of papers on these alternate topics, highlights the narrowness of focus on the recent burst of retrofitting publications seen this last decade. Candidate areas for further study, though touched on by the odd paper but in need of deeper examination, include ventilation, insulation, building envelopes, materials, air quality, weather protection and occupant behaviour.

Moreover, there are clearly many additional themes that would be worthwhile but have not evidently been pursued in residential retrofit. Given the continuing interest in climate change and the unlikeliness that rising energy costs and global warming concerns will abate anytime soon, these themes can be expected to be taken up presently. Indeed, we would commend interested researchers to champion any of the following areas to improve our understanding of how best to conduct residential retrofitting. Suggested emerging themes are:

- Government policy as it relates to retrofitting
- Retrofitting market and supply chains
- Comparisons in practice across national boundaries
- Comparisons across residential building types
- Comparisons across types and modes of retrofitting intervention
- Comparisons of proprietary system performance benchmarks
- Cost-benefit analysis of before and after retrofitting scenarios
- Examination of conduciveness on initial residential designs to absorb retrofits
- Materials, integrated systems and real-time performance measuring
- Assessment of existing energy rating systems to absorb and rate retrofitting
- Occupant values and preferences regarding optimisation of outcomes
- Impact of retrofitting strategies to mitigate effects of climate change

6. Conclusions

Climate change and global warming have been identified as the crisis of our times. The consensus appears to be that carbon emissions must be reduced drastically and soon if catastrophe is averted. Much of the focus regarding efforts to secure a sustainable future neglect recognising that it is the built environment and the building industry that contributes the most to pollution, waste, landfill, resource depletion and, above all, energy consumption and carbon emissions. Moreover, to the extent that the harm done by the built environment is acknowledged, most research has focused on the role played by commercial and industrial buildings. These are typically showcased as 'green buildings', all part of corporate efforts to promote 'virtue signalling' and 'social responsibility'. However, the bulk of the built environment is not commercial but residential. Moreover, while transitioning to more sustainable construction in new residential builds is both necessary and commendable, the inescapable reality is that the existing residential building stock continues to pose the greatest sustainability challenge. Specifically, the existing stock of residential buildings and their proclivity to consume huge amounts of energy and generate vast volumes of CO_2 pose the most significant single barrier to the sustainability goals of the world's nations.

Short of tearing down existing buildings, which of course is no solution at all, what then is to be done regarding the existing stock of residential buildings worldwide. The answer is to retrofit them and do so with the view of reducing energy consumption, either through efficiency gains or waste minimisation. There is significant research available in this realm and some critical publication outlets that support the mission. However, to date, the extent, scope and quality of that research has not been mapped, documented or described. This study delivers on that deficit. It presents a scientometric analysis of the research so far conducted in retrofitting residential buildings. It finds that the research theme has only recently gained any significant traction, with a significant publication surge evident from only 2017. Moreover, it reveals that while there are many potential research themes—from occupant behaviour to energy performance certification—the bulk of research is very circumspect in scope, being primarily limited to studies exploring trade-offs between energy usage reduction strategies thermal comfort. Finally, this study notes that there are only a limited number of academics working in this area, with most citations arising from only a few journals, of which 'Energy and Buildings' stand out as premiere.

This study has its limitations; specifically, it is only exploratory and general in nature and scope. More detailed analysis and appraisal of study findings in this field would be a logical next step to be undertaken.

Author Contributions: M.A. was responsible for literature review, data collection, software use, methodology, critical analysis, visualization and writing. I.M. was responsible for conceptualization, methodology, critical analysis and writing. M.B.L. was responsible for identifying the importance and relevance of this field of research. All authors have read and agreed to the published version of the manuscript.

Funding: This research received no external funding.

Institutional Review Board Statement: Not applicable.

Informed Consent Statement: Not applicable.

Conflicts of Interest: The authors declare that there is no conflict of interest.

References

1. McKibben, W. Winning Slowly Is the Same as Losing. Rolling Stone, 2017. Available online: https://www.rollingstone.com/politics/news/bill-mckibben-winning-slowly-is-the-same-as-losing-w512967 (accessed on 5 March 2021).
2. Ortiz, O.; Castells, F.; Sonnemann, G. Sustainability in the construction industry: A review of recent developments based on LCA. *Constr. Build. Mater.* **2009**, *23*, 28–39. [CrossRef]
3. Martek, I.; Hosseini, M.R.; Shrestha, A.; Zavadskas, E.K.; Seaton, S. The Sustainability Narrative in Contemporary Architecture: Falling Short of Building a Sustainable Future. *Sustainability* **2018**, *10*, 981. [CrossRef]
4. Masanauskas, J. *Melbourne's Continuing Population Boom Means Another 720,000 Homes Will Be Needed by 2031*; Herals Sun: Melbourne, VIC, Australia, 2014.
5. Australian Sustainable Built Environment Council. *A National Framework for Residential Ratings: Discussion Paper. 2016*; ASBEC: Darlinghurst, NSW, Australia, 2010.
6. Aye, L.; Hes, D. Green Building Rating System Scores for Building Reuse. *J. Green Build.* **2012**, *7*, 105–112. [CrossRef]

7. Building Energy Efficiency Task Group. *Building Energy Rating Schemes: Assessing Issues and Impacts*; International Partnership for Energy Efficiency Cooperation: Paris, France, 2014.
8. Jackson, S. A summary of urban assessment tools for application in Australia. In *Environment Design Guide*; Australian Institute of Architects: Potts Point, NSW, Australia, 2016.
9. Doan, D.T. A critical compariosn of green building rating systems. *Build. Environ.* **2017**, *123*, 243–260. [CrossRef]
10. Green Building Council of Australia. Available online: https://new.gbca.org.au/ (accessed on 8 April 2021).
11. Rashid, M.; Spreckelmeyer, K.; Angrisano, N. Green buildings: Environmental awareness and corporate image. *J. Corp. Real Estate* **2012**, *14*, 21–49. [CrossRef]
12. Tam, V.W.Y.; Zeng, S.X. Sustainable Performance Indicators for Australian Residential Buildings. *J. Leg. Aff. Disput. Resolut. Eng. Constr.* **2013**, *5*, 168–179. [CrossRef]
13. Harvey, L.D.D. Reducing energy use in the buildings sector: Measures, costs, and examples. *Energy Effic.* **2009**, *2*, 139–163. [CrossRef]
14. Hughes, P.J.; Shonder, J.A. The Evaluation of A 4000-Home Geothermal Heat Pump Retrofit at Fort Polk, Louisiana: Final Report. *OSTI* **1998**. [CrossRef]
15. Bell, M.; Lowe, R. Energy efficient modernisation of housing: A UK case study. *Energy Build.* **2000**, *32*, 267–280. [CrossRef]
16. Viridén + Partner. 2020. Available online: https://www.viriden-partner.ch/plus-nullenergiehaeuser (accessed on 9 April 2021).
17. Francisco, P.W.; Palmiter, L.; Davis, B. Modeling the thermal distribution efficiency of ducts: Comparisons to measured results. *Energy Build.* **1998**, *28*, 287–297. [CrossRef]
18. Voss, K. Solar energy in building renovation—Results and experience of international demonstration buildings. *Energy Build.* **2000**, *32*, 291–302. [CrossRef]
19. Chen, S.; Guan, J.; Levine, M.D.; Haiying, L.; Yowargana, P. Evaluation on Retrofit of One Existing Residential Building in North China: Energy Saving, Environmental and Economic Benefits. *Procedia Eng.* **2015**, *121*, 3–10. [CrossRef]
20. Friedman, C.; Becker, N.; Erell, E. Retrofitting residential building envelopes for energy efficiency: Motivations of individual homeowners in Israel. *J. Environ. Plan. Manag.* **2017**, *61*, 1805–1827. [CrossRef]
21. Aghili, N.; Amirkhani, M. SEM-PLS Approach to Green Building. *Encyclopedia* **2021**, *1*, 472–481. [CrossRef]
22. Amirkhani, M.; Garcia-Hansen, V.; Isoardi, G.; Allan, A. Innovative window design strategy to reduce negative lighting interventions in office buildings. *Energy Build.* **2018**, *179*, 253–263. [CrossRef]
23. Amirkhani, M. *Innovative Integrated Window Design with Electric Lighting Design System to Reduce Lighting Intervention in Office Buildings*; Queensland University of Technology: Brisbane, QLD, Australia, 2018.
24. Eom, J. China's building energy demand: Long-term implications from a detailed assessment. *Energy* **2012**, *46*, 405–419. [CrossRef]
25. OECD. *Cooperation, O.F.E. and Development, Energy Balances of Non-OECD Countries 2015*; OECD: Paris, France, 2015.
26. Union European. *EU Energy in Figures: Statistical Pocketbook 2018*; Publications Office of the European Union: Luxembourg, 2018.
27. Eichhammer, W.; Fleiter, T.; Schlomann, B.; Faberi, S.; Fioretto, M.; Piccioni, N.; Lechtenböhmer, S.; Schüring, A.; Resch, G. *Study on the Energy Savings Potentials in EU Member States, Candidate Countries and EEA Countries: Final Report for the European Commission Directorate-General Energy and Transport 2009*; EU Energy Poverty Observatory: Brussels, Belgium, 2009.
28. Paiho, S.; Seppä, I.P.; Jimenez, C. An energetic analysis of a multifunctional façade system for energy efficient retrofitting of residential buildings in cold climates of Finland and Russia. *Sustain. Cities Soc.* **2015**, *15*, 75–85. [CrossRef]
29. AE Smith. *City of Melbourne, 1200 Buildings: Melbourne Report Survey*; AE Smith: Brisbane, QLD, Australia, 2015.
30. Shah, S. *Sustainable Refurbishment*; Wiley-Blackwell: Hoboken, NJ, USA, 2012.
31. Zou, P.X. Retrofitting public buildings for energy and water efficiency. In *Sustainable Built Environment*; National Research Centre: Boulder, CO, USA, 2016.
32. Volt, J.; Fabbri, M.; Groote, M. Understanding Potential User Needs. In *A Survey Analysis of the Markets for Individual Building Renovation Roadmaps in Bulgaria, Poland and Portugal*; iBRoad: Europe, 2018; p. 106.
33. Web of Science. 2020. Available online: https://clarivate.com/webofsciencegroup/solutions/web-of-science/ (accessed on 2 February 2021).
34. Scopus. 2020. Available online: https://www.elsevier.com/solutions/scopus?dgcid=RN_AGCM_Sourced_300005030 (accessed on 2 February 2021).
35. Li, J.; Goerlandt, F.; Reniers, G. An overview of scientometric mapping for the safety science community: Methods, tools, and framework. *Saf. Sci.* **2021**, *134*, 105093. [CrossRef]
36. Chaomei, C. Science Mapping: A Systematic Review of the Literature. *J. Data Inf. Sci.* **2017**, *2*, 1–40.
37. Small, H.; Rorvig, M.E.; Lunin, L.F. Visualizing science by citation mapping. *J. Am. Soc. Inf. Sci.* **1999**, *50*, 799–813. [CrossRef]
38. Cobo, M.J.; Lopez-Herrera, A.G.; Herrera-Viedma, E.; Herrera, F. An approach for detecting, quantifying, and visualizing the evolution of a research field: A practical application to the Fuzzy Sets Theory field. *J. Inf.* **2011**, *5*, 146–166. [CrossRef]
39. Merschbrock, C.; Hosseini, M.R.; Martek, I.; Arashpour, M.; Mignone, G. Collaborative Role of Sociotechnical Components in BIM-Based Construction Networks in Two Hospitals. *J. Manag. Eng.* **2018**, *34*, 05018006. [CrossRef]
40. Cobo, M.; López-Herrera, A.; Herrera-Viedma, E.; Herrera, F. Science mapping software tools: Review, analysis, and cooperative study among tools. *J. Am. Soc. Inf. Sci. Technol.* **2011**, *62*, 1382–1402. [CrossRef]
41. Pritchard, A. Statistical bibliography or bibliometrics. *J. Doc.* **1969**, *25*, 348–349.

42. Henderson, M.; Shurville, S.; Fernstrom, K. The Quantitative Crunch: The Impact of Bibliometric Research Quality Assessment Exercises on Academic Development at Small Conferences. *Campus-Wide Inf. Syst.* **2009**, *26*, 149–167. [CrossRef]
43. Antonio-Rafael, R.-R.; José, R.-N. Changes in the Intellectual Structure of Strategic Management Research: A Bibliometric Study of the "Strategic Management Journal", 1980–2000. *Strateg. Manag. J.* **2004**, *25*, 981–1004.
44. Broadus, R.N. Toward a definition of "bibliometrics". *Scientometrics* **1987**, *12*, 373–379. [CrossRef]
45. Martín-Martín, A.; Malea, E.O.; López-Cózar, E.D. A novel method for depicting academic disciplines through Google Scholar Citations: The case of Bibliometrics. *Scientometrics* **2017**, *114*, 1251–1273. [CrossRef]
46. Nalimov, V.V.; Mulchenko, Z.M. *Measurement of Science. Study of the Development of Science as an Information Process*; National Technical Information Service: Springfield, VA, USA, 1971.
47. Hood, W.W.; Wilson, C.S. The Literature of Bibliometrics, Scientometrics, and Informetrics. *Scientometrics* **2001**, *52*, 291–314. [CrossRef]
48. Tague-Sutcliffe, J. An introduction to informetrics. *Inf. Process. Manag.* **1992**, *28*, 1–3. [CrossRef]
49. Morris, S.A.; Martens, B.V.D.V. Mapping research specialties. *Annu. Rev. Inf. Sci. Technol.* **2009**, *42*, 213–295. [CrossRef]
50. Nacke, O.I. Ein neuer Name für eine neue Disziplin. *Nachr. Dokum.* **1979**, *30*, 219.
51. Bar-Ilan, J. *Beyond Citations: Scholars' Visibility on the Social Web*; Cornell University: Ithaca, NY, USA, 2012.
52. Zhao, X. A scientometric review of global BIM research: Analysis and visualization. *Autom. Constr.* **2017**, *80*, 37–47. [CrossRef]
53. He, Q.; Wang, G.; Luo, L.; Shi, Q.; Xie, J.; Meng, X. Mapping the managerial areas of Building Information Modeling (BIM) using scientometric analysis. *Int. J. Proj. Manag.* **2017**, *35*, 670–685. [CrossRef]
54. Ren, R.; Hu, W.; Dong, J.; Sun, B.; Chen, Y.; Chen, Z. A Systematic Literature Review of Green and Sustainable Logistics: Bibliometric Analysis, Research Trend and Knowledge Taxonomy. *Int. J. Environ. Res. Public Health* **2019**, *17*, 261. [CrossRef] [PubMed]
55. Wang, Z.-Y.; Li, G.; Li, C.-Y.; Li, A. Research on the semantic-based co-word analysis. *Scientometrics* **2011**, *90*, 855–875. [CrossRef]
56. Leung, X.Y.; Sun, J.; Bai, B. Bibliometrics of social media research: A co-citation and co-word analysis. *Int. J. Hosp. Manag.* **2017**, *66*, 35–45. [CrossRef]
57. Small, H. Co-citation in the scientific literature: A new measure of the relationship between two documents. *J. Am. Soc. Inf. Sci.* **1973**, *24*, 265–269. [CrossRef]
58. Shiau, W.-L.; Dwivedi, Y.K.; Yang, H.S. Co-citation and cluster analyses of extant literature on social networks. *Int. J. Inf. Manag.* **2017**, *37*, 390–399. [CrossRef]
59. Yang, S.; Wang, F. Visualizing information science: Author direct citation analysis in China and around the world. *J. Inf.* **2015**, *9*, 208–225. [CrossRef]
60. Van Eck, N.J.; Waltman, L. Citation-based clustering of publications using CitNetExplorer and VOSviewer. *Scientometrics* **2017**, *111*, 1053–1070. [CrossRef] [PubMed]
61. Chen, C. CiteSpace II: Detecting and visualizing emerging trends and transient patterns in scientific literature. *J. Am. Soc. Inf. Sci. Technol.* **2006**, *57*, 359–377. [CrossRef]
62. Chen, C.; Ibekwe-SanJuan, F.; Hou, J. The Structure and Dynamics of Co-Citation Clusters: A Multiple-Perspective Co-Citation Analysis. *J. Am. Soc. Inf. Sci. Technol.* **2010**, *61*, 1386–1409. [CrossRef]
63. Vliet, G.C. *Retrofitting of an Apartment Building for Solar Heating, Cooling and Hot Water*; Regulatory Processes of the US Department of Energy: Dearborn, MI, USA, 1979; pp. 387–401.
64. Ding, Y.; Rousseau, R.; Wolfram, D. (Eds.) *Measuring Scholarly Impact Methods and Practice*, 1st ed.; Springer International Publishing: Cham, Switzerland, 2014.
65. Zhang, P.; Liu, Y.; Yu, F.-X.; Wu, G.-W.; Li, M.-Y.; Wang, Z.; Ding, H.-Y.; Wang, L.-X.; Zhao, K.-X.; Zhang, Z.-Y.; et al. Hierarchical integrated processing of reward-related regions in obese males: A graph-theoretical-based study. *Appetite* **2021**, *159*, 105055. [CrossRef] [PubMed]
66. Opsahl, T.; Agneessens, F.; Skvoretz, J. Node centrality in weighted networks: Generalizing degree and shortest paths. *Soc. Netw.* **2010**, *32*, 245–251. [CrossRef]
67. Newman, M.E.J. Modularity and community structure in networks. *Proc. Natl. Acad. Sci. USA* **2006**, *103*, 8577–8582. [CrossRef]
68. Rousseeuw, P.J. Silhouettes: A graphical aid to the interpretation and validation of cluster analysis. *J. Comput. Appl. Math.* **1987**, *20*, 53–65. [CrossRef]
69. Deb, K. A fast and elitist multiobjective genetic algorithm: NSGA-II. *IEEE Trans. Evol. Comput.* **2002**, *6*, 182–197. [CrossRef]
70. Noris, F.; Adamkiewicz, G.; Delp, W.W.; Hotchi, T.; Russell, M.; Singer, B.C.; Spears, M.; Vermeer, K.; Fisk, W.J. Indoor environmental quality benefits of apartment energy retrofits. *Build. Environ.* **2013**, *68*, 170–178. [CrossRef]
71. Asadi, E.; da Silva, M.C.G.; Antunes, C.H.; Dias, L.; Glicksman, L. Multi-objective optimization for building retrofit: A model using genetic algorithm and artificial neural network and an application. *Energy Build.* **2014**, *81*, 444–456. [CrossRef]
72. Ma, Z.; Cooper, P.; Daly, D.; Ledo, L. Existing building retrofits: Methodology and state-of-the-art. *Energy Build.* **2012**, *55*, 889–902. [CrossRef]
73. Kavgic, M.; Mavrogianni, A.; Mumovic, D.; Summerfield, A.; Stevanovic, Z.; Djurovic-Petrovic, M. A review of bottom-up building stock models for energy consumption in the residential sector. *Build. Environ.* **2010**, *45*, 1683–1697. [CrossRef]
74. Ballarini, I.; Corgnati, S.P.; Corrado, V. Use of reference buildings to assess the energy saving potentials of the residential building stock: The experience of TABULA project. *Energy Policy* **2014**, *68*, 273–284. [CrossRef]

75. Nic, R.; Mark, J. Combining Top-Down and Bottom-Up Approaches To Energy-Economy Modeling Using Discrete Choice Methods. *Energy J. Camb. Mass.* **2005**, *26*, 83–106.
76. Nair, G.; Gustavsson, L.; Mahapatra, K. Factors influencing energy efficiency investments in existing Swedish residential buildings. *Energy Policy* **2010**, *38*, 2956–2963. [CrossRef]
77. Dowson, M.; Poole, A.; Harrison, D.; Susman, G. Domestic UK retrofit challenge: Barriers, incentives and current performance leading into the Green Deal. *Energy Policy* **2012**, *50*, 294–305. [CrossRef]
78. Sunikka-Blank, M.; Galvin, R. Introducing the prebound effect: The gap between performance and actual energy consumption. *Build. Res. Inf.* **2012**, *40*, 260–273. [CrossRef]
79. Mata, É.; Kalagasidis, A.S.; Johnsson, F. Energy usage and technical potential for energy saving measures in the Swedish residential building stock. *Energy Policy* **2013**, *55*, 404–414. [CrossRef]
80. Waltman, L.; Eck, N.J. A new methodology for constructing a publication-level classification system of science. *J. Am. Soc. Inf. Sci. Technol.* **2012**, *63*, 2378–2392. [CrossRef]
81. Shibata, N.; Kajikawa, Y.; Takeda, Y.; Matsushima, K. Detecting emerging research fronts based on topological measures in citation networks of scientific publications. *Technovation* **2008**, *28*, 758–775. [CrossRef]

energies

Review

Radiant Conditioning Retrofitting for Residential Buildings

Hung Q. Do [1], Mark B. Luther [1,*], Mehdi Amirkhani [2], Zheng Wang [1] and Igor Martek [1]

1 School of Architecture and Built Environment, Deakin University, Geelong, VIC 3220, Australia; qhdo@deakin.edu.au (H.Q.D.); wangzheng@deakin.edu.au (Z.W.); igor.martek@deakin.edu.au (I.M.)
2 UniSA Online, Science Technology Engineering and Mathematics (STEM), University of South Australia, Adelaide, SA 5000, Australia; mehdi.amirkhani@unisa.edu.au
* Correspondence: mark.luther@deakin.edu.au

Abstract: In order to achieve Australia's greenhouse gas emissions reduction targets, a majority of the existing residential building stock in Australia will require retrofitting in favour of energy-efficient solutions. This paper considers retrofitting for conditioning to be one of the most straightforward and offers the greatest potential to deliver significant comfort and energy-saving results. Radiant conditioning systems are not new, yet some game-changing innovations have taken place over the last decade that may require an entire paradigm shift in the manner we condition our buildings. The reiteration of the principle 'thermally active systems' suggests that our buildings need to accommodate these systems into the fabric of building components. However, extremely few products and/or innovative solutions for doing such seem to be provided by the industry. We seem incompetent with solutions that are not costing the Earth, insulating, lightweight, and offering an instant response time to conditioning. We still have the concept embedded in our minds that radiative systems consist of heavy 'combat' construction with time lags of a day or two and that they are very costly to implement, especially if we are to retrofit a project. The purpose of this paper is to rectify and change our understanding of radiant systems, namely through a review of the existing technology and its recent advancements. It intends to introduce the fact that radiant systems can become highly reactive, responsive, and thermally dynamic conditioning systems. Lightweight radiant systems can be 40% more energy-efficient than common air conditioners and can respond in less than 15 min rather than in the hours required of heavy radiant systems. Thus, an insulated, lightweight radiant system is ideal for retrofitting residential buildings. Furthermore, this paper supports and introduces various systems suited to retrofitting a residential building with hydronic radiant systems.

Keywords: radiant system; conditioning system; residential building; retrofitting; thermal comfort

Citation: Do, H.Q.; Luther, M.B.; Amirkhani, M.; Wang, Z.; Martek, I. Radiant Conditioning Retrofitting for Residential Buildings. *Energies* **2022**, *15*, 449. https://doi.org/10.3390/en15020449

Academic Editor: Fabrizio Ascione

Received: 21 November 2021
Accepted: 1 January 2022
Published: 10 January 2022

1. Introduction: Retrofitting Conditioning Systems in Residential Buildings

1.1. The Benefits of Retrofitting Condition Systems

Australia has set a target of achieving a 26–28% reduction in greenhouse gas emissions by 2030 from the 2005 levels, and 25% of this targeted reduction needs to be achieved by improving energy efficiency in the building sector [1]. Meanwhile, Australia was slow in adopting energy-efficient building regulations, leading to many housing projects that were built without any energy efficiency standards [1]. This has led to a gigantic need for retrofitting residential housing in Australia with new energy-efficient technologies. To bring forth maximum effectiveness, the retrofitting should be low cost and easy to be implemented [2]. Aside from meeting the sustainable national target, energy-efficient retrofitting housing can benefit occupants with lower energy costs and improved well-being.

Ma et al. [3] reviewed studies on the most modern methods of improving residential building thermal and energy performance. Among the studies on residential building retrofitting, envelope system refurbishment and mechanical retrofitting prove most popular. Several attempts have also been made to improve the envelope of old residential housing in Australia. For example, Wilkinson and Feitosa [2] want to retrofit Australian housing with green roofs, while Bulut et al. [1] improved windows with secondary glazing. Jamil et al. [4] looked at the internal system and installation material used in ceilings to stabilise the thermal conditions of residential housing. In most cases, research papers on housing retrofitting focus on improving the building envelope system as a whole or components of the envelope such as the wall, roof, glazing, or shading systems. This is the mainstream and follows the basic principle of the retrofitting process, reducing passive energy transfer between the buildings and the external environment.

However, this research paper considers a hitherto neglected approach to retrofitting: conditioning retrofitting. The conditioning systems, without a doubt, are a critical component in modern residential buildings and, yet, one of the most problematic. In most states in Australia, air conditioners cause a severe problem to the electricity network, as they are the main reason for peak energy consumption [5]. The energy usage of air conditioners is the biggest, making up 40% of the total energy consumption in Australian housing [5]. Meanwhile, even a small adjustment to the control systems of commonly used conditioning systems can significantly save energy use and cost [6]. Additionally, while new housing projects in Australia have to meet a high energy efficiency standard and benefit from new technologies, older residential houses built before 1980 do not [5]. This results in poorer energy performance, especially for conditioning systems in older residential houses in Australia [5]. Additionally, Willand, Maller, and Ridley [7] conducted a study on the retrofitting of residential buildings in Melbourne and concluded that even small retrofits could bring forth notable improvements in residential building energy and thermal performance, reducing energy costs [7]. Hence, there is a need to retrofit the conditioning systems in Australian residential buildings, especially those built prior to 1980.

Over the past 100 years, convective systems have become the primary condition system in modern buildings worldwide with the development of mechanical ventilation. These systems in which air is used as the heat transfer medium for conditioning required spaces make up the Heating, Ventilating, and Air-conditioning (HVAC) industry [8]. While these conditioning systems can be successful for thermally heating and cooling, they are not necessarily the most economical and practical principle in producing thermal comfort [8,9]. For example, decreasing the cooling temperature by 1 °C will increase the cooling energy consumption by about 8–10% [10]. Miriel et al. [11] argue that, in the cooling mode, radiation accounts for about two-thirds of the heat transfer between the human body and room environment, while, in the heating mode, 80% of the total heat transfer is radiation. In short, radiant systems replace convection heat transferring between a human body and the surrounding environment with radiation. Additionally, many research papers have proven the merits of radiant systems over traditional HVAC in energy and comparable thermal comfort performances [12]. These improved performances and advantages over HVACs have promoted radiant systems in most building sectors, including residential, commercial, education, and even industrial buildings of all sizes [13].

1.2. Comparison of Energy Consumption of Radiant Systems with Existing HVAC Systems

While mechanical conditioning systems have become fundamental in ensuring thermal comfort in modern buildings, conventional HVAC systems are highly energy-intensive [14]. The reason for this low performance is the inefficient heat-transferring mechanism that traditional air conditioners rely on [8]. Conventional air conditioning systems use air as the heat-transferring medium and convection as the primary heat exchanging mode [13]. According to Moe [8], low-density air is a poor heat-transferring medium, while only about 27% of the human body heat transfer is via convection. Conventional HVAC systems are based on disadvantageous elements that provide high thermal and energy performance setbacks. Additionally, air conditioning systems with large supply air volumes are also burdened by several limitations, such as noise, drafts, and poor temperature gradients [10,14].

Moreover, the heat-transferring mode of radiant systems has been shown to be superior. For example, Feustel and Stetiu [15], after running a numerical computer model, concluded that radiant cooling ceilings offer advantages over air conditioning systems, despite their limited capacity. They suggested that hydronic radiant systems can significantly reduce the amount of energy used in transporting heat and air by using water as a heat-transferring medium, resulting in a 40% more energy-efficient system.

Imanari and Omori [16] conducted a numerical simulation comparing the annual energy consumption between two systems used by an office building in Tokyo, Japan. The results showed 10% energy savings for the radiant ceiling. These savings confirmed that radiant systems in the context of office buildings are proven to be financially viable, with the payback period being between 1 and 17 years [16]. The US Technology Department has published three technical support documents recommending radiant conditioning systems as an alternative to HVAC, quoting savings of up to 50% in energy consumption across small, medium, and large office buildings [12]. Hao et al. [17] investigated the performance of radiant cooling ceilings combined with displacement ventilation and desiccant dehumidification, using simulations in the context of hot, humid summers in Beijing, China. The radiant cooling system was compared to a conventional convective system, utilising a mere 8% total energy consumption. That is, the radiant system generated savings of up to 68.5% of chiller and 39.0% of fan energy systems. Hence, on the whole, the radiant system proved to be more energy-efficient than the convective system, even when the radiant system consumed 25.6% more energy for pumping and about four times more energy for the boiler.

2. Research Methodology

2.1. Literature Review

This study is conducted based on a literature review of radiant systems. It confirms their potential as a preferred approach to the conditioning retrofitting of residential buildings. As mentioned, retrofitting residential buildings is an emergent field of study [3]. However, most studies on this subject have been conducted using simulations or instrumental experiments. For example, Jamil et al. [4] utilised simulations and instrumental experiments, while the studies conducted by Bulut et al. [1] and Wilkinson and Feitosa [2] were only conducted using instrumental experiments. Nevertheless, the knowledge in this field is rather well-developed [3]. Radiant systems have also been comprehensively studied since the 1960s [18]. Moreover, these studies have been reviewed in several high-quality publications over the last six years. Despite that, the potential for radiant systems to service the conditioning retrofitting of residential buildings has yet to be comprehensively appraised.

A review study was conducted by Rhee and Kim on radiant systems, revealing 50 years of knowledge on radiant systems since the 1960s [18]. Rhee, Olesen, and Kim [19] and Karmann, Schilevel of thermal comfortavon, and Bauman [12] also provided valuable overviews on the body of knowledge on radiant systems. The review studies on retrofitting residential housing by Ma et al. [3] also had similar values categorising the methods of housing retrofitting. Thus, a literature review serves as an appropriate initial approach in clarifying the application of radiant systems for the retrofitting of residential buildings. This literature review study can also lay the foundation for prototype development and further instrumental experiments.

As such, this research paper will investigate the capabilities of various types of radiant systems and the potential optimum application and control of hydronic radiant systems in retrofitting residential buildings. It also proposes an innovative hydronic radiant system based on the literature review. This research paper explains the principle of thermal comfort and the nature behind the benefit of radiant conditioning for housing project owners and conditioning systems manufacturers, as well as architects. Based on such knowledge, existing convective conditioning systems in residential housing can be supplemented and/or be retrofitted with radiant condition systems for better thermal and energy efficiency results.

2.2. Review Method

This research paper was primarily conducted with qualitative data gathered from previous publications since the 1960s. This was also the period applied in the aforementioned high-quality review radiant systems paper of Rhee and Kim [18]. A breakthrough research paper in radiant cooling was published in Australia during this period. The material was searched on Google Scholar linked with our university library. The keywords applied in this research were "residential building energy consumption", "energy retrofitting residential buildings", "radiant conditioning", "radiant cooling and heating", "residential air conditioning systems", "radiant panel", "thermally active building systems", and "mean radiant temperature". Several searches were done with Australian context or keywords such as "thermal comfort", "control", "measuring", "condensation prevention", and "ventilation".

Two hundred and twenty-four research papers were gathered and catalogued according to their research topics and contents. Around 80 articles were marked as highly relevant, containing valuable information used in this study. Recently, published papers and review papers have been prioritised and used to trace back to earlier research papers. The REHVA Handbook Number 7, EN ISO 7730, and ASHRAE 55 2017 are critical essential standards and guidelines for thermal comfort and conditioning principles used in this research.

2.3. Research Design and Findings

The research approach of this paper was to provide and explain the existing conditioning systems at present, which are primarily based upon convective conditioning. Therefore, it was essential to introduce and consider the mechanisms of conditioning and differences from those that provide radiative conditioning. To clarify and evaluate different conditioning mechanisms, the principle of thermal comfort was revisited and discussed; specifically, the role of the mean radiant temperature (MRT) in regard to comfort was reviewed. Additionally, this paper focused on stand-alone houses that make up the majority of Australian housing.

A review of the various types of hydronic systems is briefly presented in terms of the past problems in conditioning control, and the practicality of installation in retrofitting is considered. Based upon the collected information of recently introduced lightweight products, the capillary hydronic conditioning system is pursued. Preliminary empirical testing and actual installations are presented. A relatively new principle of hydronic cooling from the ceiling is introduced, with its associated concern of condensation. Finally, innovation and breakthrough discovery from the 1960s is revisited as a modularised hydronic panel system.

The data gathered from the previous studies are categorised and analysed to shed light on the salient concerns. What is the difference between new radiant systems and conditioning systems commonly used in residential buildings in Australia? By what mechanisms do radiant systems provide thermal comfort? What are the types of radiant systems, and how well do they perform? What are their merits and limitations? Answering these questions allows a determination on the most appropriate evident configuration of radiant systems in the retrofitting of residential buildings.

3. Common Conditioning Systems in Australian Houses

As mentioned above, air conditioners are the most commonly used systems providing thermal comfort for residential housing in various world regions, such as China and the USA, while the demand for air conditioners is dramatically increasing in China, India, Indonesia, and Brazil [20]. The most frequently applied conditioning systems within the residential sector in Australia include the window-mounted or wall-mounted direct expansion A/C unit (Figure 1) [5]. These have traditionally provided both heating and air-conditioning (cooling) of a single room or zone within a house. Over the last two decades, these systems have been gradually replaced by split systems that consist of an external heat pump compressor and an internal wall-mounted fan (cassette) (Figure 2). The split system is also one of Australia's most widely used conditioning systems [5]. Less popular systems in Australia's residential buildings are ducted (centralised conditioning systems), evaporative conditioners, and portable conditioners [5].

Figure 1. Wall-mounted direct expansion A/C. Source: First author.

Figure 2. Split systems air conditioner: (**A**) Indoor unit—Evaporator; (**B**) Outdoor unit—Condenser. Source: First author.

Other popular systems, used in heating only, are wall-mounted gas burners (Figure 3A) and completely ducted HVAC systems. It is not common to have both heating and cooling introduced within an HVAC ducted system. Further to these heating approaches, radiant panels, both hydronic and electric (Figure 3B,C), have recently become popular. Hydronic systems often apply a gas or diesel-fuelled boiler. It is important to note that these conditioning systems primarily condition occupants under the principle of convection [21,22]. In other words, they condition the air in a space.

In the past and, again, more recently, residential buildings in Australia have applied electric floor heating (Figure 4). These systems use a mat of electric heating cable that runs underneath the finished floor, turning the floor into a thermally active surface that radiates heat.

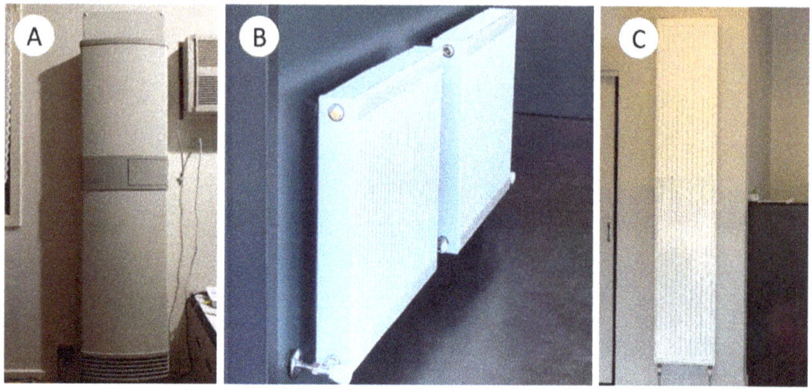

Figure 3. (**A**) A convective heater. (**B**) A hydronic radiator. (**C**) A modern hydronic radiator panel. Source: First author.

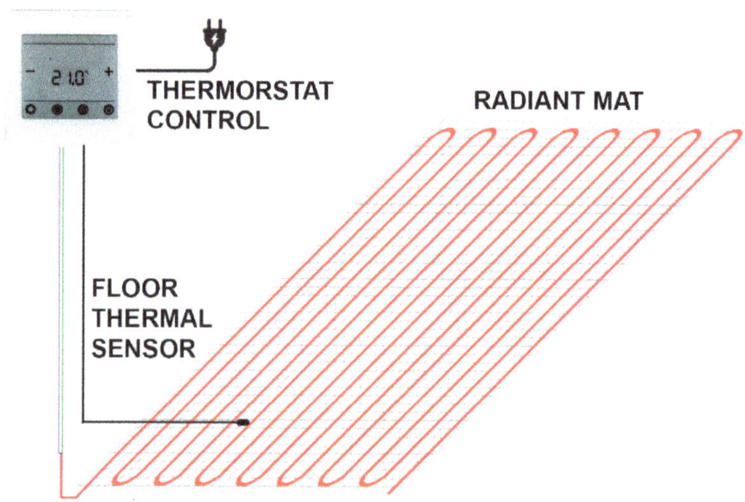

Figure 4. An electric floor heating system. Source: First author.

A radiant system mainly utilises the radiant heating transferring mechanism rather than air temperature conditioning. Heat can be radiated directly between the occupants and the radiant surfaces. Hydronic radiant systems use water as the heat transfer medium from the heat sources or sink to their radiant surfaces. A hydronic radiant system pumps hot or cold water through a series of pipes embedded inside or attached to the wall, floor, or ceiling [23] (Figure 5). This will turn the wall, floor, or ceiling surfaces into "thermally active surfaces" for radiating heat or absorbing heat [8]. The radiant surfaces will also have a heat conduction effect on the air in contact with them and thus promote convective heat transfer. For Australian housing, which mainly consists of separated houses [24], the commonly used heat sources can be a gas boiler or heat pump. Figure 5 shows a radiant heating and cooling system with a heat pump as the heat source and sink. Additionally, the heat pump can run on renewable energy technologies such as solar panels.

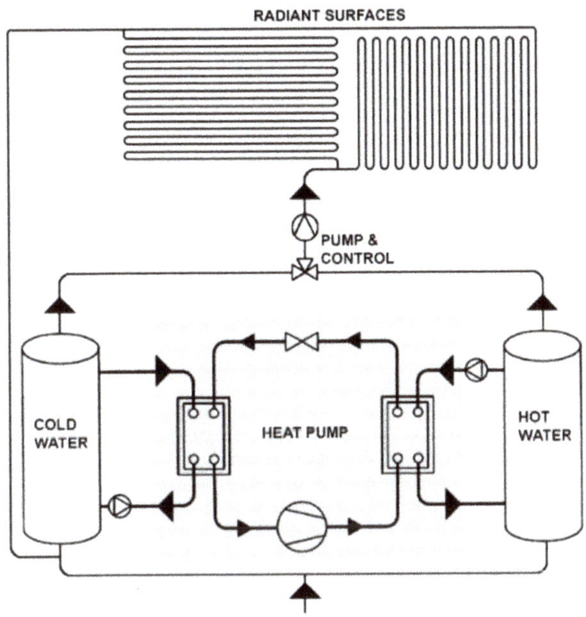

Figure 5. Radiant system diagram. Source: Fourth author.

The capillary tube mat is a recently introduced configuration of hydronic radiant systems. This type of radiant system has more flexibility, simplifies installation, and localises energy performance compared to HVAC systems or other types of radiators [25–27]. Radiant systems have become ubiquitous in Northeast Asia and Europe [23]. While, in the past, radiant systems were not yet widespread in North America, recently, the market for radiant systems in this region has shown rapid growth [23]. There has been no significant research on Australia's radiant systems in the last few decades [18]. In general, radiant systems can be considered superior to convective systems, both in terms of thermal and energy performances [12]. Furthermore, Australian houses exhibit far more air leakage than European houses. This leakage's effectiveness inherently jeopardises convective conditioning systems. Hence, the prospect of replacing conventional HVAC with radiant systems could lead to better thermal comfort and reduce energy costs.

4. Thermal Comfort

Thermal comfort can be defined as "the condition of the mind in which satisfaction is expressed with the thermal environment" [28]. However, several other definitions for thermal comfort have been pointed out and even debates around how thermal comfort can be defined [29]. The problem is that the mechanism of being "thermally comfortable" is highly complicated and affected by various factors, such as physiology, psychology, and behaviour [30]. Hence, several comfort models are used for different environments [31].

According to Fanger [32], thermal comfort can be quantitatively measured using six parameters, namely air velocity (Av), mean radiation temperature (MRT), air temperature (dry-bulb temperature—DB), air relative humidity (RH), metabolic rate (met), and clothing insulation (clo). Combining these parameters indicates thermal comfort in a controlled and conditioned environment. The level of thermal comfort (or discomfort) of a group of people is rated via predicted mean vote (PMV) index scaling from −3 (cold) to +3 (hot) [33]. Along with PMV, the index of the predicted percentage of dissatisfied (PPD) indicates the proportion of thermally dissatisfied people in the group. This principle makes up the classic PMV (Fanger) thermal comfort model shown in Figure 6.

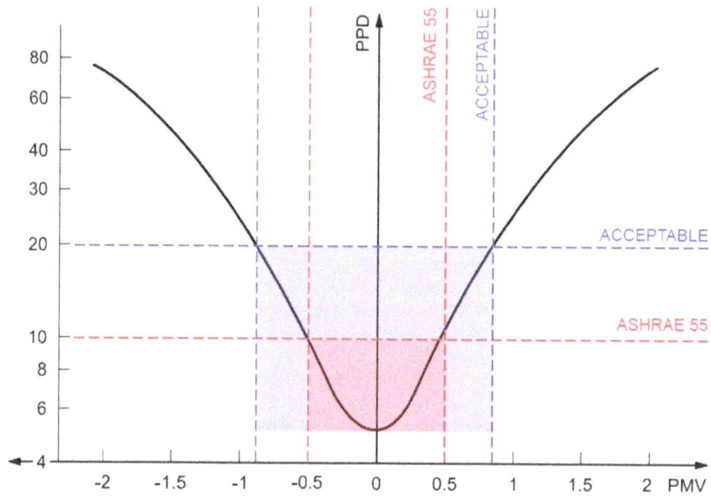

Figure 6. PMV—Fanger thermal comfort models. Source: Redrawn by the first author.

As the most classical thermal comfort model, the PMV thermal comfort model is commonly used in standards or guidelines such as ISO 7730, ASHRAE 55 (standard of The American Society of Heating, Refrigerating and Air-Conditioning Engineers) [28], or the REHVA (Federation of European Heating, Ventilation, and Air Conditioning Associations) Guidebook Number 7 [34]. The PMV thermal comfort model is well-suited and commonly used to assess thermal comfort in conditioned office spaces [35]. Loveday et al. [36] conducted an experiment proving that the PMV model accurately assesses thermal comfort in a radiant conditioned environment and concluded that a radiant conditioned space is similar to an air-conditioned space. According to ASHRAE 55, a thermal environment with an acceptable comfort level would have a PMV between −0.5 and +0.5, resulting in a PPD of less than 10% (Figure 6) [28]. However, the Fanger PMV comfort model can be considered conservative [37], and acquiring a PPD of less than 10% is not common [28]. Hence, the PPD of less than 20% (PMV between −0.8 and +0.8) is also welcomed as an acceptable level of thermal comfort (Figure 6) [28,37].

Mean Radiant Temperature (MRT)

MRT is a crucial parameter determining thermal comfort [14,28,38]. It is defined as "the uniform surface temperature of an imaginary black enclosure in which the radiation from the occupant equals the radiant heat transfer in the actual non-uniform enclosure" [39]. Stefan Boltzmann defined MRT as "a temperature value to evaluate the total radiation flux over the human body" [40]. MRT is the relationship between indoor surface temperature, direct and scattering solar radiation power, view factor toward transparent and translucent surfaces, and solar radiation receiving area of the occupant body [41]. MRT is the most complex parameter of thermal comfort, often difficult to control, calculate, and measure [41,42]. Yet, it is important to note that radiant systems function on the principle of MRT, not air temperature (dry-bulb). This is the crucial problem in radiative system control, which functions on the control principle of a thermostatic sensor (air temperature).

MRT was assumed to be equal to the air temperature in many early research papers in space conditioning [39,43,44]. However, this is not the case on many occasions, especially when solar heat gain is introduced [28,45]. In residential buildings, access to sunlight is a fundamental requirement. Thus, MRT will not be equal to the air temperature in residential buildings. Instead, MRT is a better index to represent occupants' thermal sensation than air temperature [38,46]. Hence, in retrofitting residential buildings, MRT control should be prioritised.

Sui and Zhang [10] investigated the relationship between MRT and other thermal comfort parameters, namely air temperature, air velocity, and air humidity. The research proved the significant influence of MRT on thermal comfort. Sui and Zhang [10] illustrated that thermal comfort can be achieved even at high air temperatures with a low MRT. This allows the concept of high air temperature cooling and low air temperature heating to occur in a room. Air temperature and MRT are the two most critical aspects determining thermal comfort [34]. Radiant conditioning focuses on changing MRT via the internal surface temperature rather than changing the air temperature. Hence, cooling can be achieved even with high air temperatures, while heating can be provided with low air temperatures [34]. Additionally, Sui and Zhang [10] argued that, while a low air velocity means reductions in convection and evaporation, MRT can play a significant role to improve thermal comfort. Their research also indicated that the higher the humidity, the lower the MRT.

Figure 7 charts an investigation in achieving thermal comfort performed by Luther, Tokede, and Lui [37], applying similar principles as Sui and Zhang [10]. The circle and the triangle represent thermally unsatisfactory conditions, and there are three options to reach the acceptable level of thermal comfort (PPD = 20%). Option A raises the air temperature by about 3.5 °C, which can increase the conditioning energy cost by 30–45% [25]. Option B increases the MRT by 4.5–23.5 °C (for the circle) and provides heating while the air temperature remains at 20.5 °C. This can be achieved with the radiant surface's temperature at 30–35 °C at a minimum cost [25]. Hence, the radiant system is more energy-efficient. Nevertheless, option C raises the air temperature and MRT, the shortest path, although the energy consumption will be higher than option B.

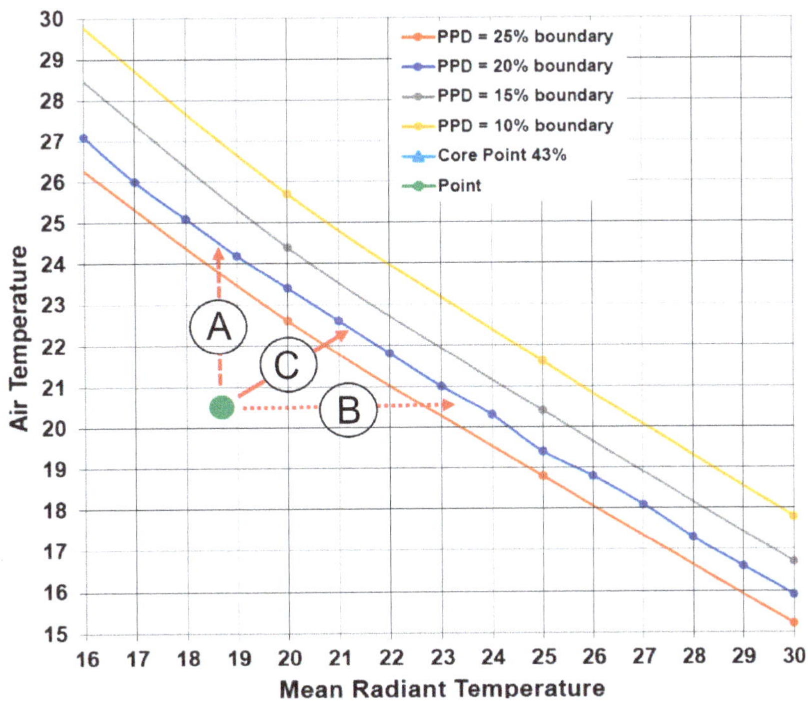

Figure 7. Air temperature vs. mean radiant temperature concerning PPD [25]. Source: Redrawn by the second author.

In a uniform air temperature environment, nonuniform MRT can cause thermal discomfort [36]. In turn, with uniform MRT, local thermal discomfort can be avoided even with high air temperature gradients [36]. Hence, MRT has more effect on human thermal comfort than the air temperature itself [47].

MRT can be calculated using the surface's temperature and view factor (angle factor) between an occupant and internal surfaces. The equation is [48]:

$$T_r^4 = T_1^4 \times F_{p-1} + T_2^4 \times F_{p-2} + \ldots + T_N^4 \times F_{p-N} \quad (1)$$

where

T_r = mean radiant temperature, K;

T_N = surface temperature of surface N, K;

F_{p-N} = view factor (angle factor) between a person and surface N.

The view factor is the proportion of energy radiated from a surface and received by another surface [49]. The view factor represents the radiant heat exchanges between two surfaces [50]. The total view factor toward an occupant in a room equals 1 [34]. The view factor is affected by the surface shapes, angle, and distance between the two surfaces [49]. Hence, in thermal comfort, the view factor depends on the dimensions of the radiant surfaces and the distance to the occupant. The view factor for a person in a room also depends on the posture and position [51]. Although determining the view factor is highly complicated, two conventional methods of approximately determining the view factor's value are the Fanger-Rizzo method and the Nusselt Analog method [52]. While the Fanger-Rizzo method is used in fundamental documents ASHRAE and the REHVA Handbook [34,48], the Nusselt method is not. This is an important factor in designing radiant systems for retrofitting residential buildings, especially where the systems can be mounted for maximum effectiveness.

5. Hydronic Radiant Systems

5.1. Types of Hydronic Radiant Systems

The Federation of European Heating, Ventilation, and Air Conditioning Associations (REHVA) Guidebook Number 7 classifies radiant conditioning systems into three types based on how the piping is installed [34]. Table 1 shows the three types of radiant systems, including:

1. Embedded surface system (ESS), with the tubing installed in a wall, floor, or ceiling layer but isolated from the main building structure.
2. Radiant panel (RP), where the tubing runs within prefabricated lightweight panels.
3. Thermal active building system (TABS), with the tube installed directly inside the building structure.

While the ESS and RP can be used in new constructions and retrofitting projects, TABS can only be used in new buildings. Two tube types are currently used in radiation systems, including small diameter tubes (metal or plastic tubes) and polyethylene capillary tube mats [53]. Hence, the standard ISO-11855 listed seven types of radiant systems [25]. In this category, the ESS is divided into four subtypes depending on the tubing position (type A–C), type F with a capillary tube, and type G with a wooden structure, while type E is identical to TABS.

Table 1 illustrates three different radiant hydronic systems. The thermal active building system (TABS) is the early radiant system that dates back to the 1930s [23]. It considers one substantial thermal mass that requires conditioning. This system can be relatively costly to run, as it stores a great deal of energy within the building fabric and is not reactive to sudden climate changes. TABS also does not show a specific insulation layer, and the hydronic tubing is located deep inside the structure. This results in potential thermal and energy performance reductions. Several research papers have revealed the problems of a heavy system design like TABS. As the sentinel load is stored and released by the thermal mass, the properties of the structure, such as thickness or thermal conductivity of the slab material, have a significant impact on the heat transfer process and thermal performance of this heavy system [54]. This heavy thermal mass results in a slow reaction time, potential overheating, and low energy efficiency [18]. While the thermal mass can be advantageous in maintaining the temperature of radiant surfaces, receiving an extra external load like solar heat gain can cause overheating or large temperature swings [55]. Heavy thermal mass can also reduce the efficiency of radiant conditioning systems, as energy is wasted when space is not occupied or wasted on a structure mass rather than contributing directly to the conditioning of the required environment [56]. In terms of reaction time, an experiment with radiant slab floor done by Merabtine et al. [57] showed that the system surface needs about 4 h to be heated up from 20 °C to 29 °C. Hence, this heavy system requires an advanced control system involving the water flow rate, inlet water temperature, and reaction time [18]. Additionally, the slow reaction time can increase the risk of condensation in the cooling mode if the space's humidity increases faster than the system response [11].

Table 1. Types of hydronic radiant systems [11]. Source: Redrawn by the first author.

System Types	Schematic	Classify
Thermal active building system (TABS)	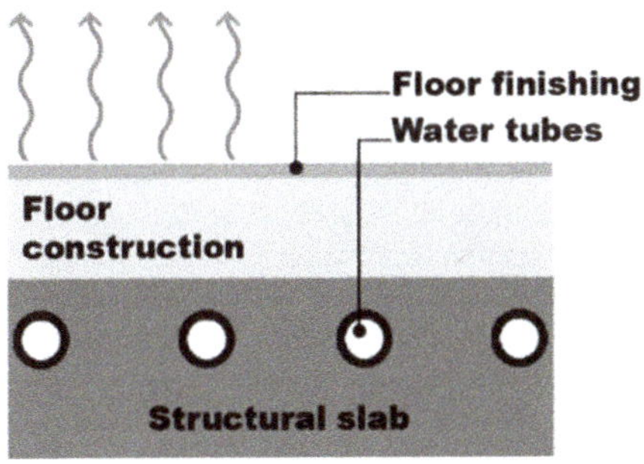	Heavyweight radiant system
Embedded surface system (ESS)	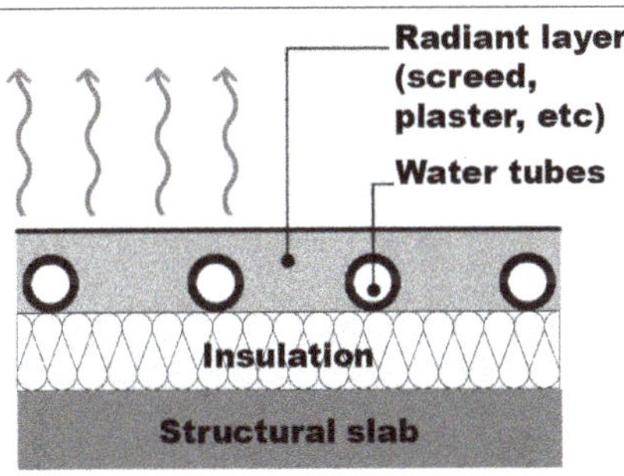	Medium-weight radiant system
Radiant panel (RP)	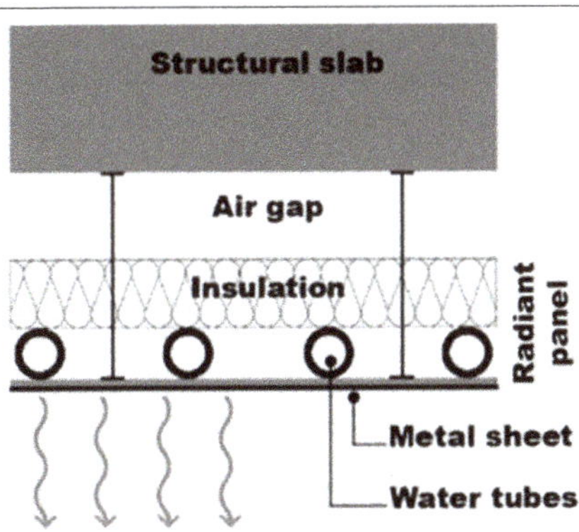	Lightweight radiant system

In contrast to the thermal active building system (TABS), the embedded surface system (ESS) has smaller thermal mass components. The radiant hydronic tubes are isolated from the main structure mass and are closer to the room surface. The ESS is a lighter system than TABS, with less thermal mass resulting in lower thermal inertia [11], which means a faster reaction time and less energy waste. The system is also isolated from the main structure. The insulation layer is crucial in directing the heat transfer [11], preventing energy waste.

Finally, the radiant panel (RP) is the lightest, with the radiant hydronic tube wholly isolated from any considerable thermal mass. RP systems are relatively new and improved lightweight, reactive systems [11]. The panels can be integrated into a ceiling system [58], which allows concentration on conditioning the environment with the radiant tube isolated from a heavy thermal mass. Miriel et al. [11] experimented with a radiant cooling ceiling panel with metal tubes resulting in a low responsive time, although the time was not explicitly stated. The commonly used radiant ceiling has the top surface insulated, while the other, downward surface is used for heat transfer [13]. The insulation layer on top of the system can also consider acoustic absorption [11]. An experiment showed that a radiant panel system with tube mats had less than 15-min reaction times [53].

5.2. Radiant Panel Systems

Piping plays a critical role in radiant panel (RP) performance. The pipe diameter, space between pipes, and thermal resistance of the pipe material significantly impact the surface temperature of radiant systems [54]. Besides the water temperature, in heating, bigger pipes with low thermal resistance and less space between them result in higher surface temperatures [54]. In short, more tube surface area and less material thermal resistance mean better surface temperature and more efficient heat transfer. There are two main piping, including small-bore tubes (about ø20-mm metal or plastic tube) and polyethylene capillary tube mats used in RP systems. Zhao et al. [26] experimented on an embedded heating floor with a ø20-mm plastic tube and a capillary mat gypsum heating panel system. The results showed that the capillary mat system can provide thermal comfort with inlet water of about 8 °C lower and consume 45% less energy than panel systems with small-bore tubes. Additionally, the temperature difference between the inlet and outlet water in the capillary tube system and the small-bore tubes system is around 5 °C [26]. According to Miriel and Serres [11], polyethylene capillary tube mats provide flexibility in design and installation. Additionally, the distance between tubes and the heat resistance of the tubes and filling material in capillary tube radiant systems significantly influence the system's thermal capacity [27]. Capillary radiant systems are also considered more flexible and responsive than metallic tubes, although the capillary tubes can encounter blockage due to the small tube diameter [11].

Consulting with our partner in the industry, KPW Solutions, the recommended water flow rated for normal radiant systems with small-bore tubes (usually ø20 mm) is 3–5 L/min for every circuit and will be higher for the capillary tube mat. Additionally, the water cycling rate for the system should be at least 40 times/h. This is necessary for ensuring uniform radiant surfaces.

Figure 8 shows a capillary radiant ceiling system being installed in a residential project in Victoria. The capillary mats are installed on the existing plasterboard ceiling (area A), and the top is covered using batt insulation (area B). The upper insulation ensures the downward stream of heat transfer, while the capillary tube enhances the thermal performance. The flexible and lightweight polyethylene capillary tube mat works well in this context of a confined and complex ceiling. The flexibility of the capillary mat allows installation with relative ease in various situations, making this type of piping ideal for retrofitting projects.

(**a**) (**b**)

Figure 8. Capillary tube installation on a ceiling: (**a**) Capillary mats on existing plasterboard ceiling; (**b**) Batt insulation cover the top of Capillary mats. Source: KPW solutions.

Figure 9 shows an EES installed on a radiant floor system in a residential project in Victoria. The system's section is shown in Figure 10. The capillary tube mats are utilised to reduce the response time that is negatively affected by the concrete screed and load-distributing layer needed for a solid floor system. Underneath the load-distributing layer is a solid insulation layer that prevents heat from reaching the structure mass directing heat to the top floor surface. The total thickness of this system is only 40–50 mm instead of the 120–150 mm of a thermally active building system (TABS).

Figure 9. Capillary tube installation on a floor. Source: KPW Solutions.

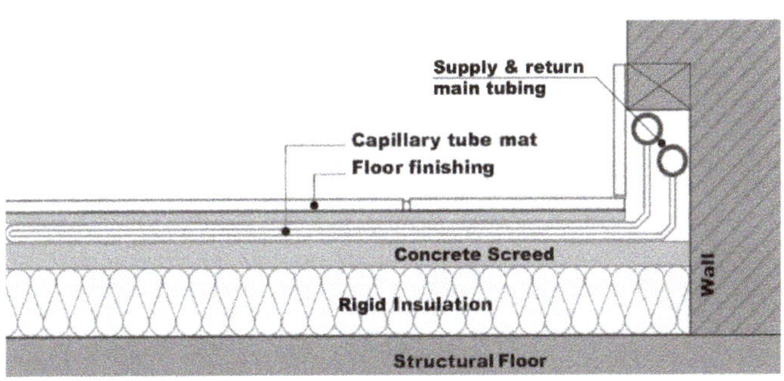

Figure 10. Capillary tube radiant floor. Source: BEKA Klima-Komfort, redrawn by the first author.

5.3. A Prototype for a Lightweight and Dynamic Radiant Cooling System

Based on the lessons learned from the previous review, we propose an RP system with polyethylene capillary tube mats (Figure 11). Each panel has a capillary tube mat attached to a solid insulation board (Figure 11A). The insulation board has a layer of aluminium foil in contact with the capillary tube to increase the heat-directing effect (Figure 11B). The proposed finishing surface of a panel can be a perforated metal sheet or plasterboard that is in contact with the tube mat (Figure 11C). This finishing surface ensures the architectural appearance of the panel and the efficiency of radiant heat transfer. This lightweight and modular system integrates with the ceiling system (Figure 11D) or wall cover. This design focuses on solving two problems of radiant systems in retrofitting, non-modularity and system installation. Our system is lightweight, modular, and allows rapid system installation.

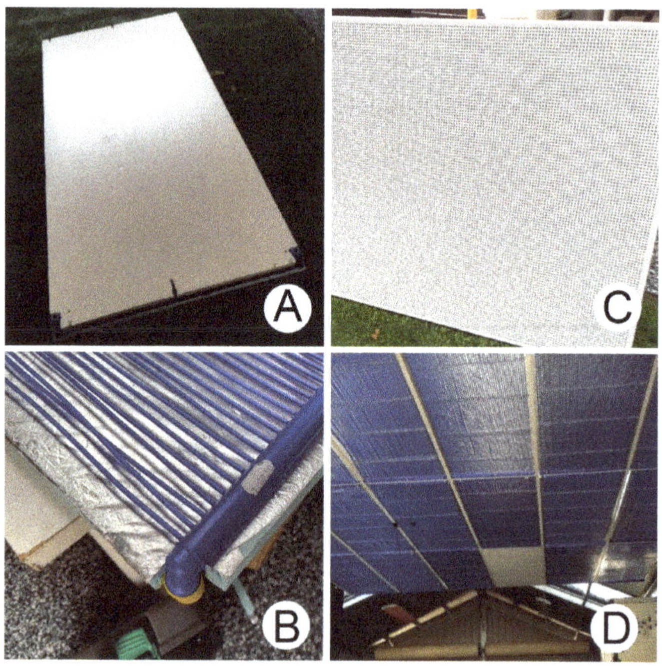

Figure 11. Capillary radiant panel system prototype: (**A**) Insulation board at the back; (**B**) Aluminium foil in contact with tube mat; (**C**) Perforated metal sheet at the front; (**D**) integrated into a ceiling panel. Source: First and second authors.

This prototype is currently under testing, indicating promising results. An initial demo heating operation was conducted. The water flow rate can reach about 14 L/min, leading to a 95 times/h water cycling rate. The system's response time (from turning on to a stable surface temperature) is only about 15 min. However, we face a nonuniform surface temperature problem when the difference between the inlet and outlet water temperatures is too high (at about 20 °C). Due to limited data gathered from the initial test, further testing will be conducted

A diagram of the complete system components is shown in Figure 12. The design follows the principal diagram for the radiant system shown in Figure 5 and includes several pumping solutions to ensure adequate water flow, reduce the water leaks, and prevent blocking caused by air bubbles. The pressure tank is used to take the pressure changes in the system as the water temperature changes leading to changes in the water volume. The vacuum relief valves are used to trap the bubbles and prevent blocking. At the same time, these valves can release pressure if the water pressure in the system increases to a critical level. A pressure gauge is used for monitoring the water pressure. The system functions correctly without leaking if the pressure is stable between 1–3 Bar (14.5–43.5 psi).

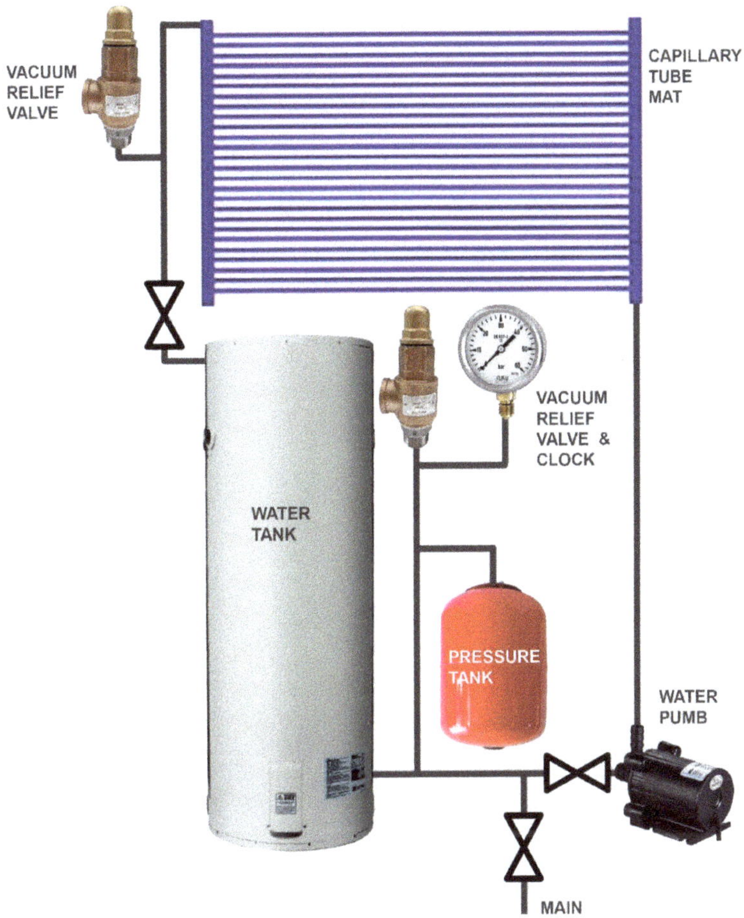

Figure 12. Capillary radiant panel system diagram. Source: KPW Solutions.

5.4. Mounting Position for Hydronic Radiant Systems

Sui and Zhang [10] claimed that a radiant cooling floor can provide the same level of thermal comfort as a radiant ceiling. Olesen [59] explained that having a radiant cooling system mounted on the floor can be more advantageous than the ceiling due to the close distance between a floor and the occupants, leading to a better view factor (angle factor). The view factor (angle factor) is a fundamental aspect of radiant heat transfer representing the radiant heat exchanges between two surfaces [50]. A better view factor means more efficient heat transfer.

Typically, the floor has the highest view factor compared to the wall and ceiling due to the close and uniform distance to the occupants (Figure 13A), resulting in the best heat exchange coefficient. Hence, the floor should be preferred. However, the efficiency of floor-mounting radiant systems can be obstructed by furniture or carpet covering the floor radiant surfaces [60]. Fontana [61] experimented with the impact of furniture coverage on the thermal performance of a radiant heating floor. The results showed that furniture can significantly impact the heat flux between a radiant floor and the subject environment, resulting in up to a 30% reduction in heat transfer. Hence, furniture coverage must be considered when designing radiant systems [61].

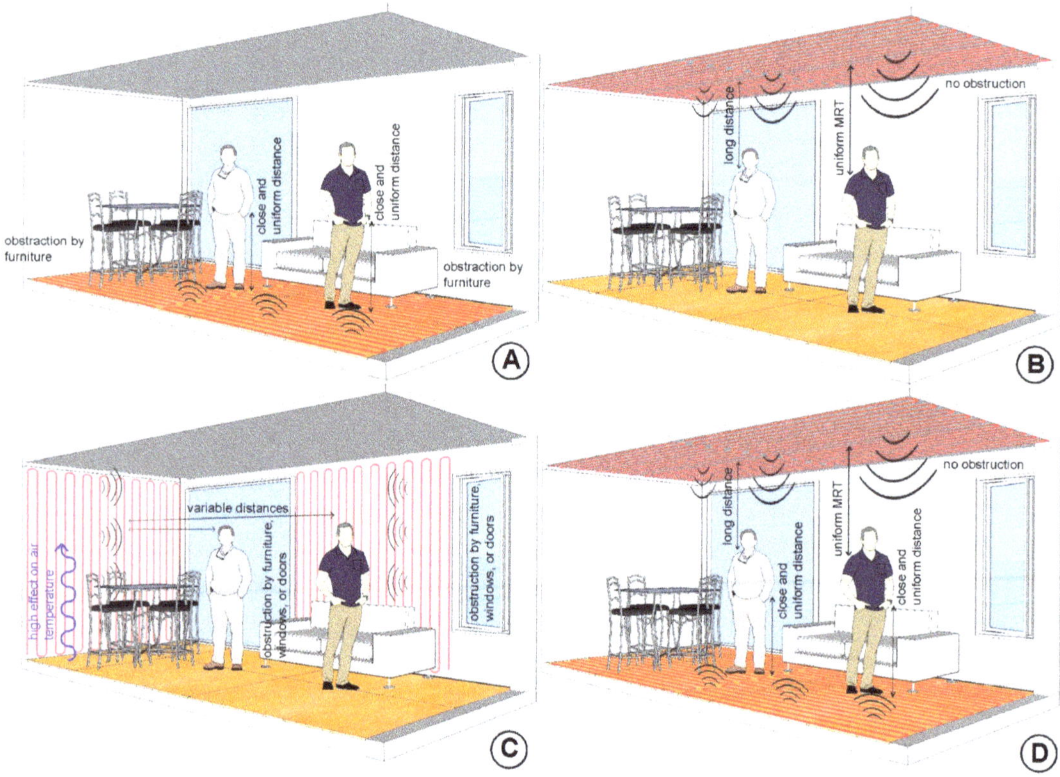

Figure 13. (**A**) Floor-mounted radiant systems; (**B**) Ceiling-mounted radiant systems; (**C**) Wall-mounted radiant systems; (**D**) Floor and Ceiling-mounted radiant systems. Source: First author.

It is claimed that the ceiling is the least-favourable mounting location for radiant systems [58]. The reason can be the usual long-distance from the system to the occupants, leading to low view factors. However, some clear advantages of a radiant ceiling can be seen, such as uniform MRT and not being obstructed by objects (Figure 13B). In fact, empirical results from a residential project indicated that a radiative ceiling, during cooling, kept the polished concrete floor cool within a 4-m floor-to-ceiling height. While this setup will not be advantageous for heating, it will still function and work with the assistance of air movement. Several research papers have suggested that the floor is not the optimal mounting position for radiant systems. Bojić et al. [58] compared the performance of radiant heating panels in four different positions. The results showed that, while mounting heating panels on the ceiling yielded the worst thermal performance, heating panels on both floor and ceiling was the best option for thermal comfort (Figure 13D), followed by a heating wall and floor. Karabay, Arıcı, and Sandık [60] conducted a similar experiment comparing a heating wall and floor, with the results also favouring a heating wall due to the better effect on room air temperature (Figure 13C) while not being affected by furniture coverage. Having said that, while conducting a design for radiant walls, we found that radiant wall systems can still be affected by furniture coverage, while doors and windows can render radiant walls unfeasible on some occasions. The distance between a radiant wall and the occupants can vary in terms of view factor (Figure 13C), leading to a nonuniform view factor and MRT. A question can be raised about how this nonuniform MRT can cause local thermal discomfort.

There are also debates on the effectiveness and efficiency of a heating ceiling and cooling floor. In fact, standard ISO 7730 shows that a heating ceiling can easily cause thermal discomfort [33]. Additionally, in EN 1264, the heat exchange coefficients of the heating ceiling and the cooling floor are similarly low, showing their limitation and ineffectiveness compared to other mounting positions [62]. However, as the heating ceiling and cooling floor are mentioned in standard ISO 7730, EN 1264, and the REHVA Handbook, it is still considered a viable option.

Overall, wall, ceiling, and floor-mounting radiant systems can provide adequate thermal comfort. This proves the feasibility of radiant system application. Although, in some buildings, it is not possible to install radiant panels at their preferred location, the radiant system with its flexibility can still be applied to other, less favourable locations and perform adequately.

6. Hydronic Radiant System Control

6.1. Instrumentation for Control

The study conducted by Sui and Zhang [10] showed the relationship between MRT and other measurable factors of thermal comfort and provided a basic mechanism in controlling radiant systems. Firstly, air temperature is not a significant factor, as a radiant system can provide cooling with a high air temperature and heating with a low air temperature [34]. Secondly, the lower the MRT, the lower the air velocity should be to maintain thermal comfort. However, it is not practical to reduce the air velocity to 0 m/s. Thirdly, the lower the MRT, the higher the humidity should be to maintain thermal comfort [10]. However, high air humidity increases the risk of condensation, which is the most problematic for radiant cooling systems. Interestingly, the recommended humidity for TABS in cooling mode is only 40% [34]. Hence, the principle of controlling a radiant system can be summarised by keeping the air velocity low, keeping the humidity in control, and, mainly, adjusting the MRT and adjusting the air temperature if necessary.

The flow rate of water and the temperature of the inlet water significantly impact the hydronic radiant system thermal performance [63]. In particular, a steady and fast flow rate will ensure a uniform heat flux on the radiant surface and better thermal performance, while the water temperature is the most vital aspect ensuring the thermal and energy performance of the system [64]. Hence, to properly function, the water flow rate should be fast enough to provide a uniform temperature across the radiant surfaces, while the water temperature needs to be adjusted to maintain thermal comfort and control the energy consumption appropriately. There are two flow rate control mechanisms: on–off and variable control speed. However, flow rate control faces several disadvantages, such as a nonuniform radiant surface temperature, high fluctuations in the room temperature, and even noise make it far inferior to the more precise temperature control [65]. Nevertheless, the flow rate control method can be used in an individual room via a simple on–off control system linked with a room thermostat [59]. Commonly used TABS radiant systems operate with a stable flow rate while adjusting the water temperature [66].

As the water temperature is the most critical aspect in controlling radiant systems, most of the proposed control strategies focus on optimising the water temperature supply to the spaces [65–68]. Olesen [69] pointed out that the set point for the water temperature (average of inlet and return water) needs to be adjusted based on the outside temperature and specific building type regarding mass, heat loss, and external and internal gain. The control mechanism recommended in the REHVA Handbook for central control is based on the average water temperature (between the inlet and returned water) [34]. The European standard UNE EN1264 and the REHVA Handbook dictate the maximum capacity of the radiant system in both cooling and heating (Table 2).

Table 2. Maximum capacity of radiant systems. Source: REHVA Handbook [34]. Source: Remade by the third author.

	Total Heat Exchange (W/m^2 K)		Acceptable Surface Temperature (°C)		Maximum Capacity (W/m^2)	
	Heating	Cooling	Maximum Heating	Minimum Cooling	Heating	Cooling
Floor occupied zone	11	7	29	19	99	42
Floor perimeter zone	11	7	35	19	165	42
Wall	8	8	~40	17	160	72
Ceiling	6	11	~27	17	42	99

Valves control the flow rate and inlet water temperature in hydronic systems. For example, Miriel et al. [11] used a three-way valve to control the inlet water temperature for cooling and heating while keeping the flow rate stable. The valve control is still the primary control for even modern hydronic radiant systems [18,25,34]. A simple control mechanism used in early radiant systems is adjusting the inlet water temperature and flow rate via manual valve control [34]. Radiant systems can also be controlled automatically using the data provided by sensors [34]. These sensors collect the operative temperature data (MRT and air temperature) in a conditioned room and use the data on the control system and valves. The data on the inlet and return water temperatures and the mass flow rate allows the heat transfer between the system and the conditioned space to be calculated and controlled, resulting in improved thermal and energy performance [34].

While the traditional thermostat systems control the conditioning systems using dry-bulb temperature, the thermal comfort method applies the PMV index [70]. Meanwhile, controlling HVAC systems with only dry-bulb temperatures is inadequate and can benefit from PMV calculations [71]. As the radiant conditioned spaces are similar to air-conditioned spaces in terms of thermal comfort, a PMV calculation should be the controlling index for radiant systems [36].

Modern thermal imaging technologies can potentially aid in automatic control for radiant systems. Infrared thermal cameras can provide detailed data on the room surface temperatures and the location of heat sources [72]. Infrared thermal cameras allow data on MRT to be collected on a real-time basis and provide a precise calculation of the thermal comfort levels (PMV and PPD) [70]. This infrared thermal imaging can provide data to the control systems of HVACs, resulting in about a 0.4 °C difference in conditioned air temperature compared to traditional thermostat systems [70]. This can be equal to 3.2–4% cooling load savings. Zampetti, Arnesano, and Revel [71] also tested similar real-time thermal comfort monitoring systems to control heaters, resulting in 17% energy savings and better thermal comfort. Therefore, thermal imaging technologies, together with thermal comfort calculation methods, should be considered for hydronic radiant systems.

6.2. Problem with Thermal Mass and Controlling Strategy

The main aim of an optimal control strategy for the radiant systems is achieving peak thermal comfort and energy performance during operation [68]. This means maintaining an adequate level of thermal comfort when required while keeping energy consumption and operation costs low [34]. As some types of radiant systems are challenging to control and not reactive, the control strategy plays a more critical role in ensuring performance [56]. In this case, radiant systems with high thermal masses proved to be the most challenging [25,66]. The thermal mass can effectively reduce the peak conditioning load and close the gap between the day and night conditioning loads [73]. However, this also leads to a slow response time that challenges the system design and energy evaluation [74]. Therefore, high thermal mass radiant systems require efficient controlling strategies [66,68,73]. The REHVA Handbook recommends using two controlling levels for high thermal mass radiant systems, namely central and individual controls [34]. The room core air temperature to set the water temperature and individual control to adjust the flow rate in each room [34]. On the other hand, Raftery et al. [66] promoted zone control alone, as each thermal zone in a building may have different inlet water temperatures, average water temperatures, or flow rates.

The traditional control strategy for radiant systems is the continuous mode in which the heat flux between the system and the room is monitored, and the individual on–off control will respond accordingly to maintain a comfortable range of the room temperature [75]. However, this strategy has drawbacks, such as a slow response to the external environment and high risk of discomfort [75]. Hence, advanced control strategies were developed considering high thermal mass systems [74]. Romaní, de Gracia, and Cabeza [74] mentioned improved control strategies such as pulse width modulation, model predictive control, adaptive and predictive controls, and gain scheduling control. These advanced control strategies use data from weather forecasts or previous operations to set the control variables and parameters [74].

On the contrary, radiant systems with a low thermal mass, such as radiant panels, are more energy-efficient [34], more responsive, and easier to control and, hence, do not need a complex control strategy to properly function [25,74]. This is a significant advantage of radiant panels over other high thermal mass radiant systems. Nevertheless, an improved control strategy remains a viable option for retrofitting existing high thermal mass systems.

7. Radiant Cooling and Condensation Prevention

7.1. Radiant Cooling

The idea of hydronic radiant cooling systems has been developed and faces several difficulties. In the 1930s, Dr Oscar Faber successfully applied a cooling capability to a radiant floor system installed in the Bank of England building [23]. However, the early hydronic chilled ceiling system developed in the 1930s was unsuccessful due to the high condensation risk. Condensation remains one of the most significant problems of radiant cooling systems and prevented radiant cooling from being widely utilised for several decades [76]. Condensation can also reduce the air quality of a conditioned space and reduce the thermal efficiency of the radiant cooling systems [63]. Nevertheless, new knowledge and technology allow the problem to be avoided, allowing radiant cooling systems to be constantly improved. Hence, the interest in radiant cooling has recently returned, and radiant cooling systems have been used in several modern buildings, even in humid climates such as Singapore or Thailand [19].

As mentioned above, the condensation risk has been the biggest issue for radiant cooling systems from the very beginning in the 1930s [23]. Moreover, nearly 90 years later, Teitelbaum et al. [76] suggested that the condensation risk caused by the low surface temperature of the radiant surface is one of the main reasons preventing radiant cooling systems from being commonly used, especially in hot, humid climates. Morse [77] pointed out that the dew point temperature for condensation is high in this type of climate, while the radiant cooling surface must be operated at a low temperature to balance the high air temperature and humidity. However, condensation will occur if the temperature of the cooling surfaces is lower than the dew point. This makes setting up radiant cooling for hot, humid climates highly complicated. Additionally, the condensation risk for ceiling cooling is significantly higher than wall cooling or floor cooling [78], while ceiling cooling is the most common concept mentioned in radiant cooling research papers.

7.2. Preventing Condensation

Many studies have been conducted to solve this condensation problem of radiant cooling, focusing on hot, humid climates. The standard strategy is to combine radiant cooling with a mechanical ventilation system capable of dehumidification [17,79–81]. This method aims to reduce the relative humidity and dew point temperature, allowing radiant cooling to be operated at the required temperature. Some research papers have been conducted to improve the efficiency of dehumidification ventilation. For example, Binghooth and Zainal [80] and Hao et al. [17] used desiccant for dehumidification in ventilation systems to solve the problem in Malaysia's hot, humid climate conditions, similar to Beijing, China. Both research papers concluded that condensation could be prevented, along with finding an energy-saving potential. Kim and Leibundgut [81] instead used hydraulic airbox units to reduce the moisture in the supply air, ending with good results.

In a conditioned environment, air humidity mostly comes from air infiltration, and when using dehumidification ventilation, it is essential to give the system enough time to reduce the interior air humidity [79,82]. Zhang and Niu [79] investigated the problem and recommended a strategy to prevent air infiltration and dehumidify the internal air at least one hour before the radiant cooling operation. Meanwhile, Ge, Xiao, and Wang [82] used a network of sensors and control systems to collect data on the outdoor air quality and calculated an optimal radiant cooling strategy. The resulted in a significant reduction in the dehumidification time to about half an hour.

Another approach to reducing the condensation risk is to prevent condensation droplets from forming using the material surface property. Yin and Wang [63] found that radiant cooling systems with gypsum surfaces have less condensation risks than metal and plastic surfaces. Tang et al. [83] looked at the condition of forming droplets on the surfaces of radiant cooling panels and proposed a superhydrophobic treatment to the radiant surface of the cooling ceiling. The result was that the condensation droplets become too small to be noticed. However, this method can only make condensation negligible rather than solving the problem.

Surprisingly, a breakthrough was made in the 1960s to tackle the condensation problem of radiant cooling in hot, humid climates [76]. Morse [77] proposed a radiant cooling system that could operate below the room dew point and make radiant cooling viable in hot, humid climates. The principle was to isolate the cooling panel within a high humidity environment while allowing material radiation to operate freely. The inventor used a sheet of polyethylene, a material transparent to radiation, to cover the cooling surface and isolate it from the high humid air. The result was a competent radiant system that had almost no impact on the air temperature or humidity. Hence, this system could be operated without any limitation on ventilation. Unfortunately, the principal did not receive much attention. After nearly half a century, Teitelbaum et al. [76] applied this principle to build a radiant cooling system prototype and conducted an instrumental experiment. A polyethylene membrane as the cover sheet material was used. The membrane allowed more than 80% of the radiation to pass by, ensuring the system's thermal performance. The system was tested at an open-air bus stop in Singapore's hot, humid climate and proved that thermal comfort could be achieved even with a high metabolic rate and with no condensation risk. The results also showed that the membrane-assisted radiant cooling system could be operated at temperatures below the dew point and could even be used in naturally ventilated environments.

7.3. Proposing a Design for a Condensation Resisted Radiant Panel System

Figure 14 illustrates a proposed cooling radiant panel (RP) system inspired by the design of Morse [76]. Figure 15 shows the components of the panel. The system utilises capillary tube mats to reduce the responding time and improve the thermal performance. The rigid insulation boards can provide a heat direction effect, as well as form the frame for the panel. The moulded aluminium sheet behaves like a radiant plate and enhances the heat direction effect. The panel is sealed by a membrane (EPDM, PP, or HDPE). This membrane allows most of the radiation from the environment to pass through and be absorbed by the tube mat and the aluminium sheet while isolating them from the environment. This design aims at providing a lightweight, dynamic, and condensation-resistant radiant panel system. This system is expected to be a solution for radiant cooling for retrofitting and new projects in hot, humid climates. In a future work, a prototype will be built and tested.

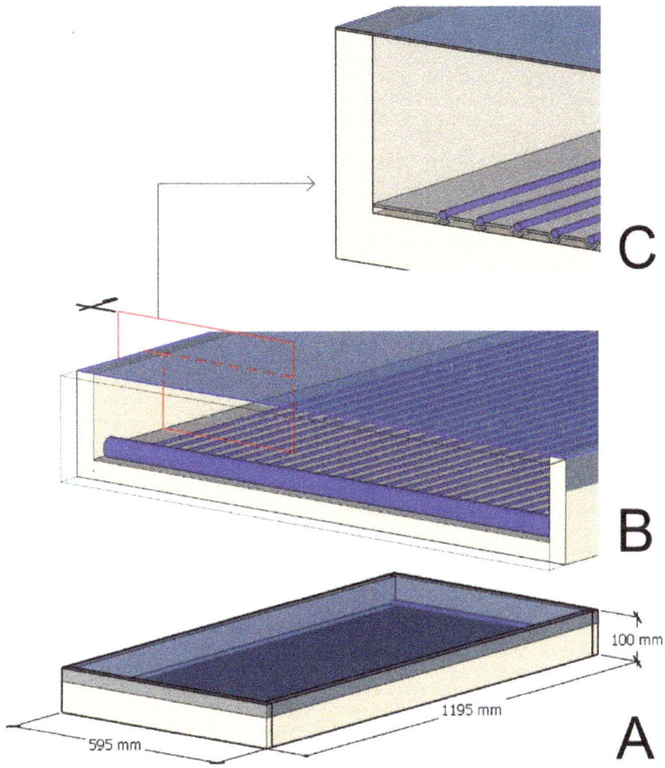

Figure 14. A condensation-resistant radiant cooling panel: (**A**) A complete panel and total dimensions; (**B**) Panel section; (**C**) Panel section detail. Source: First author.

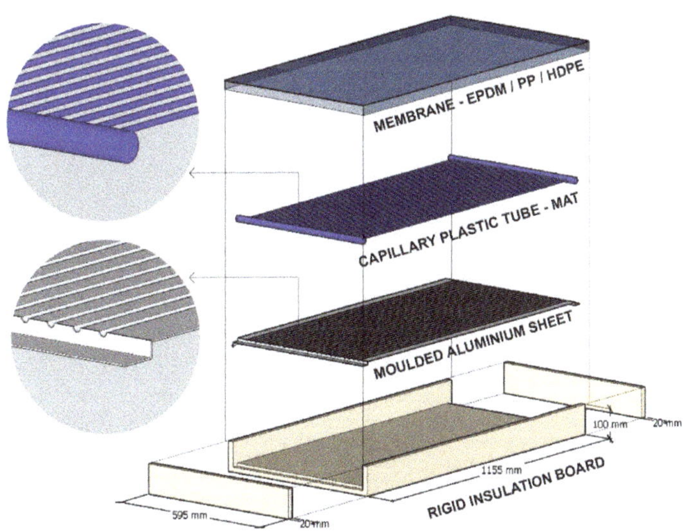

Figure 15. Components of a condensation-resistant radiant cooling panel. Source: First author.

8. Conclusion: Why Retrofit Residential Housing with the Radiant System?

8.1. General Advantages of Radiant System

It is now well-known among researchers that radiant systems offer high thermal performance, energy efficiency, and more [10,18,64,69]. For example, Rhee, Olesen, and Kim [19] conducted an informative literature review on the radiant system and concluded that hydronic radiant systems could offer several advantages: reducing the discomfort caused by drafts, a low noise level, and good energy performance, as well as design flexibility. The reduction in air velocity and the risk of a cold draft are key aspects in radiant systems' superior thermal performance over HVAC systems [16,84]. Specifically, radiant systems offer several other advantages, including [10]:

1. creating a uniform thermal environment, reducing vertical temperature gradients and local discomfort;
2. requiring less air supply, leading to a reduction in the discomfort caused by drafts;
3. a decrease in the energy load for transporting the air supply;
4. using water as a more efficient heat-carrying medium can significantly decrease the energy load for cooling transport;
5. incorporating MRT, which the human body senses the most, resulting in a more efficient energy performance;
6. cooling can be provided with relatively high water temperatures; hence, unconditioned water temperatures can be used on some occasions;
7. quieter operations.

8.2. Limitation of Radiant Systems and Recent Improvements

Despite the merits mentioned earlier, radiant cooling systems still have several limitations. According to the authors of Reference [16], the high price, difficulties with control, condensation risk, and system operational life are issues of radiant systems. Similarly, Sui and Zhang [10] pointed out that problems such as indoor thermal environment design, system control, and anticondensation need to be solved for radiant cooling systems in residential buildings. Serageldin et al. [13] repeated some of the main issues with hydronic systems: condensation risk, difficulty in control, cold draught, high initial cost, and small capacity. Additionally, local discomfort can occur in a radiant heating environment caused by the distance between the human body and the radiant surface [18].

There is also complexity in installing and utilising radiant systems. As the radiant system alone cannot supply fresh air to the conditioned space, there should always be an air supply system [85]. This, however, may not be the case for the leaky windows and façade of old Australian residential buildings, where a high infiltration rate exists. This contributes to the implementation of radiant systems and hinders radiant systems as secondary conditioning options [64]. The pumping water system required for hydronic radiant systems can also be considered complicated, contributing to installation [86]. A major problem with radiant floor systems is that the radiant heat transfer is affected by furniture coverage and carpets [60].

8.3. The Potential Solution of Lightweight Capillary Tube Radiant Systems

The above review also highlights the advantage of lightweight radiant panel systems over heavier systems such as TABS or ESS. In general, lightweight radiant systems:

1. do not require complicated control systems;
2. are more flexible and easier to install than their heavier counterparts;
3. are more dynamic and respond faster;
4. are relatively cost-effective.

A thermal dynamic and fast response time of radiant panel systems can be further enhanced with capillary tube mats.

These properties, along with the advantages mentioned above of radiant systems in general, make radiant panel systems ideal for the retrofitting residential buildings and even other types of buildings, such as commercial or industrial. However, further studies are needed to confirm this. Hence, experimentation is being conducted.

Author Contributions: The paper was a collaborative effort between the authors. H.Q.D. is the main contributor to the initial draft and revised the following drafts. Z.W. contributed to the method section and experiments. M.B.L., M.A., and I.M. contributed equally to conceptualisation, project administration, writing the initial paper, and reviewing and editing the final drafts. All authors have read and agreed to the published version of the manuscript.

Funding: This research received no external funding.

Institutional Review Board Statement: Not applicable.

Informed Consent Statement: Not applicable.

Data Availability Statement: Not applicable.

Acknowledgments: Our thanks to Klaus Wanke, Christopher Wanke, and KPW Solutions Company for helping us develop our radiant systems prototype and providing precious empirical information on radiant system operations. Peter Horan and Richard Buldys are also acknowledged for helping us set up the demo operation for our prototype.

Conflicts of Interest: The authors declare no conflict of interest.

References

1. Bulut, M.B.; Wilkinson, S.; Khan, A.; Jin, X.-H.; Lee, C.L. Thermal performance of retrofitted secondary glazed windows in residential buildings–two cases from Australia. *Smart Sustain. Built Environ.* **2021**. [CrossRef]
2. Wilkinson, S.; Feitosa, R.C. Retrofitting housing with lightweight green roof technology in Sydney, Australia, and Rio de Janeiro, Brazil. *Sustainability* **2015**, *7*, 1081–1098. [CrossRef]
3. Ma, Z.; Cooper, P.; Daly, D.; Ledo, L. Existing building retrofits: Methodology and state-of-the-art. *Energy Build.* **2012**, *55*, 889–902. [CrossRef]
4. Jamil, H.; Alam, M.; Sanjayan, J.; Wilson, J. Investigation of PCM as retrofitting option to enhance occupant thermal comfort in a modern residential building. *Energy Build.* **2016**, *133*, 217–229. [CrossRef]
5. Goldsworthy, M. Towards a residential air-conditioner usage model for Australia. *Energies* **2017**, *10*, 1256. [CrossRef]
6. Simshauser, P.; Nelson, T. The outlook for residential electricity prices in Australia's National Electricity Market in 2020. *Electr. J.* **2013**, *26*, 66–83.
7. Willand, N.; Maller, C.; Ridley, I. Addressing health and equity in residential low carbon transitions–Insights from a pragmatic retrofit evaluation in Australia. *Energy Res. Soc. Sci.* **2019**, *53*, 68–84. [CrossRef]
8. Moe, K. *Thermally Active Surfaces in Architecture*; Princeton Architectural Press: New York, NY, USA, 2010.
9. de Dear, R.; Zhang, F. Why the Perfect Office Temperature Is a Myth. Available online: https://www.sydney.edu.au/news-opinion/news/2019/03/14/why-the-perfect-office-temperature-is-a-myth.html (accessed on 20 July 2021).
10. Sui, X.; Zhang, X. Analysis on combinations of indoor thermal microclimate parameters in radiant cooled residential buildings and drawing of new thermal comfort charts. *Build. Serv. Eng. Res. Technol.* **2016**, *37*, 66–84. [CrossRef]
11. Miriel, J.; Serres, L.; Trombe, A. Radiant ceiling panel heating–cooling systems: Experimental and simulated study of the performances, thermal comfort and energy consumptions. *Appl. Therm. Eng.* **2002**, *22*, 1861–1873. [CrossRef]
12. Karmann, C.; Schiavon, S.; Bauman, F. Thermal comfort in buildings using radiant vs. all-air systems: A critical literature review. *Build. Environ.* **2017**, *111*, 123–131. [CrossRef]
13. Serageldin, A.A.; Ye, M.; Radwan, A.; Sato, H.; Nagano, K. Numerical investigation of the thermal performance of a radiant ceiling cooling panel with segmented concave surfaces. *J. Build. Eng.* **2021**, *42*, 102450. [CrossRef]
14. Catalina, T.; Virgone, J.; Kuznik, F. Evaluation of thermal comfort using combined CFD and experimentation study in a test room equipped with a cooling ceiling. *Build. Environ.* **2009**, *44*, 1740–1750. [CrossRef]
15. Feustel, H.E.; Stetiu, C. Hydronic radiant cooling—Preliminary assessment. *Energy Build.* **1995**, *22*, 193–205. [CrossRef]
16. Imanari, T.; Omori, T.; Bogaki, K. Thermal comfort and energy consumption of the radiant ceiling panel system.: Comparison with the conventional all-air system. *Energy Build.* **1999**, *30*, 167–175. [CrossRef]
17. Hao, X.; Zhang, G.; Chen, Y.; Zou, S.; Moschandreas, D.J. A combined system of chilled ceiling, displacement ventilation and desiccant dehumidification. *Build. Environ.* **2007**, *42*, 3298–3308. [CrossRef]
18. Rhee, K.-N.; Kim, K.W. A 50 year review of basic and applied research in radiant heating and cooling systems for the built environment. *Build. Environ.* **2015**, *91*, 166–190. [CrossRef]
19. Rhee, K.-N.; Olesen, B.W.; Kim, K.W. Ten questions about radiant heating and cooling systems. *Build. Environ.* **2017**, *112*, 367–381. [CrossRef]
20. Debnath, K.B.; Jenkins, D.P.; Patidar, S.; Peacock, A.D. Understanding residential occupant cooling behaviour through electricity consumption in warm-humid climate. *Buildings* **2020**, *10*, 78. [CrossRef]

21. Pedersen, T.H.; Hedegaard, R.E.; Kristensen, K.F.; Gadgaard, B.; Petersen, S. The effect of including hydronic radiator dynamics in model predictive control of space heating. *Energy Build.* **2019**, *183*, 772–784. [CrossRef]
22. Embaye, M.; Al-Dadah, R.; Mahmoud, S. Thermal performance of hydronic radiator with flow pulsation–Numerical investigation. *Appl. Therm. Eng.* **2015**, *80*, 109–117. [CrossRef]
23. Bean, R.; Olesen, B.W.; Kim, K.W. Part 2: History of radiant heating & cooling systems. *Ashrae J.* **2010**, *52*, 50.
24. Australian Bureau of Statistics. Year Book Australia, 2009–10. Available online: https://www.abs.gov.au/AUSSTATS/abs@.nsf/Lookup/566B88038EED654CCA25773700169C79 (accessed on 10 December 2021).
25. Feng, J.D. *Design and Control of Hydronic Radiant Cooling Systems*; University of California: Oakland, CA, USA, 2014.
26. Zhao, M.; Kang, W.; Luo, X.; Yu, C.W.; Meng, X.; Gu, Z. Performance comparison of capillary mat radiant and floor radiant heating systems assisted by an air source heat pump in a residential building. *Indoor Built Environ.* **2017**, *26*, 1292–1304. [CrossRef]
27. Ding, P.; Li, Y.; Long, E.; Zhang, Y.; Liu, Q. Study on heating capacity and heat loss of capillary radiant floor heating systems. *Appl. Therm. Eng.* **2020**, *165*, 114618. [CrossRef]
28. ASHRAE. *ASHRAE Standard 55-2017: Thermal Environment Conditions for Human Occupancy*; American Society of Heating, Refrigerating and Air Conditioning Engineers Inc.: Atlanta, GA, USA, 2017.
29. Bessoudo, M.; Tzempelikos, A.; Athienitis, A.; Zmeureanu, R. Indoor thermal environmental conditions near glazed facades with shading devices–Part I: Experiments and building thermal model. *Build. Environ.* **2010**, *45*, 2506–2516. [CrossRef]
30. Anderson, T.; Luther, M. Designing for thermal comfort near a glazed exterior wall. *Archit. Sci. Rev.* **2012**, *55*, 186–195. [CrossRef]
31. Zhao, Q.; Lian, Z.; Lai, D. Thermal Comfort models and their developments: A review. *Energy Built Environ.* **2021**, *2*, 21–33. [CrossRef]
32. Fanger, P.O. Assessment of man's thermal comfort in practice. *Occup. Environ. Med.* **1973**, *30*, 313–324. [CrossRef]
33. International Organization for Standardization. *ISO 7730 2005-11-15 Ergonomics of the Thermal Environment: Analytical Determination and Interpretation of Thermal Comfort Using Calculation of the PMV and PPD Indices and Local Thermal Comfort Criteria*; ISO: Geneva, Switzerland, 2005.
34. Babiak, J.; Olesen, B.W.; Petras, D. *Low Temperature Heating and High Temperature Cooling: REHVA GUIDEBOOK No 7*; REHVA: Ixelles, Belgium, 2007.
35. Fanger, P.O.; Toftum, J. Extension of the PMV model to non-air-conditioned buildings in warm climates. *Energy Build.* **2002**, *34*, 533–536. [CrossRef]
36. Loveday, D.; Parsons, K.; Taki, A.; Hodder, S. Displacement ventilation environments with chilled ceilings: Thermal comfort design within the context of the BS EN ISO7730 versus adaptive debate. *Energy Build.* **2002**, *34*, 573–579. [CrossRef]
37. Luther, M.; Tokede, O.; Liu, C. Applying a comfort model to building performance analysis. *Archit. Sci. Rev.* **2020**, *63*, 481–493. [CrossRef]
38. Hua Ge, P. An infrared sphere method to measure mean radiant temperature. *ASHRAE Trans.* **2013**, *119*, 1F.
39. Chung, J.D.; Hong, H.; Yoo, H. Analysis on the impact of mean radiant temperature for the thermal comfort of underfloor air distribution systems. *Energy Build.* **2010**, *42*, 2353–2359. [CrossRef]
40. Chen, Y.-C.; Lin, T.-P.; Matzarakis, A. Comparison of mean radiant temperature from field experiment and modelling: A case study in Freiburg, Germany. *Theor. Appl. Climatol.* **2014**, *118*, 535–551. [CrossRef]
41. La Gennusa, M.; Nucara, A.; Rizzo, G.; Scaccianoce, G. The calculation of the mean radiant temperature of a subject exposed to the solar radiation—A generalised algorithm. *Build. Environ.* **2005**, *40*, 367–375. [CrossRef]
42. Kántor, N.; Unger, J. The most problematic variable in the course of human-biometeorological comfort assessment—The mean radiant temperature. *Cent. Eur. J. Geosci.* **2011**, *3*, 90–100. [CrossRef]
43. Walikewitz, N.; Jänicke, B.; Langner, M.; Meier, F.; Endlicher, W. The difference between the mean radiant temperature and the air temperature within indoor environments: A case study during summer conditions. *Build. Environ.* **2015**, *84*, 151–161. [CrossRef]
44. Chaudhuri, T.; Soh, Y.C.; Bose, S.; Xie, L.; Li, H. On assuming Mean Radiant Temperature equal to air temperature during PMV-based thermal comfort study in air-conditioned buildings. In Proceedings of the IECON 2016-42nd Annual Conference of the IEEE Industrial Electronics Society, Florence, Italy, 23–26 October 2016; pp. 7065–7070.
45. Chaiyapinunt, S.; Phueakphongsuriya, B.; Mongkornsaksit, K.; Khomporn, N. Performance rating of glass windows and glass windows with films in aspect of thermal comfort and heat transmission. *Energy Build.* **2005**, *37*, 725–738. [CrossRef]
46. Alfano, F.R.d.A.; Dell'Isola, M.; Palella, B.I.; Riccio, G.; Russi, A. On the measurement of the mean radiant temperature and its influence on the indoor thermal environment assessment. *Build. Environ.* **2013**, *63*, 79–88. [CrossRef]
47. Barnaby, C.S.; Pedersen, C.O. Develop a Radiant System Module for the Simulation and Analysis of Spaces and Systems. *ASHRAE Trans.* **2015**, *121*, 364–374.
48. ASHREA. *ASHREA Handbook Fundamentals, SI ed.*; American Society of Heating, Refrigerating and Air-Conditioning Engineers, Inc.: Atlanta, GA, USA, 2021.
49. Vujičić, M.; Lavery, N.; Brown, S. View factor calculation using the Monte Carlo method and numerical sensitivity. *Commun. Numer. Methods Eng.* **2006**, *22*, 197–203. [CrossRef]
50. Lee, D.-S.; Kim, E.-J.; Cho, Y.-H.; Kang, J.-W.; Jo, J.-H. A field study on application of infrared thermography for estimating mean radiant temperatures in large stadiums. *Energy Build.* **2019**, *202*, 109360. [CrossRef]
51. Olesen, B.W. *Thermal Comfort*; Bruel & Kjaer: Nærum, Denmark, 1982; Volume 2.

52. Guo, H.; Aviv, D.; Loyola, M.; Teitelbaum, E.; Houchois, N.; Meggers, F. On the understanding of the mean radiant temperature within both the indoor and outdoor environment, a critical review. *Renew. Sustain. Energy Rev.* **2020**, *117*, 109207. [CrossRef]
53. Díaz, N.F.; Cuevas, C. Testing and thermal modeling of radiant panels systems as commissioning tool. *Energy Convers. Manag.* **2010**, *51*, 2663–2677. [CrossRef]
54. Liu, Y.; Wang, D.; Liu, J. Study on heat transfer process for in-slab heating floor. *Build. Environ.* **2012**, *54*, 77–85. [CrossRef]
55. Athienitis, A.K. Investigation of thermal performance of a passive solar building with floor radiant heating. *Sol. Energy* **1997**, *61*, 337–345. [CrossRef]
56. Wang, Z.; Luo, M.; Geng, Y.; Lin, B.; Zhu, Y. A model to compare convective and radiant heating systems for intermittent space heating. *Appl. Energy* **2018**, *215*, 211–226. [CrossRef]
57. Merabtine, A.; Mokraoui, S.; Kheiri, A.; Dars, A.; Hawila, A.A.-w. Experimental and multidimensional numerical analysis of the thermal behavior of an anhydrite radiant slab floor heating system: A multi-objective sensitivity study. *Energy Build.* **2018**, *174*, 619–634. [CrossRef]
58. Bojić, M.; Cvetković, D.; Marjanović, V.; Blagojević, M.; Djordjević, Z. Performances of low temperature radiant heating systems. *Energy Build.* **2013**, *61*, 233–238. [CrossRef]
59. Olesen, B. Radiant floor cooling systems. *Ashrae J.* **2008**, *50*, 16–22.
60. Karabay, H.; Arıcı, M.; Sandık, M. A numerical investigation of fluid flow and heat transfer inside a room for floor heating and wall heating systems. *Energy Build.* **2013**, *67*, 471–478. [CrossRef]
61. Fontana, L. Thermal performance of radiant heating floors in furnished enclosed spaces. *Appl. Therm. Eng.* **2011**, *31*, 1547–1555. [CrossRef]
62. BS EN 1264-5:2021. *Water Based Surface Embedded Heating and Cooling Systems—Part 5: Determination of the Thermal Output for Wall and Ceiling Heating and for Floor, Wall and Ceiling Cooling*; BSI Standards Limited: London, UK, 2021.
63. Yin, Y.; Wang, R.; Zhai, X.; Ishugah, T. Experimental investigation on the heat transfer performance and water condensation phenomenon of radiant cooling panels. *Build. Environ.* **2014**, *71*, 15–23. [CrossRef]
64. Tye-Gingras, M.; Gosselin, L. Comfort and energy consumption of hydronic heating radiant ceilings and walls based on CFD analysis. *Build. Environ.* **2012**, *54*, 1–13. [CrossRef]
65. Ryu, S.-R.; Jae-Han, L.; Yeo, M.-S.; Kwang-Woo, K. A study on the control methods for radiant floor heating and cooling system in residential building. *ASHRAE Trans.* **2004**, *110*, 106.
66. Raftery, P.; Duarte, C.; Schiavon, S.; Bauman, F. A New Control Strategy for High Thermal Mass Radiant Systems. 2017. Available online: https://escholarship.org/content/qt5tz4n92b/qt5tz4n92b.pdf (accessed on 8 October 2021).
67. Karlsson, H.; Hagentoft, C.-E. Application of model based predictive control for water-based floor heating in low energy residential buildings. *Build. Environ.* **2011**, *46*, 556–569. [CrossRef]
68. Joe, J.; Karava, P. A model predictive control strategy to optimize the performance of radiant floor heating and cooling systems in office buildings. *Appl. Energy* **2019**, *245*, 65–77. [CrossRef]
69. Olesen, B.W. Radiant floor heating in theory and practice. *ASHRAE J.* **2002**, *44*, 19–26.
70. Revel, G.; Arnesano, M.; Pietroni, F. An innovative low cost IR system for real-time measurement of human thermal comfort. In Proceedings of the IAQ2013 Conference, Vancouver, BC, Canada, 15–18 October 2013.
71. Zampetti, L.; Arnesano, M.; Revel, G. Experimental testing of a system for the energy-efficient sub-zonal heating management in indoor environments based on PMV. *Energy Build.* **2018**, *166*, 229–238. [CrossRef]
72. Revel, G.; Sabbatini, E.; Arnesano, M. Development and experimental evaluation of a thermography measurement system for real-time monitoring of comfort and heat rate exchange in the built environment. *Meas. Sci. Technol.* **2012**, *23*, 035005. [CrossRef]
73. Lehmann, B.; Dorer, V.; Koschenz, M. Application range of thermally activated building systems tabs. *Energy Build.* **2007**, *39*, 593–598. [CrossRef]
74. Romaní, J.; de Gracia, A.; Cabeza, L.F. Simulation and control of thermally activated building systems (TABS). *Energy Build.* **2016**, *127*, 22–42. [CrossRef]
75. Cho, S.; Zaheer-Uddin, M. Predictive control of intermittently operated radiant floor heating systems. *Energy Convers. Manag.* **2003**, *44*, 1333–1342. [CrossRef]
76. Teitelbaum, E.; Chen, K.W.; Aviv, D.; Bradford, K.; Ruefenacht, L.; Sheppard, D.; Teitelbaum, M.; Meggers, F.; Pantelic, J.; Rysanek, A. Membrane-assisted radiant cooling for expanding thermal comfort zones globally without air conditioning. *Proc. Natl. Acad. Sci. USA* **2020**, *117*, 21162–21169. [CrossRef] [PubMed]
77. Morse, R. Radiant cooling. *Archit. Sci. Rev.* **1963**, *6*, 50–53. [CrossRef]
78. Tang, H.; Liu, X.-H.; Jiang, Y. Theoretical and experimental study of condensation rates on radiant cooling surfaces in humid air. *Build. Environ.* **2016**, *97*, 1–10. [CrossRef]
79. Zhang, L.; Niu, J. Indoor humidity behaviors associated with decoupled cooling in hot and humid climates. *Build. Environ.* **2003**, *38*, 99–107. [CrossRef]
80. Binghooth, A.; Zainal, Z. Performance of desiccant dehumidification with hydronic radiant cooling system in hot humid climates. *Energy Build.* **2012**, *51*, 1–5. [CrossRef]
81. Kim, M.K.; Leibundgut, H. Advanced Airbox cooling and dehumidification system connected with a chilled ceiling panel in series adapted to hot and humid climates. *Energy Build.* **2014**, *85*, 72–78. [CrossRef]

82. Ge, G.; Xiao, F.; Wang, S. Neural network based prediction method for preventing condensation in chilled ceiling systems. *Energy Build.* **2012**, *45*, 290–298. [CrossRef]
83. Tang, H.; Liu, X.-H.; Li, H.; Zhou, Y.; Jiang, Y. Study on the reduction of condensation risks on the radiant cooling ceiling with superhydrophobic treatment. *Build. Environ.* **2016**, *100*, 135–144. [CrossRef]
84. Kitagawa, K.; Komoda, N.; Hayano, H.; Tanabe, S.-i. Effect of humidity and small air movement on thermal comfort under a radiant cooling ceiling by subjective experiments. *Energy Build.* **1999**, *30*, 185–193. [CrossRef]
85. Loveday, D.L.; Parsons, K.C.; Taki, A.H.; Hodder, S.G. Designing for thermal comfort in combined chilled ceiling/displacement ventilation environments/Discussion. *ASHRAE Trans.* **1998**, *104*, 901.
86. Peng, P.; Gong, G.; Mei, X.; Liu, J.; Wu, F. Investigation on thermal comfort of air carrying energy radiant air-conditioning system in south-central China. *Energy Build.* **2019**, *182*, 51–60. [CrossRef]

energies

Article

Guidance on Implementing Renewable Energy Systems in Australian Homes

Peter Horan, Mark B. Luther * and Hong Xian Li

School of Architecture and Built Environment, Deakin University, Geelong, VIC 3220, Australia; peter@deakin.edu.au (P.H.); hong.li@deakin.edu.au (H.X.L.)
* Correspondence: luther@deakin.edu.au

Abstract: The purpose of this paper is to examine several real house cases as renewable energy resources are installed. It is an empirical study, based on first principles applied to measured data. In the first case presented, a PV solar system has been installed and a hybrid vehicle purchased. Battery storage is being considered. Smart Meter data (provided in Victoria, Australia) measures the electrical energy flowing to and from the grid in each half hour. Missing is the story about what the house is generating and what its energy requirements are through each half hour interval. We apply actual (on site) solar PV data to this study, resolving the unknown energy flows. Analysing energy flow has revealed that there are five fundamental quantities which determine performance, namely energy load, energy import, energy harvesting, energy export and energy storage. As a function of PV size these quantities depend on four parameters, easily derivable from the Smart Meter data, namely the house load, the night-time house load (no PV generation), the rating of the solar PV system and the tariffs charged. This reveals most of the information for providing advice on PV array size and whether to install a battery. An important discovery is that a battery, no matter what size, needs a PV system large enough to charge it during the winter months. The analysis is extended to two more houses located within 5 km for which detailed solar data is unavailable.

Keywords: solar PV; battery; costs; CO_2 emissions; electric vehicle

Citation: Horan, P.; Luther, M.B.; Li, H.X. Guidance on Implementing Renewable Energy Systems in Australian Homes. *Energies* **2021**, *14*, 2666. https://doi.org/10.3390/en14092666

Academic Editors: George Kosmadakis and Carlo Renno

Received: 24 March 2021
Accepted: 29 April 2021
Published: 6 May 2021

1. Introduction

The application of renewable energy systems to Australian houses is not new, yet, it is doubtful that many owners could tell you why their selections of size of solar PV, batteries, solar hot water or electric vehicles were made. In fact, most residents have blindly left the decision making to others without any real economic or environmental understanding of what the combination of renewable energy systems implies for themselves as well as for the global environment. This paper is about a journey taken to convert an existing house into one using renewable energy. Furthermore, it is about discoveries and achievements through several stages of conversion from grid dependent to self-generated energy, as follows:

1. A house and car (petrol fuelled) with no renewable energy except a solar gas boosted water heater;
2. A house with a 3 kW Solar PV system and a plug-in hybrid (battery/petrol) car; and
3. Given the above, is a battery (and if so, what size) a good investment?

In fact, the above systems are those contemplated by many homeowners and system designers when considering what to include in a system using renewable energy. There is little guidance for the layman about integrating technologies as to what system sizes are recommended, or in fact are reasonable, for a particular homeowner's energy use and which considers their patterns of use. Guidance that is available from system installers tends to be derived in an opaque manner.

This paper presents a method readily available to the homeowner for the analysis of electrical energy consumption which predicts performance when such technologies are installed in an existing household. While the method is applied here, at first, to a specific household, it is not limited to this case and can be expanded to households elsewhere. Two more cases are analysed, one with a solar PV system that has a high feed-in tariff, and another household without any solar. This demonstrates a way of arriving at informed decisions for the selection and sizing of renewable energy products and implementations that are justified by data.

Underlying this, almost all Victorian households are monitored by "Smart Meters" which record electrical energy, both imported and exported, every half hour. In the absence of household electricity generation, this reveals a homeowner's distinctive patterns of electrical energy use. Unfortunately, if a solar PV system is installed, it does not reveal what happens "behind the meter", namely, the solar PV harvest and its contribution to the house load. For a complete picture, data from the PV system is required. However, even in the absence of detailed data from the solar PV system, Smart Meter data also provides the information to size a PV system, or alternatively extend it, and to determine whether energy diversion to a battery or thermal storage is viable. It may even assist at revisiting the existing heating and/or cooling system(s) installed as well as its fuel source for conditioning.

2. A Review of Current Work

Solar-powered homes are receiving increasing attention worldwide as a means of using renewable energy to reduce fossil fuel consumption and the environmental burden of residential buildings [1–3]. In Australia, private solar photovoltaic installations have been growing at a significant rate, both in capacity, more than doubling since 2009, and in number [4]. That paper reviewed the residential solar PV policy between 2001 and 2012 in Australia and identified that more than 2300 MWp of residential PV were installed during that period as one of the results of policy incentives [3].

Feed-in Tariffs (FiTs) as well as significant rebate incentives, have been the primary driving force for the growth of solar PV homes in the last decade; however, FiTs continue to change over time and from state to state in Australia, ranging between AUD 0.66/kWh and AUD 0.05/kWh [3]. Apart from the politics, the environment of supplier tariffs and plans for grid electricity, especially when they involve solar PV, is complex [5].

Nicholls et al. [6] conducted comprehensive lifecycle economic and environmental analysis for rooftop PVs in Australia and found that the cost payback time ranges from 11 to more than 25 years in Australia. With the current rebate to homeowners, the dramatic fall in PV system costs and the increase in electrical costs (doubling in 10 years) this payback has decreased significantly with anecdotal 6–7 years payback [3]. More recent analyses for some homeowners indicates that paybacks of even 4 years are possible.

It has been reported recently that the growth of battery installations is on an exponential rise [7]; indeed, in 2019, Australians installed almost 23,000 home battery systems with a capacity of 233 MWh [8]. When considering the application of batteries in households, the relations between the two parties, the consumer and the distribution network operators, need to be considered. Each party is affected differently. Increasing demand on the low voltage network due to changes in consumer behaviour will force network operators to change the way in which they manage and reinforce the network [9]. Ren et al. [10] have modelled the impact of PV battery systems on energy consumption and electricity bills under various tariffs in several Australian cities. No doubt, there are advantages for the distributor when the electrical feed-in is time-of-day regulated in order to smooth the output gains to the network. Both parties prosper when government develops a feed-in tariff incentive for PV systems with battery storage that can provide this balance [3].

The intermittent nature of solar PV output and the mismatch between customer-sited energy load profiles suggests that battery storage may be a means to maximise savings [11–13]. Figure 1 is an example of such a period of mismatch for the first house investigated. Through battery storage systems the variable output of PV to the grid is minimised and anticipates better control [3,11,14].

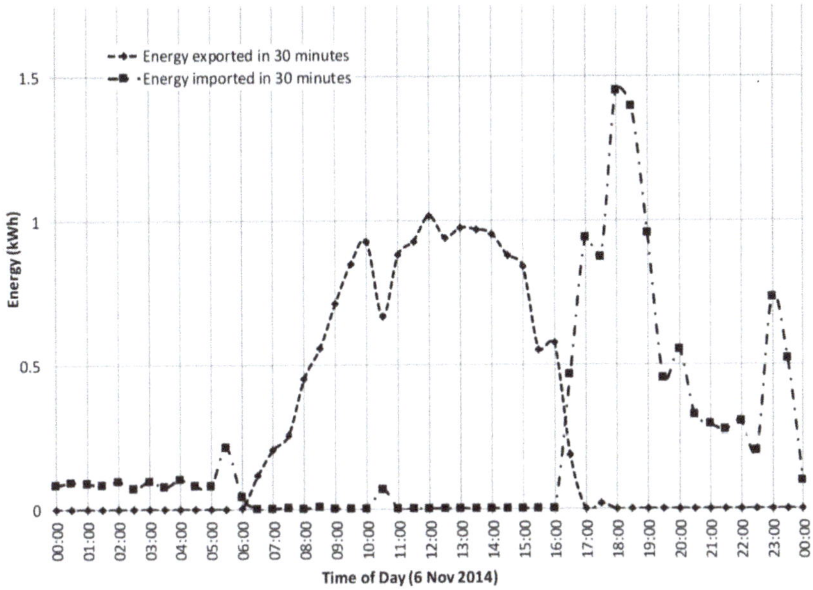

Figure 1. The mismatch of exported and imported energy.

Hassan et al. [11] provides such a circumstance where real data from a solar PV home is used to study by simulation the benefit of battery storage over a 30 min interval dataset for a full year. However, it is not clear whether the data came from individual homes or is an average of multiple projects. As reported in [15], a feed-in-tariff scheme that favours the installation of battery storage to maximise self-consumption is already in place in Germany. There is a growing need for affordable decentralised energy storage and the market is expected to grow rapidly.

Two different methods are available for energy storage in residential buildings, including onsite battery storage and through an electric vehicle (EV). Ren et al. [10] have analysed the impacts of integrated PV battery systems, including net present value (NPV) and peak demand for various tariff scenarios in several Australian cities for the studied cases. Horan and Luther [14] monitored and analysed the energy performance of an EV-integrated solar home and compared the actual energy performance of the home before and after the solar PV and EV were integrated.

Forrest et al. [16] compared the impacts of immediate charging, smart charging, and vehicle-to-grid charging on the scale of infrastructure required to meet renewable utilisation targets, and highlighted the importance of intelligent plug-in electric vehicle charging. Sarabi et al. [17] explored the potential of plug-in electric vehicles for vehicle to grid charging, based on statistical data; this research provides a full picture for the utilisation of electric vehicles.

Sizing battery storage for grid connected residential solar PV applications has been studied extensively. Research conducted on optimising battery sizing for residential projects is relevant and apparent [11,13,18–20]. Most of the existing research deals with the prediction of energy generation, its time-of-use and load calculation. Even the computation of available solar energy harvested by a PV system of a particular size is considered.

While some work uses existing 30 minute interval data to build an optimal model, the data is often not directly used to calculate and anticipate load time-of-use [21]. Relevant to the work presented here, that paper presents algorithms for determining optimal battery sizing, for a particular PV array and based upon time-of-use. It considers the dynamic use of collected energy, periods of mismatch as well as missed opportunities for storage.

Battery size is an important factor for onsite energy storage in solar homes. As there is a considerable difference between the tariffs to supply electricity from the grid and to feed electricity into the grid, this is an important topic for solar homes. There are several estimating tools for solar homes regarding solar PV sizing and onsite battery sizing. The calculator [22] offers a method to calculate solar PV size based on annual electricity consumption and solar hours per day at a specific location to achieve net zero electrical emissions. Solar Calculator [23] uses a house postal code, average bill, roof design, solar system size, and battery size, to conduct a comprehensive analysis. Aichhorn et al. [19] highlighted the benefits of using energy storage for residential buildings and conducted the economic analysis to determine battery size based on simulation. Weniger et al. [18] analysed residential PV battery systems to examine the system sizing. Simulation and sensitivity analysis were conducted to identify appropriate system configurations. This research concluded that PV systems with batteries would not only be profitable but also the most economical solution from the long-term perspectives. Hassan et al. [11] studied the techno-economic benefits of battery storage coupled to photovoltaic (PV) generation system; in this research, simulation, optimisation, and sensitivity analysis are conducted to analyse the benefits considering different electricity tariffs and battery storage. In addition to the growth in popularity of solar homes, another concern is the impact of solar homes on the electricity grid. Agnew and Dargusch [24] explored the impact of residential solar and storage on the electrical distribution network.

Given the fact that there exists a considerable price difference between supplying electricity from the grid and feeding electricity into the grid, local energy storage with appropriate battery size is an emerging topic for solar homes. While metering data provides valuable information for deciding on energy storage, the research on using metering data, collected in the absence of energy storage, has not been extensively conducted. In addressing this identified research gap, this paper presents research on using Smart Meter data, to determine the impact of energy storage on cost.

Another significant aspect of the literature is in regard to battery type, recharging cycles, deterioration and lifespan [9,12,17,19]. The affordability of battery storage critically depends on capital cost and life span. Estimating battery life span and optimising battery management to increase life span is difficult [21]. Furthermore, determining the optimal size of a battery according to its optimal daily use is a challenging decision. All the factors as well as their time-of-occurrence need to be considered: charging and discharging, depth-of-discharge, minimum and maximum cycling limits, battery currents and battery temperature. In [21], Muenzel et al. have studied Li-ion battery cycling and discovered that cycles that fully charge and discharge them degrades these batteries more than small more frequent cycles.

Our work is distinct because it is based on an existing PV system with actual generation and user patterns for a house with a hybrid EV rather than averaged or modelled data. It utilises a year of Smart Meter data, solar PV data and EV charging data. The Smart Meter, installed by the electricity provider, records the quantity of electricity supplied from the grid and fed into the grid at half-hourly intervals. Applying first principles to actual data is worthwhile because averages and models may hide unexpected behaviour. For example, in summer, the maximum and minimum daily energy harvested was 21.2 kWh on 11 December 2018 and 5 kWh two days later. In winter 2019, the maximum was 9.17 kWh on 7 May and the minimum was 2.17 kWh on 10 May.

The main assumptions in this analysis are that the battery is charged only from the solar PV system and not from the grid and that energy is drawn from the battery before drawing energy from the grid [14].

The use of Smart Meters in grid connected systems is the norm for most households in the state of Victoria, Australia. Smart Meter data can be used to determine the effects of Solar PV system size and battery storage on a particular household according to its user patterns. As the immediate past two years of Smart Meter data can be downloaded from the network supplier, our method can be applied to any household considering adding battery storage [14].

In addition to Smart Meter data, solar data is required to calculate the impact of PV system size. The resulting analysis shows the effect of both PV system size and battery size on a particular household and its energy use and may suggest alternatives to manage excess electricity and other means of energy storage.

The question of whether to invest in solar PV or batteries or some mix, depends on one's aims. Many domestic installers aim to reduce their carbon footprint, so a PV system which achieves a "net zero energy" goal may be chosen. Other installers aim to reduce or eliminate the recurrent cost of energy, in which case a larger PV system would be the choice. Some users may wish to shift the time of peak load, provide solar power throughout the night or exploit off-peak grid supply rates, so batteries would be required. In country areas, some may aim to work "off-grid", so a still larger PV system and battery installation would be the choice.

In this paper, based on grid-connected houses, we investigate the relationship between imported and exported grid energy, solar PV system size, battery size, house load and night-time load, and energy costs based on a simple flat tariff. When a house does not have solar energy, or a record of solar energy is not available, we used solar data from a nearby house as a proxy. This study shows how the various elements of the system interact to map out the landscape of domestic solar PV systems in an understandable way for the homeowner.

It is not the purpose of this paper to replace the expert advice available from qualified installers, but rather to enable meaningful discussions between parties. We ignored, for example, energy losses in inverters and cables and round-trip losses in batteries. Furthermore, we did not account for variable tariffs and we assumed that batteries charge only from the solar PV system and feed only the house. Charging batteries from and discharging them to the grid is not considered. Nevertheless, the analyses done reveal how the parts of an installation interact and relate to each other in an understandable way.

3. The Move towards Going Solar

When first considering solar PV, finding uses for solar energy when it exceeds the load, rather than passively exporting it to the grid, was also considered. The outcome was a decision to purchase a plug-in hybrid electrical vehicle as well as installing a 3 kW solar photovoltaic (PV) system. The additional load of charging the vehicle, being compensated by the solar harvest, did not affect the annual grid demand or cost of electricity. As a result, comparing the years before and after installation, about 4 MWh of electrical energy was generated and AUD 1300 saved by reducing petrol purchases by over 80%. Overall, estimated CO_2 emissions from the household were reduced by 54% [5].

For this case, we present an analysis of data relating to energy use within the house. In Victoria, Australia, electrical data consumption is available from the provider for the entire year for each half-hour using Smart Meters, which were installed in 2013. Data from the vehicle was recorded before and after each charging event and when the car was refuelled. As a result, the data was available to do a before and after analysis of energy generation and consumption. Additional instrumentation records the energy generation from the solar PV and charging energy of the vehicle [5].

After installing solar PV and a hybrid car, the next questions to consider are the impact of the PV system size and whether a battery for energy storage is a viable option. Here we quantify the relationships between system size and harvested solar energy and energy imported from and exported to the grid. We also estimate how energy costs fall as system size grows as we explore the effect of introducing batteries of various size into the system. Again, this was accomplished using Smart Meter data from the power supply company and solar data recorded from the PV system controller. The conclusion reached is that increasing PV system size is much more effective than installing batteries, (at the moment) both in terms of energy and cost [14].

Finally, based on the work performed thus far, we consider its application to other households: one, a household that has a solar PV system; and one that has no solar PV. Based on the Smart Meter data from each house, and solar data gathered at the first household, we conclude that suitable guidance can be provided to individual households for solar PV and battery sizing based upon their energy usage patterns.

3.1. Working with Smart Meter and Solar Data

Energy related to a house are the grid energy and, if a PV system is installed, solar energy. Therefore, in a time interval, n:

$$E_H(n) = E_G(n) + E_S(n) \tag{1}$$

where:

- E_H is the house load;
- E_G is the grid energy, positive if imported into the house, and negative if exported; and
- E_S is the harvested solar energy.

For billing purposes, Smart Meter data separates imported and exported energy and records them separately:

$$E_G(n) = E_I(n) - E_E(n) \tag{2}$$

where:

- E_I is the energy imported from the grid; and
- E_E is the energy exported to the grid.

In terms of Smart Meter data, the house load is:

$$E_H(n) = E_I(n) - E_E(n) + E_S(n) \tag{3}$$

Note that the imported energy and exported energy flow on the same wire, and at any instant, either the exported energy is zero when energy is being imported from the grid, or else the imported energy is zero when energy is being exported. Furthermore, in the absence of a solar PV system, there is no exported energy so that E_S and E_E are zero. Smart Meter data totalise energy in a particular half hour interval, so that, when the solar harvest and load are almost equal, grid energy can be imported for some subintervals and exported for the remainder. Therefore, the Smart Meter data can show energy both being imported and exported in the whole 30 minute interval.

3.2. The House Prior to Any Renewable Energy Installations

The original house was built in 2005 to achieve good summer and winter performance by using double glazing, insulation in ceiling, walls and floor, eaves over windows and external blinds. As it was built on fill (of a former gully), a lightweight timber and brick veneer building supported by piles engaged with footings and brick piers was used. Therefore, the building incorporated no major concrete slab to provide thermal storage. A gas-boosted solar hot water system was the only renewable energy technology incorporated.

The energy used by the house and car for 12 months prior to installing renewable energy is listed in Table 1. This includes the energy used by the car as well, anticipating the installation of the hybrid electric vehicle [5].

Table 1. Energy use prior to installing solar panels (May 2013–April 2014).

		Energy in	Payment
House	Electricity Grid	3806 kWh	AUD 1524.26
	Gas	3842 kWh	AUD 479.58
	Solar HWS	Up to 1694 kWh available [23]	
Car (9696 km)	Petrol	10,685 kWh (1113 L × 9.6 kWh/L)	AUD 1679.03
		Total running cost:	AUD 3682.87

3.2.1. The House with Solar PV and a Hybrid Car

This work began with considerations arising when contemplating a solar PV system. In particular, there would be times when energy harvested by the system would exceed the house load and would be exported. As the feed-in tariff was low, the question to be asked was whether there would be uses for this excess harvested energy which would be of benefit, either monetarily or environmentally? Could the imported energy be reduced by the excess of another period? Water heating was ruled out, because a gas-boosted solar hot water system was already installed; air conditioning was deemed unnecessary because of the good house performance; and battery storage was too expensive. However, replacing a vehicle was necessary at the time and a plug-in hybrid vehicle was purchased.

A 3.0 kW photovoltaic system was switched on 1 July 2014 and harvested 3.94 MWh of energy to 30 June 2015 or 10.8 kWh/day. The system has 12 solar panels rated at 250 W each with a microinverter generating 240 V AC power, which reports to the system controller. (In hindsight, due to the experimental nature of this household case, it was a wise decision to select panels with independent microinverters. This means that each panel converts its DC power into 240 AC, allowing for easy future expansion of the solar array without the need to purchase a new stand-alone inverter). The total active area is 20 m^2, so the harvested power density averages 22.8 W/m^2 over a year. (The power density of 22.8 W/m^2 is found by averaging over a whole year including night-time when the solar power is zero). This figure is 1.7% of the solar constant of about 1350 W/m^2 [25]. It is also 12% of the annual averaged direct normal radiation of 193 W/m^2 as reported, for example, in [26]. Some of the difference is caused by the innate efficiency limits of silicon solar panels and the angle of installation, flat on the roof of about 5° incline. The power density is consistent with (of the same order as) that reported for the United Kingdom in [27].

The annual energy flows to and from the house and hybrid car are listed in Table 2 and illustrated in Figure 2. Costs are also listed. Energy from five sources, the grid, the solar panels, the gas supply, petrol supply and the solar hot water service, flows to the house and the car. The incoming electrical energy of almost 8 MWh is shared almost equally between the grid and the solar harvest. The house takes about 4 MWh, the car, 1.9 MWh, and the balance of about 2 MWh is in excess of demand from time to time during the day and is exported.

Table 2. Energy use and costs with solar panels and plug-in hybrid vehicle.

		Energy in (kWh)	Payments	Energy out (kWh)	Receipts
House	Electricity grid	4017	AUD 1641.98	2047	AUD 145.20
	Solar panels	3941			
	Gas	3573	AUD 544.56	3573	
	Solar HWS	1694		1694	
	Electricity use			4019	
Car	Petrol (191 L)	1833	AUD 250.90	1833	
(10,519 km)	Electricity			1892	
	Total		AUD 2437.44		AUD 145.20
	Total energy (kWh)	15,058		15,058	
	Net running cost		AUD 2292.24		

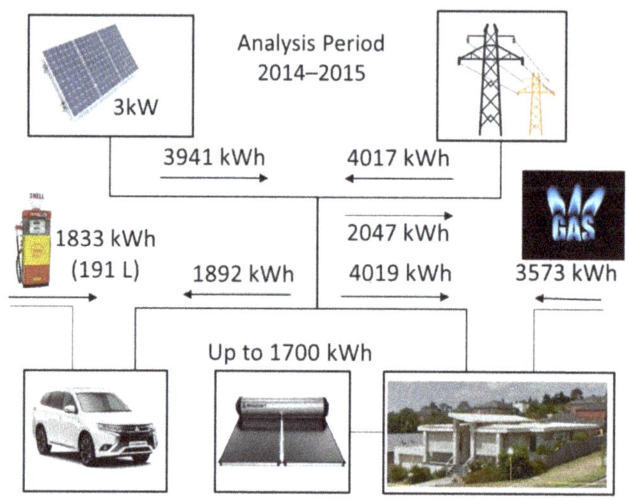

Figure 2. Household energy flows (July 2014–June 2015).

3.2.2. Varying the Solar PV System Size

How do the energy flows change when the solar PV system size changes? Scaling the solar energy by a factor k, new values for the imported and exported energy are:

$$E'_G(n) = E_H(n) - kE_S(n) \tag{4}$$

Splitting the grid energy into imported and exported components

$$E'_I(n) = E_H(n) - kE_S(n) \tag{5}$$

when $E_H > kE_S$; and:

$$E'_E(n) = kE_S(n) - E_H(n) \tag{6}$$

otherwise.

Summing Equations (4)–(6) over the year, the grid energy is:

$$\sum_n E'_G(n) = \sum_n E_H(n) - k\sum_n E_s(n) \tag{7}$$

The imported energy when the load exceeds the solar harvest, $E_H > kE_S$ is:

$$\sum_n E'_I(n) = \sum_n E_H(n) - k\sum_n E_s(n) \tag{8}$$

The exported energy when the harvest exceeds the load, is:

$$\sum_n E'_E(n) = \mathrm{k}\sum_n E_S(n) - \sum_n E_H(n) \tag{9}$$

Splitting the annual grid energy into imported and exported components in Equation (7):

$$\sum_n E'_I(n) - \sum_n E'_E(n) = \sum_n E_H(n) - k\sum_n E_s(n) \tag{10}$$

Therefore, the modified totals for the imported and exported energy can be determined for different system sizes by varying the value of the scale factor, k.

Figure 3 shows the result. The harvested solar energy changes linearly, from zero through 4.0 MWh at the nominal 3 kW system size. The imported grid energy falls from the zero solar estimated house load of 5.4 MWh, through 4 MWh drawn with the installed 3 kW PV system to 2.94 MWh with a 12 kW system. With a very large PV system, such that it cancels all imported energy during the day, the night-time energy of 2.1 MWh must still be imported [14]. The procedure for constructing Figure 3 is listed in Appendix A. Figure 3 also shows the net zero electrical energy point when the annual exported energy equals the imported energy.

Rearranging Equation (10):

$$\sum_n E_H(n) - \sum_n E'_I(n) = k\sum_n E_S(n) - \sum_n E'_E(n) \tag{11}$$

In other words, whatever the PV system size, the difference between the load energy and the imported energy is always equal to the difference between the harvested solar energy and the excess energy exported independent of the solar system size.

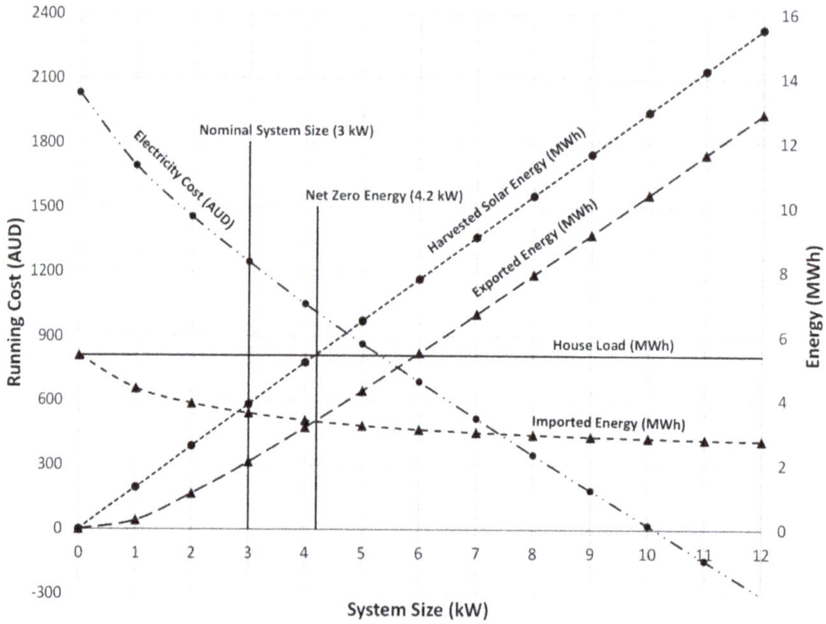

Figure 3. House 1: Energy flows and cost of electricity as a function of PV system size.

It is important to note that Figure 3 is constructed using the Smart Meter data, the solar PV data and the house load as described in Appendix A. By summing the imported energy when no solar energy is being generated at night, the lower limit on the imported energy can be calculated. Additionally, by recognising that when the solar system is small, all the solar energy feeds the load, completely displacing the imported energy; this means that the derivative (slope) of the imported energy at zero size is equal and opposite to the derivate of the harvested solar energy.

The cost of electricity, also shown in Figure 3, can then be calculated from the tariff (AUD 0.30/kWh flat rate; tariffs are for the year 2014–2015), the feed-in tariff (AUD 0.14/kWh) and the supply charge (AUD 1.14/day). For a small solar system, the harvested energy displaces imported energy at the import tariff. A large system harvests energy which, if greater than the house load, is exported at the feed-in tariff. A 1 kW PV system creates an annual saving AUD 339 and at today's prices would cost about AUD 1000. Therefore, its payback period is just under 3 years. A 12 kW system saves AUD 2334.85 annually; if it costs AUD 12,000, it has a payback period of just over 5 years.

3.2.3. CO_2 Equivalent Emissions

Table 3 summarises the CO_2 equivalent emissions of the house and car for the 3 kW case. The estimated emissions increase when the plug-in hybrid is used without installing solar PV. This result applies in Victoria because the emissions factor for brown coal is so high [28]. (At time of writing, the reported emissions intensity of electricity in Victoria has fallen by 13% [29]). However, with solar PV, emissions are reduced by 65%. On the other hand, the hybrid vehicle reduces petrol consumption by 84% and a pure electric vehicle would use no petrol at all.

Table 3. Summary of CO_2 equivalent emissions (July 2013–June 2015).

	Before Solar		Hybrid Only		Solar PV Only		Solar and Hybrid	
		CO_2-e		**CO_2-e**		**CO_2-e**		**CO_2-e**
Grid (net)	3806 kWh	4301 kg	5911 kWh	7090 kg	78 kWh	88 kg	1970 kWh	2226 kg
Gas	3842 kWh	711 kg	3573 kWh	661 kg	3573 kWh	661 kg	3573 kWh	661 kg
Petrol	1113 L	2451 kg	191 L	421 kg	1207 L	2658 kg	191 L	421 kg
	Total	7463 kg	**Total**	8172 kg	Total	3407 kg	**Total**	3308 kg

3.2.4. Working with Batteries

At any time, the solar harvest exceeds the load of the house and it is in excess; or, the load exceeds the harvest and there is a shortfall. A battery stores energy for later use by charging from the excess solar energy or by importing energy from the grid. For this analysis, we exclude the case of charging from the grid, so the battery is charged when solar energy is in excess and provides energy when there is a shortfall, either at night or during the day, when the load exceeds the solar harvest. If the battery is fully charged, excess solar is exported to the grid; and, when discharged, the shortfall is imported (Figure 4).

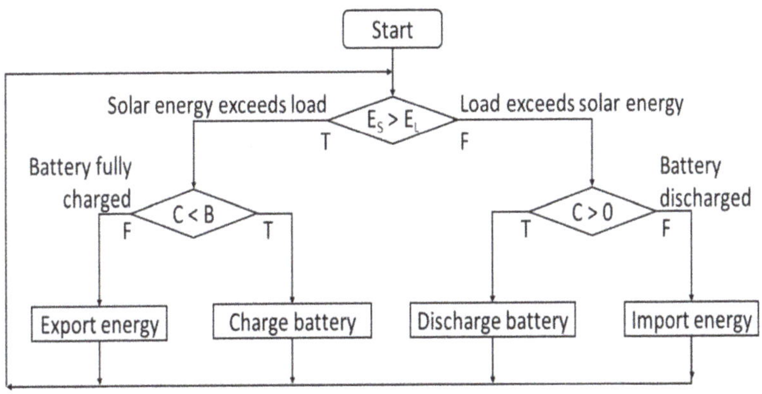

Figure 4. Simple battery control (solar charging only).

Battery storage can increase savings by reducing grid imports and exports of energy which arise from a mismatch of the solar harvest time and the energy usage time [11,12]. Figure 1 is an example of a period of gross mismatch for the house investigated, when an electric vehicle is charged from 16:15. By using battery storage systems the variable output of PV to the grid is minimised and anticipates better control [3,5,11].

As shown in [14], a battery reduces the energy imported and exported depending on its size. In Figure 5, the effect of introducing a battery, as a function of its size, and for several PV system sizes is shown. These graphs are constructed as described in Appendix B. As the PV system size is increased, the imported energy drops and the exported energy increases. In each case, as the battery size grows, the energies fall from their initial values. Storing excess energy occurs during the day, when the solar energy exceeds the load. Savings are created by replacing the night-time energy drawn from the grid with energy drawn from the battery. Drawing energy from the battery will also occur during the day whenever the load is greater than the harvested solar energy. Savings from using a battery are limited. For example, a 12 kW PV system feeding a 14 kWh battery (the capacity of a Tesla Powerwall II) saves about AUD 600 annually.

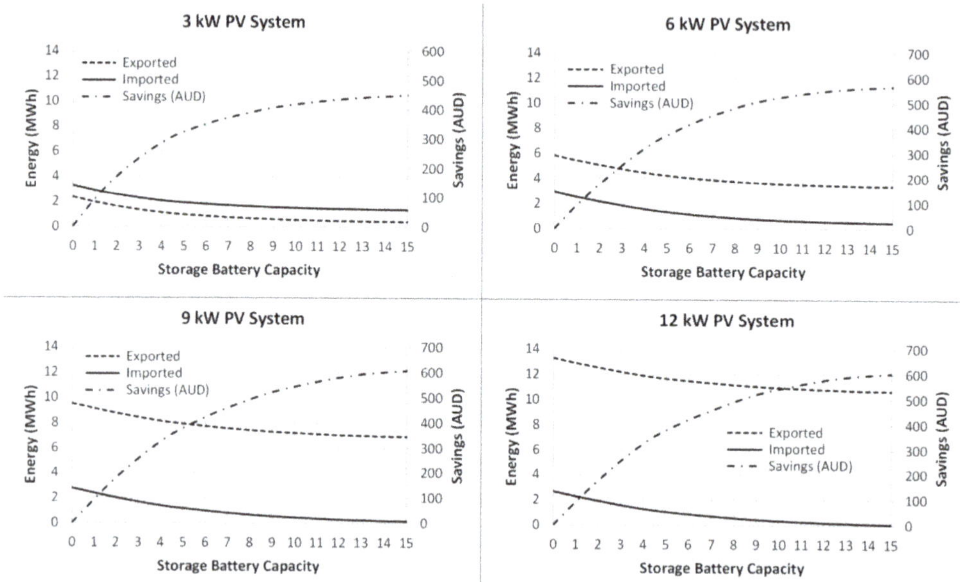

Figure 5. Energy flows and cost savings as a function of battery size and PV system size.

With a sufficiently large solar PV system, energy imported from the grid during the day will be minimised, in the limit, to zero. Without a battery, any excess solar energy will be exported. A battery will reduce the exported energy provided it is not fully charged. A large battery coupled with a large solar PV system can reduce the imported energy at all times to zero, allowing off-grid operation. Off-grid operation requires a battery large enough to support the house load over several days in mid-winter. However, then a sufficiently large enough solar PV system is required to charge the battery during short mid-winter days.

3.3. House 2 with a 2.25 kW Solar PV System and a AUD 0.66/kWh Feed-In Tariff

In the case of the second house, solar data was not available, only Smart Meter data. Without data on the solar harvest the effect of varying the solar system size cannot be calculated directly. However, we used the solar data from the first house as a proxy as the two houses are separated by 4.2 km and the solar intensity will be similar at both sites. This means that the effect of varying the solar system size and introducing battery storage can be calculated (Figure 6). Furthermore, this accounts for variable weather which will be similar at the two houses.

A problem with using proxy solar data is matching the two solar installations. Using nominal sizes, as here, suffers from differences in installation, such as orientation and tilt, and from differences in the actual power developed by each set of panels, independent of geometrical issues. Both systems are oriented towards north, but, at the first house, panels are tilted at about 5°, whereas, at the second, they are tilted at about 25°.We took the approach that the nominal power of each system equals its actual power, being alert for any major anomalies, such as times when the house load, calculated by subtracting the exported from the harvested energy, is negative. This is so for 3% of the readings. Increasing the proxy solar figure by 10% reduces this to 1.5%.

The point in Figure 6 referring to the subsidised AUD 0.66 feed-in tariff is the annual running cost under an old, but still active, "Premium Feed-in Tariff". This subsidy was introduced in 2009 and closed in 2011. The annual saving for this case is AUD 1504 For a PV system of 2.5 kW costing about AUD 6000 in 2011 prices, this corresponds to payback period of 4 years. However, since electricity tariffs were lower at the time, the annual saving would have been greater and the payback period less. Figure 6 also shows the net zero energy point at which the annual exported energy is equal to the annual imported energy, and that a system size greater than 8.25 kW would reduce the running cost below that achieved with the AUD 0.66 feed-in tariff.

As solar energy is not available during the night, the total imported energy can never be zero without using battery storage. Indeed, the imported energy has a horizontal asymptote of 3.11 MWh independent of any scale error arising from the foreign solar data. This figure is obtained from the imported energy when the solar energy is zero. As it depends only on imported energy, the estimate is free of scaling errors and will be a good estimate.

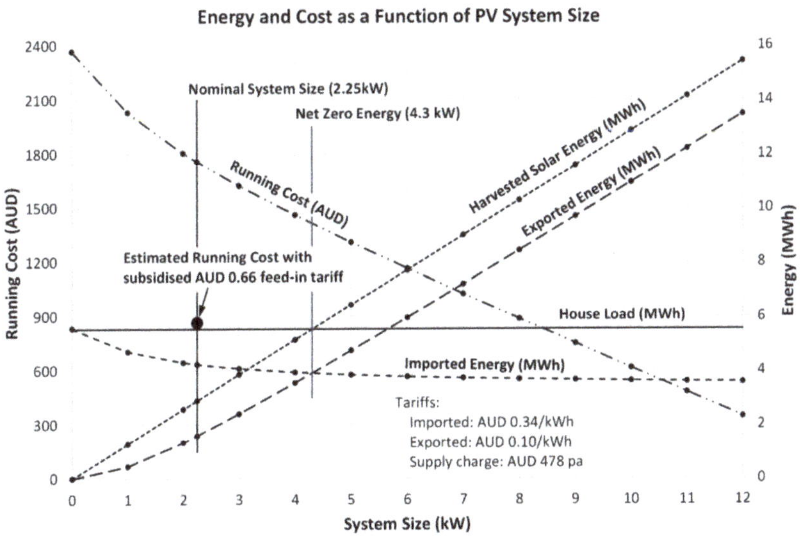

Figure 6. House 2: Energy flows and cost of electricity as a function of Solar PV system size.

To explore the question of how sensitive are readings taken from Figure 6 to errors in mismatching the target and reference solar PV system sizes consider Figure 7. This shows the modelled imported energy using both the nominal size of the proxy solar PV system and double that size. The imported energy is known and is independent of any mismatch of scale between the original and target systems. Such a mismatch will alter the estimate of the house load, which is the value of energy when the PV system size is assumed to be zero. Using the proxy solar energy unscaled, the house load is 5.53 MWh. If the proxy solar energy is doubled, the house load is estimated to be 8.42 MWh. Therefore, the actual house load is very sensitive to scaling errors.

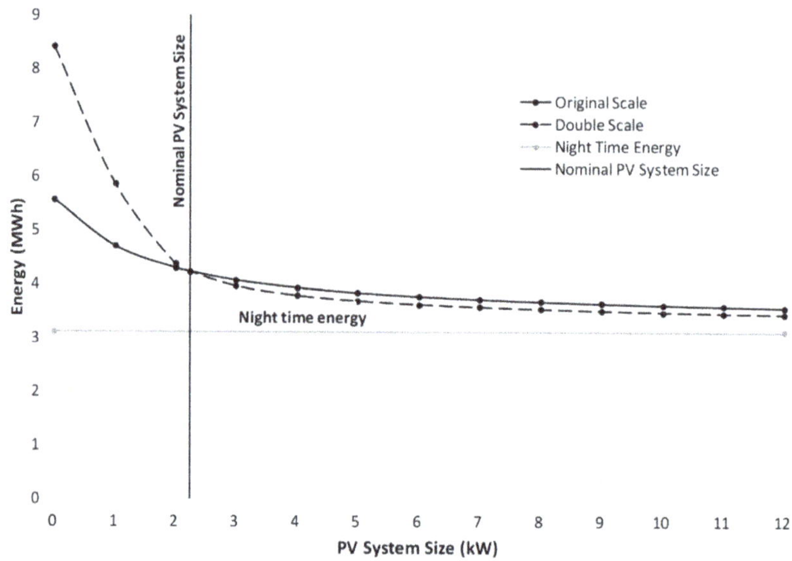

Figure 7. The effect of doubling the proxy solar data on the imported energy.

However, in the case of the target house, with a 2.25 kW PV system, the imported energy is 4.23 MWh, and the night-time energy is 3.11 MWh as noted above. Therefore, the imported energy curve is confined between the imported energy at nominal size and the night-time energy. It is tightly controlled and a good estimate of imported energy. That is, for PV systems larger than the nominal, error arising from using the proxy solar data to estimate the imported energy is small.

On the other hand, the curve shows that estimates below the known, nominal size, are very sensitive to mismatches between the actual and proxy solar data. However, reducing the size of a solar PV system is probably not of interest in normal situations. The answers users would most likely seek is to the effect of increasing the size of a solar PV system.

3.3.1. CO_2 Equivalent Emissions

Table 4 lists the net grid energy for House 2, which shows that the CO_2-e is large. This is because the installed solar system is much smaller than that required to achieve net-zero energy. Figure 6 shows that a system of about 4.2 kW is required to achieve this. However, because the subsidy would no longer be applied, the annual running cost would rise by about AUD 500.

Table 4. House 2: CO_2 equivalent emissions (2018).

	Energy	CO_2-e
Grid (net)	2637 kWh	2822 kg
Gas (est)	4863 kWh	900 kg
Diesel (est.) 25,000 km	2750 L	7483 kg
	Total	11,205 kg

3.3.2. Introducing Batteries

As Figure 7 shows, for House 2, the energy drawn at night in 2018 is 3.11 MWh. Figure 8 shows the total annual imported and exported energy and cost savings as a function of battery size for various sizes of the solar PV system. The maximum possible saving is about AUD 700 with a 15 kW battery and a 12 kW PV system. A Tesla Powerwall II would save about AUD 600.

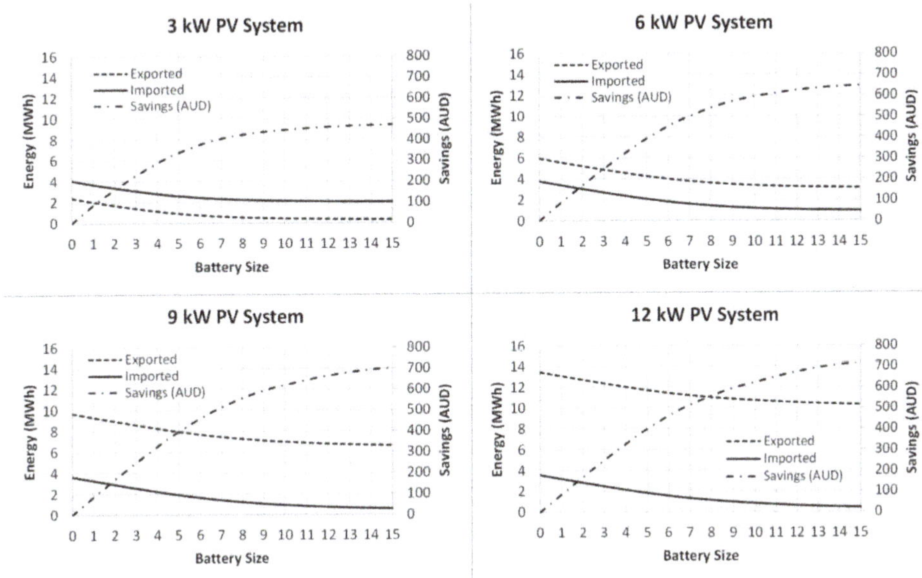

Figure 8. House 2: Energy flows and cost savings as a function of battery size.

The maximum possible savings resulting from installing a battery is higher in the second house than the first, which can be traced to a higher load overall, but mainly at night.

3.4. House 3 with No Solar PV System

Information for this house is available only from the Smart Meter data. This means that the load to be served is known, but the night-time energy cannot be deduced from the Smart Meter data because there is no solar energy data. There are two ways to estimate the night-time component, by using the proxy solar data or by using times for sunrise and sunset at the house location.

Figure 9 shows the energy flows and cost of electricity for House 3 which has no solar PV system. Nevertheless, it is similar to Figures 3 and 6. The parameters which set the curves for this house are the house load (4.6 MWh), the night-time load (2.1 MWh), the ratio of annual energy harvest to PV system size (1.3 MWh/kW), the grid tariff (AUD 0.341/kWh) and the feed-in tariff (AUD 0.106/kWh).

In a house with no solar PV, the nominal size is zero and the curve of imported energy is confined between the total house load and the night-time house load. A solar PV system size of 3.5 kW will achieve annual net-zero emissions, and a 12 kW system will reduce running cost to zero.

Estimating night-time load is a problem without solar data. Here, the solar data from House 1 is used as a proxy and can be used as a mask to select the load when solar energy is unavailable. Otherwise, times of sunrise and sunset at the house can be used.

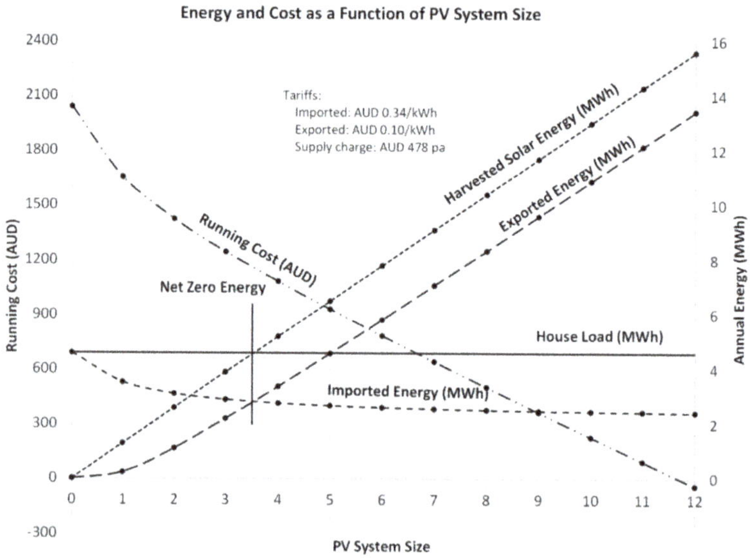

Figure 9. House 3: Energy flows and cost of electricity.

3.4.1. Introducing Batteries

Batteries in House 3 have characteristics similar to House 1 because the night-time energy is similar (Figure 10).

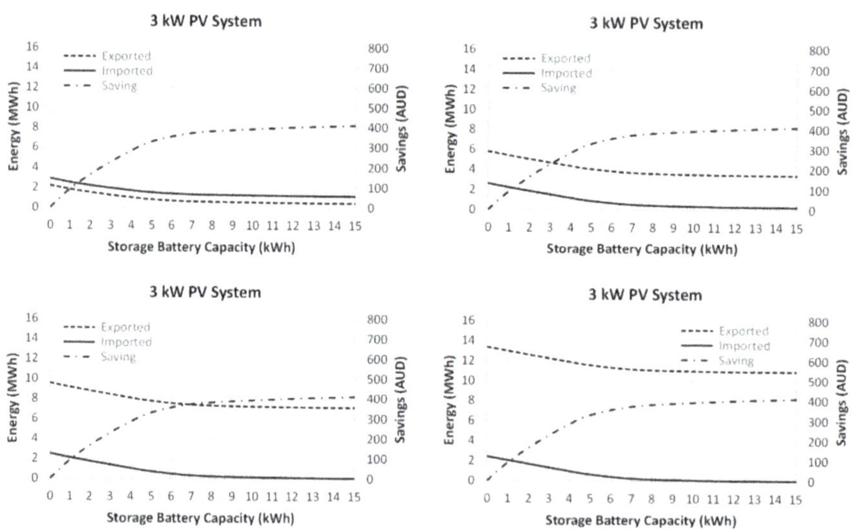

Figure 10. House 3: Energy flows and cost savings as a function of battery size.

3.4.2. CO_2 Equivalent Emissions

Table 5 lists the net grid energy and CO_2-e for House 3. A solar PV system of about 3.5 kW would achieve net-zero grid energy, and so zero CO_2-e annually. Unfortunately, gas and petrol data became unavailable since the project began.

Table 5. House 3: CO_2 equivalent emissions (2018).

	Energy	CO_2-e
Grid (net)	4574 kWh	4711 kg
Gas	n/a	
Petrol	n/a	
	Total	4711 kg

4. Comparing the Three Houses

The data for the three houses is collated in Table 6. Houses 1 and 3 are similar, as they have a similar night-time load, but House 1 also supports charging the hybrid car. House 1 is 14 years old, built with an eye to energy performance with double glazing, and insulation of walls, ceiling and floor, supports the hybrid car at no extra cost. Not listed in the table are the avoided cost of petrol for the hybrid. The night-time load in House 2 is 50% greater. One difference may be that they are a working couple, whereas the occupants of Houses 1 and 3 are retired.

Table 6. Comparison of the three houses (2018 data).

	House 1	House 2	House 3
Grid tariff		AUD 0.341/kWh	
Feed in tariff		AUD 0.106/kWh	
Supply charge		AUD 526.05	
Annual load	5.4 MWh	5.5 MWh (est.)	4.6 MWh
Night-time load	2.1 MWh	3.1 MWh	2.1 MWh
Installed PV system	3 kW	2.25 kW	n/a
PV system size achieving net zero electrical energy	4.2 kW	4.2 kW	3.5 kW
Electricity cost without solar	AUD 2164 (est.)	AUD 2040 (est.)	AUD 2085
Electricity cost with installed PV system	AUD 1399	AUD 850 (AUD 0.66/kWh FiT) AUD 1750 (AUD 0.10/kWh FiT)	n/a
Annual saving	AUD 765	AUD 1190 (AUD 0.66/kWh FiT) AUD 290 (AUD 0.10/kWh FiT)	
Electricity cost with 6 kW PV system	AUD 904	AUD 1150 (AUD 0.10/kWh FiT)	AUD 785
Annual saving	AUD 1260	AUD 890	AUD 1300
Maximum possible saving with battery	AUD 642	AUD 827	AUD 560
Saving with 6 kWh battery	AUD 413	AUD 445	AUD 437
Total saving with 6 kW PV system and 6 kWh battery	AUD 1673	AUD 1335	AUD 1737

5. Matching the Solar PV Harvest to the House Load

The two factors to consider, first, in sizing a solar PV system and battery are the load and the solar harvest. Is the daily solar harvest sufficient to support the load for the whole day, in particular at night when there is no solar energy available? Figure 11 shows the daily harvest from a 10 kW PV system and the load for House 1. Because of the latitude of the house, the harvest from a PV system peaks in the summer months and falls in the winter months. The shorter days and the panels flat on a 5° sloping roof emphasise the variation. From about the beginning of September to the beginning of April, the solar harvest is usually sufficient to support the load, the excess being exported to the grid. In the remaining months, the solar harvest would be insufficient on a number of days. When the harvested energy is insufficient, energy is imported from the grid.

The 10 kW PV system covers 109 days of the 153 from 1 April to 31 August. A larger PV system will make up for some of the remaining days. For example, if the minimum harvest on a particular day is 10 kWh, but the load is 20 kWh, the system size would need to be doubled to compensate. A 20 kW PV system will cover an extra 33 days for a total of 144 of the 153 days. The extra capacity of 10 kW covers one third of the days covered by the first 10 kW and so is less economic.

Figure 11 does not show that the energy harvested must be stored in a battery for later use at night. In Figure 12, the house load has been filtered to exclude the load energy when solar energy is being harvested; what remains is the night load, averaging 8.4 kWh a day. The imported energy, without a battery, exceeds the night load somewhat, because some energy is imported during the day when the load exceeds the solar harvest. Employing a battery (10 kWh in Figure 12), harvested energy can be stored, so that the imported energy is significantly reduced, but not completely, saving 2.31 MWh in the year. Of course, a 10 kWh battery may reduce the night time load by, at most, 10 kWh.

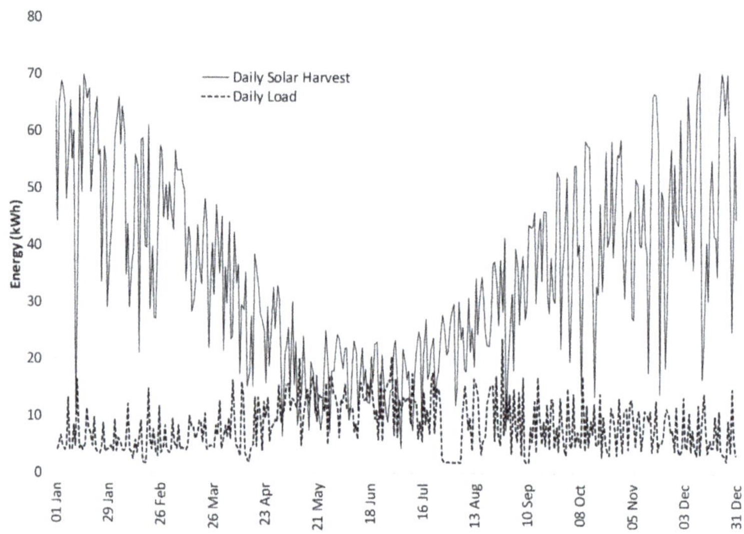

Figure 11. Daily 10 kW PV solar harvest and load for House 1.

Of interest is the higher night-time load, averaging 10.5 kWh a day, in the colder months between 1 April and 31 August. In the warmer months, the average night-time load is 6.9 kWh a day. This difference is because winter night times are longer, there is less opportunity to charge the vehicle in daytime hours and energy for lighting and driving fans associated with gas heating is higher.

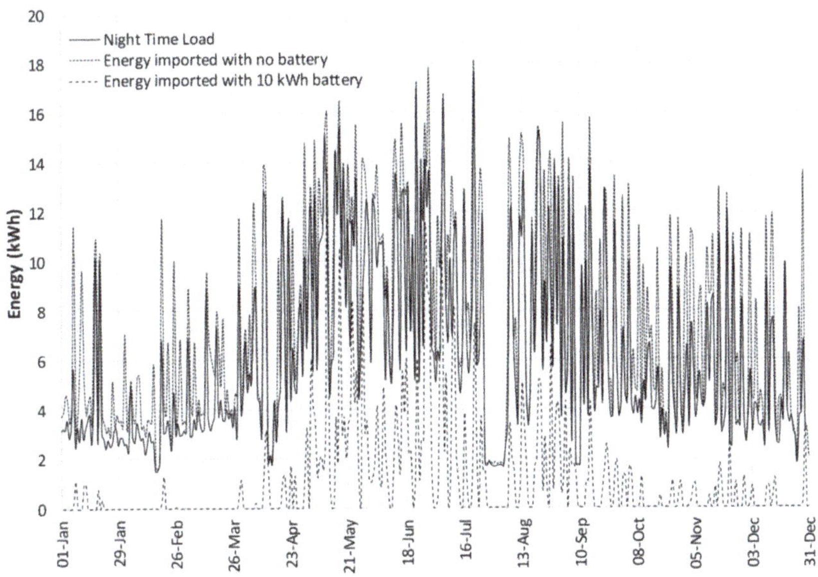

Figure 12. Night load and energy imported with and without a 10 kWh battery.

6. Guidance

Figures 3, 6 and 9 show the performance of each house as a function of PV system size. These figures depend on the house load, the night-time load, the solar PV system size, the imported and exported grid energy, and the tariffs in place. The solar PV data for House 1 was collected, but, in the absence of solar data, the harvest may be determined from data supplied by [25]. The Annual Averaged Direct Normal Radiation reported for the location of House 1 is 4.16 kWh/m^2/day. The 3 kW system occupies 20 m^2 and is about 13% efficient, so harvests 3.95 MWh per annum. This sets the gradient of the Solar Harvest line. The actual measured harvest shown in Figure 2 is 3.94 MWh.

Smart Meter data provides the values of imported and exported energy for the nominal PV system size in half hourly time intervals. The house load can be evaluated from Equation (3). By excluding Smart Meter data when solar energy is being exported, what remains is the night-time energy imported from the grid. This provides the lower limit of the imported energy. The energy costs follow as described earlier. Then the solar energy can be scaled and the imported and exported energy recalculated using Equations (8) and (9). Totalised over the whole year gives the data for Figures 3, 6 and 9 determined similarly, except that the solar data is supplied from that determined for the first house. These graphs depend directly on the data gathered, and no simulation or modelling is done, except for modelling the solar energy at House 2 and House 3 by proxy from House 1. This is a key result of this work.

The imported energy decreases from the house load to the night-time load as the PV system size is increased, and it is tightly constrained by these two load quantities as Figure 7 shows. That is, errors in PV system size when using the solar data from the first house will not be serious. For the same reason, without solar data to model half hourly variations, the annual imported energy as a function of PV system size may be modelled by fitting a curve to the house load, the night-time load and the initial gradient of the imported energy curve.

Figures 5, 8 and 10 show the cost saving achievable with a battery. This saving is limited by the night-time load, which a battery can supply after being charged by the PV system. Figure 12 shows the state of charge of the battery throughout the year. The winter performance is critical. If the battery is chosen to cater for the night-time load in winter, how big does the PV system need to be to charge it sufficiently?

The process of estimation starts from the house load. Knowing that, solar PV systems may be chosen to achieve objectives such as:

- annual net zero electrical energy use (~4.2 kW for House 1);
- net zero electrical energy use from 1 April to 31 August (~7 kW for House 1);
- annual net zero running cost without a battery (~10 kW for House 1);
- matching average winter night-time load (10.5 kWh for House 1).

The battery capacity may be chosen to lie between a lower limit such that the battery is frequently charged in winter by the solar PV system and an upper limit such that it is rarely discharged by the load overnight, but also not always fully charged at night. This latter constraint implies that the solar PV system is large enough to frequently charge the battery in winter. The average night-time load from 1 April to 31 August 2018 is 8.15 kWh with a standard deviation of 3.91 kWh. Assuming a normal distribution, a 10 kWh battery would cover 68% of the night time loads. Based on our data, a 10 kWh system will support the total load on 109 of 153 days in these months, or 71%.

7. Realisations

During the course of this work, we realised a number of factors which affect the performance of solar PV systems. First, latitude is an important element. The disparity between summer and winter performance is greatly affected by latitude. At the latitude of this study, 38°S, winter and summer harvests differ significantly, by as much as a factor of 3 or 4 (Figure 11). This is because a mid-winter day is two-thirds the length of a mid-summer day and the maximum altitude of the sun changes from 76° at midsummer to 29° at midwinter. On the other hand, in tropical latitudes, the altitude remains high all year round and the day length changes little, so there is less need to design for winter performance. For a given load, a smaller PV system is likely to suffice, and winter performance will not be so critical.

Secondly, the night-time load is significantly higher in winter, 60% in the case of House 1. This is not evident from the daily load data, but only when the night-time load itself is studied. Furthermore, the winter night-time load is the important case in considering battery size.

Thirdly, the availability of Smart Meter data is important. If a solar PV system is not installed, the Smart Meter data is the house load. It may be combined with sunrise and sunset times to extract the night-time load. Hence, combined with the efficiency of the solar panels, the performance of a new solar PV system can be calculated accurately.

Finally, if records of the solar harvest are unavailable, either because no system is installed, or because records are not available, records from a nearby property can be used as a proxy, because weather conditions are similar. It turns out that the imported energy is controlled strongly by the house load and the night-time load and so can be predicted accurately (Figure 7).

Finally, newly built houses are at a disadvantage in estimation as there is no record of load to be used in the calculation of performance as, for example, in Figure 5. Theoretical estimates of load would need to be used.

8. Advantages and Disadvantages

The advantages of the work presented are:

- The methods are based on actual data collected from Smart Meters which the power supply company uses to bill data. Our two methods are empirical, and work from first principles.
- Actual solar PV data is used if available. If not, solar data from a nearby Solar PV system may be used as a proxy. This caters for similar weather conditions at the two houses.

- The methods for calculating system performance from this data are simple. They are outlined in Apendix A and Appendix B. The methods can be implemented on a simple spreadsheet.
- The parameters required for the two methods are readily obtainable from the Smart Meter and Solar data, namely, imported energy, energy imported at night (no solar energy), exported energy, solar energy harvested and tariffs.
- No modelling is used except for scaling the sizes of the PV system and battery system.

The disadvantages of the work presented are:

- Management of heavy loads have not been considered. The power required from batteries has not been taken into account.
- Losses in the system have not been allowed for.
- Managing batteries in winter by charging from the grid when necessary has not been considered.
- Using the electric vehicle to feed energy into the house has not been considered.
- Optimisation methods have not been applied to size batteries.

9. Summary and Conclusions

This study began with House 1 when a 3 kW PV system was installed and a hybrid vehicle was purchased [13]. Since then, data has been collected from the Smart Meter, solar data from the PV system and charging data from the vehicle. It was found that the hybrid vehicle avoided the use of about 1000 L of petrol annually and the solar PV system compensated for charging the vehicle, so electricity costs did not change significantly. The financial saving was from avoiding the purchase of petrol.

Smart Meter data was used to study the energy use in Houses 2 and 3. However, as the solar energy data was not available at these houses, the data from House 1, 4.2 km distant, and so subject to much the same weather conditions, was used as a proxy. This enabled the house load, in the absence of solar data to be estimated. Then the performance of the house, as the size of the solar PV system was increased, could be estimated. There is uncertainty in estimating the house load and consequently, the energy and cost benefits from increasing the PV system size. However, the imported energy is tightly bounded above by net house load (imported less exported energy) and below by the unavoidable night-time load.

When batteries are considered, the night-time energy is readily obtainable from Smart Meter and solar data, by summing the imported energy when the solar energy is zero. Batteries, by load shifting, can be used to reduce this night-time load. However, using an electric vehicle offers a greater monetary benefit than household batteries because of the high cost of petrol. (While writing this paper, a rebate supporting batteries of up to AUD 6000 became available which has not been accounted for).

The size of the PV system and the battery depends on the house load. The annual average load in the original house is about 13 kWh/day. The average night-time load is 8.4 kWh annually, but critically, this rises to 10.5 kWh in the months from April to August. The load is expected to be far less sensitive to seasonal variation, so will be better for estimation. A 10 kWh battery, matching the average winter night-time load, will cater for most behaviours.

The 3 kW PV reference system harvests an annual average of 10.5 kWh/day. This is low because the panels are flat on a steel colorbond roof. Larger PV systems harvest more energy and significantly reduce energy costs. However, the night-time energy can be reduced only by using batteries. At present, batteries are not very cost effective, but PV systems are.

The power rating of batteries was not considered in this paper. However, it is important to consider the peak power required by the load. For example, as Figure 1 shows, 1.5 kWh is being drawn in 30 min, which is a power rating of 3 kW.

Upon reflection, several important lessons have been learned in this study and in writing this paper. These may appear obvious but need to be brought to the forefront if considering solar PV size in conjunction with batteries:

- Winter periods yield about 1/3 or less energy harvest than peak summer.
- If there is no harvested energy to store in batteries in winter for evening use, batteries are superfluous, since there could be about one-third to half of the year where no solar PV storage is provided for in the night-time hours.
- A hybrid or electric vehicle makes more economic sense than a large battery for a household, since petrol costs are avoided.
- A larger solar PV system with a smaller storage battery may prove to be more viable since there will be both excess export to the grid in summer and some storage for evening use in winter. In fact, sizing the battery for winter evening use is probably the most sensible approach given the relatively low cost and rebates at present on solar PV systems.
- Given that Smart Meter data yields the average daily energy use in winter at night, this is useful for selecting battery size.
- Private CO_2 emission reductions in the State of Victoria are achieved mainly by installing a Solar Photovoltaic System. Reducing grid supplied energy displaces the use of brown coal.
- Lastly, we used recorded data in this analysis to understand the problem from first principles. Using optimisation techniques may provide answers, but often without explanation. Our approach shows how possible answers are bounded physically.

Author Contributions: All authors contributed equally to this paper. All authors have read and agreed to the published version of the manuscript.

Funding: This research received no external funding.

Institutional Review Board Statement: Not applicable.

Informed Consent Statement: Not applicable.

Data Availability Statement: Data available on request due to restrictions e.g., privacy or ethical. The data presented in this study are available on request from the corresponding author. The Smart Meter data are not publicly available as it has been obtained from a private power distribution company using identification information obtained from private householders. The authors have no commercial association with the power distribution company. The solar PV data has been obtained from the private company which monitors the PV system.

Acknowledgments: The authors wish to acknowledge those householders who kindly provided their grid and solar data.

Conflicts of Interest: The authors declare no conflict of interest.

Appendix A. Constructing the Graphs in Figures 3, 6 and 9

The following procedure was used:

- The format of Smart Meter files is defined in [30]. They are CSV files readable by programs such as Microsoft Excel. Each line corresponds to a day with 48 readings of energy taken every half hour. There are two blocks, one of imported energy and one of exported energy. A program is used to reformat the data into lines which contain both imported and exported energy for a particular half hour. One year of data is used.
- The harvested solar energy data is downloaded from the provider and consists of readings taken every 15 min. Every consecutive pair needs to be converted into a half hourly reading.
- Both sets of data are then merged into a single file of imported, exported and solar energy for each half hour.
- The house load for each time interval is then calculated using Equation (3).
- Summing each type of data, the annual imported, exported, harvested and load energies are found.
- The harvested energy can be scaled to a number of solar PV system sizes and the resulting modified values of imported and exported energy can be calculated for each

interval using Equations (5) and (6). The new annual values of imported and exported energy are found by summing these values for each system size.
- The cost of energy can then be calculated using the tariff structure.
- The resulting data is then plotted as a function of PV system size.

Appendix B. Constructing the Graphs in Figures 5, 8 and 10

The following procedure was used (see Figure 4):

- The data for a given size of solar PV system (date and time, imported energy, exported energy, harvested energy, and load energy is used).
- The battery size is selected.
- If the harvested energy exceeds the load, during an interval, and the battery is not fully charged, the excess energy is added to the battery instead of being exported.
- If the battery is fully charged, the excess energy is exported.
- If the load exceeds the harvested energy, the shortfall is drawn from the battery.
- If the battery discharged, energy is imported from the grid.
- The exported and imported energy is totalized for the whole year
- The calculation is repeated for a range of battery sizes and the calculated annual exported and imported energy is plotted against battery size.
- The cost of energy can be calculated for each battery size and compared with the cost when the battery is absent. The difference is the saving created by the battery.

References

1. Baulch, B.; Do, T.D.; Le, T.-H. Constraints to the uptake of solar home systems in Ho Chi Minh City and some proposals for improvement. *Renew. Energy* **2018**, *118*, 245–256. [CrossRef]
2. Halder, P. Potential and economic feasibility of solar home systems implementation in Bangladesh. *Renew. Sustain. Energy Rev.* **2016**, *65*, 568–576. [CrossRef]
3. Hong, X.L.; Horan, P.; Luther, M.B.; Ahmed, T.M. Informed decision making of battery storage for solar-PV homes using smart meter data. *Energy Build.* **2019**, *198*, 491–502. [CrossRef]
4. Chapman, A.J.; McLellan, B.; Tezuka, T. Residential solar PV policy: An analysis of impacts, successes and failures in the Australian case. *Renew. Energy* **2016**, *86*, 1265–1279. [CrossRef]
5. Horan, P.; Luther, M.B. Going Solar: The ramifications. In Fifty years later: Revisiting the role of architectural science in design and practice. In Proceedings of the 50th International Conference of the Architectural Science Association, Adelaide, Australia, 7–9 December 2016; pp. 1–10.
6. Nicholls, A.; Sharma, R.; Saha, T. Financial and environmental analysis of rooftop photovoltaic installations with battery storage in Australia. *Appl. Energy* **2015**, *159*, 252–264. [CrossRef]
7. Turner, L. Energy Storage Update, Renew, 145, October 2018. pp. 18–23. Available online: https://renew.org.au/renew-magazine/solar-batteries/energy-storage-update/ (accessed on 2 May 2021).
8. Vorrath, S. Australians Installed 22,661 Home Battery Systems in 2019. 2020. Available online: https://onestepoffthegrid.com.au/australians-installed-22661-home-battery-systems-in-2019/ (accessed on 19 April 2020).
9. Rowe, M.; Yunusov, T.; Haben, S.; Holderbaum, W.; Potter, B. The Real-Time Optimisation of DNO Owned Storage Devices on the LV Network for Peak Reduction. *Energies* **2014**, *7*, 3537–3560. [CrossRef]
10. Ren, Z.; Grozev, G.; Higgins, A. Modelling impact of PV battery systems on energy consumption and bill savings of Australian houses under alternative tariff structures. *Renew. Energy* **2016**, *89*, 317–330. [CrossRef]
11. Hassan, A.S.; Cipcigan, L.; Jenkins, N. Optimal battery storage operation for PV systems with tariff incentives. *Appl. Energy* **2017**, *203*, 422–441. [CrossRef]
12. Hoppmann, J.; Volland, J.; Schmidt, T.S.; Hoffmann, V.H. The economic viability of battery storage for residential solar photovoltaic systems–A review and a simulation model. *Renew. Sustain. Energy Rev.* **2014**, *39*, 1101–1118. [CrossRef]
13. Muenzel, V.; Mareels, I.; De Hoog, J.; Vishwanath, A.; Kalyanaraman, S.; Gort, A. PV generation and demand mismatch: Evaluating the potential of residential storage. In Proceedings of the 2015 IEEE Power & Energy Society Innovative Smart Grid Technologies Conference (ISGT), Washington, DC, USA, 18–20 February 2015; pp. 1–5.
14. Horan, P.; Luther, M.B. How big should my battery be? In *ASA 2018: Engaging Architectural Science: Meeting the Challenges of Higher Density, Proceedings of the 52th International Conference of the Architectural Science Association, Melbourne, Australia, 28 November–1 December 2018*; Rajagopalan, P., Andamon, M.M., Eds.; Architectural Science Association: Melbourne, Australia, 2018; pp. 249–256. [CrossRef]
15. Tant, J.; Geth, F.; Six, D.; Tant, P.; Driesen, J. Multiobjective Battery Storage to Improve PV Integration in Residential Distribution Grids. *IEEE Trans. Sustain. Energy* **2013**, *4*, 182–191. [CrossRef]

16. Forrest, K.E.; Tarroja, B.; Zhang, L.; Shaffer, B.; Samuelsen, S. Charging a renewable future: The impact of electric vehicle charging intelligence on energy storage requirements to meet renewable portfolio standards. *J. Power Sources* **2016**, *336*, 63–74. [CrossRef]
17. Sarabi, S.; Davigny, A.; Courtecuisse, V.; Riffonneau, Y.; Robyns, B. Potential of vehicle-to-grid ancillary services considering the uncertainties in plug-in electric vehicle availability and service/localization limitations in distribution grids. *Appl. Energy* **2016**, *171*, 523–540. [CrossRef]
18. Weniger, J.; Tjaden, T.; Quaschning, V. Sizing of Residential PV Battery Systems. *Energy Procedia* **2014**, *46*, 78–87. [CrossRef]
19. Aichhorn, M.; Greenleaf, M.; Li, H.; Zheng, J. A cost effective battery sizing strategy based on a detailed battery lifetime model and an economic energy management strategy. In Proceedings of the 2012 IEEE Power and Energy Society General Meeting, San Diego, CA, USA, 22–26 July 2012; pp. 1–8. [CrossRef]
20. Xiao, H.; Pei, W.; Yang, Y.; Kong, L. Sizing of battery energy storage for micro-grid considering optimal operation management. In Proceedings of the 2014 International Conference on Power System Technology, Chengdu, China, 20–22 October 2014; pp. 3162–3169. [CrossRef]
21. Muenzel, V.; De Hoog, J.; Brazil, M.; Vishwanath, A.; Kalyanaraman, S. A Multi-Factor Battery Cycle Life Prediction Methodology for Optimal Battery Management. In Proceedings of the 2015 ACM Sixth International Conference on Future Energy Systems ACM, Bangalore, India, 14–17 July 2015; pp. 57–66.
22. Calculate Solar Array Size and Cost. Available online: http://sroeco.com/solar/calculate-solar-cost/what_size_solar_system_do_i_need/ (accessed on 11 August 2017).
23. Solar Calculator. Available online: https://solarcalculator.com.au/ (accessed on 6 April 2020).
24. Agnew, S.; Dargusch, P. Effect of residential solar and storage on centralized electricity supply systems. *Nat. Clim. Chang.* **2015**, *5*, 315–318. [CrossRef]
25. Duffie, J.A.; Beckman, W.A. *Solar Engineering of Thermal Processes*, 2nd ed.; John Wiley & Sons, Inc.: New York, NY, USA, 1991; p. 5.
26. NASA. Surface Meteorology and Solar Energy–Available Tables. Available online: https://eosweb.larc.nasa.gov/cgi-bin/sse/sse.cgi?+s01 (accessed on 28 June 2016).
27. McKay, D. *Sustainable Energy–Without the Hot Air*, 1st ed.; UTI Cambridge Ltd.: Cambridge, UK, 2009; p. 39.
28. *Department of the Environment, National Greenhouse Accounts Factors–2015*; Department of the Environment, Commonwealth of Australia: Canberra, Australia, 2015; p. 18, ISSN: 2202-333X.
29. *Department of the Environment, National Greenhouse Accounts Factors–2018*; Department of the Environment, Commonwealth of Australia: Canberra, Australia, 2018; p. 19, ISSN: 2202-333X.
30. AEMO. *Meter Data File Format Specification NEM12 and NEM13*; Version 1.06, December 2017; Australian Energy Market Operator Ltd.: Melbourne, Australia. Available online: https://www.aemo.com.au/-/media/Files/Electricity/NEM/Retail_and_Metering/Metering-Procedures/2018/MDFF-Specification-NEM12--NEM13-v106.pdf (accessed on 2 May 2021).

energies

MDPI

Article

Applying Solar PV to Heat Pump and Storage Technologies in Australian Houses

Tom Simko [1], Mark B. Luther [2,*], Hong Xian Li [2] and Peter Horan [3]

1 School of Property, Construction and Project Management, RMIT University, Melbourne, VIC 3000, Australia; tom.simko@rmit.edu.au
2 School of Architecture and Built Environment, Deakin University, Geelong, VIC 3220, Australia; hong.li@deakin.edu.au
3 Faculty of Science, Engineering and Built Environment, Deakin University, Geelong, VIC 3220, Australia; peter.horan@deakin.edu.au
* Correspondence: mark.luther@deakin.edu.au

Abstract: Innovative mechanical services coupled with renewable energy systems are crucial for achieving a net zero energy goal for houses. Conventional systems tend to be vastly oversized because they lack the means to buffer energy flows and are based on peak loads. This paper presents an approach to achieve a net zero energy goal for houses by using a solar PV system, heat pumps, and thermal and electrical storage batteries, all off-the-shelf. Constraining one part of the system and then showing how to manage energy storage and flow is a paradigm shift in sizing. The design is for a modest-sized house built in Melbourne, Australia. The output of a solar photovoltaic array drives a small-scale heat pump to heat water, buffering its energy in a thermal battery to energise a radiant space heating system. Space cooling is provided by a separate heat pump. Through energy storage in electrical and thermal batteries, it is possible to meet the electricity, heating and cooling needs of the house for the Melbourne climate with a heat pump that draws less than 1 kW. The design methodology is detailed in an appendix and can be applied to similar projects. This paper contributes to similar work worldwide that aims to reinforce innovative renewable energy driven service design.

Keywords: self-consumption; solar photovoltaics; heat pumps; thermal storage; space heating; space cooling

Citation: Simko, T.; Luther, M.B.; Li, H.X.; Horan, P. Applying Solar PV to Heat Pump and Storage Technologies in Australian Houses. *Energies* **2021**, *14*, 5480. https://doi.org/10.3390/en14175480

Academic Editor: Benedetto Nastasi

Received: 22 July 2021
Accepted: 24 August 2021
Published: 2 September 2021

1. Introduction

Over the last decade, research on the self-consumption of photovoltaic electricity within residential buildings has become a significant topic [1–5]. While Stauffer et al. do not go into great detail about their service system design, they provide a detailed theoretical explanation of a calculation for a self-consumption ratio [1]. Sánchez et al. show a more detailed service system without a battery, but they select a 1 kW input heat pump to provide for increased PV self-consumption [2]. Beck et al., in their papers on the topic, do not explain the mechanical services but rather deal with battery sizing and theoretical algorithms to predict the optimization of self-consumption [3,4]. Luthander et al. provide a review on self-consumption which deals more with PV and battery sizing. The idea of running a heat pump from the renewable generated energy and putting this into thermal storage has already proven successful [6–8]. While the types of heat pumps applied to these purposes require further understanding and investigation, they do increase self-consumption through thermal storage.

The complete decarbonization of energy will depend on using heat pumps in the residential sector. This has been acknowledged by the Association of Heat Pumps in the UK [9], as well as a report on reaching 10 million heat pumps for residential buildings by 2030 [10]. Another article presents a solar thermal assisted heat pump system in support of global emission reduction [11].

Several of the papers noted above emphasised the importance of optimized control algorithms to minimise cost, but most of the published algorithms are concerned with variable feed-in tariffs and changing the coefficient of performance (COP) of the heat pump during different stages in its operation cycle. However, in Australia, since solar energy capacity is growing to exceed present demand, the actual harvest of energy will be limited by the demand, making optimization of cost unnecessary.

Building operation is responsible for a significant portion of energy consumption and CO_2 emissions. In Australia, space conditioning and water heating account for 40% and 23%, respectively, of total household energy use [12]. However, these loads could be serviced using air-to-water heat pumps [1,2,9–11]. Such heat pumps could run on renewable solar energy PV systems to move towards net zero energy building. Feed-in tariffs are becoming a problem in Australia because the electricity grid, which is designed to deliver energy rather than receive it, is becoming overloaded. Power companies are beginning to restrict feed-ins by reducing feed-in tariffs or disallowing feed-in altogether. Consequently, some new installations, while hooked up to the grid, are not connected to supply it.

Kolkotsa et al. [13] described the net zero energy building/positive energy building (NZEB/PEB) design as a "progression from passive sustainable design". In this regard, some approaches were considered capable of improving the energy efficiency of buildings through: the improvement of the building fabric (insulation, thermal mass, phase change materials); the use of renewable energies and intelligent energy management; and the incorporation of high efficiency heating, ventilation, and air conditioning (HVAC) equipment. Building service systems that provide HVAC and domestic hot water (DHW) are central to the design of NZEBs. Fabrizio et al. [14] reviewed the technology, performance and parameters of the latest multi-energy systems for residential NZEBs. They proposed two integrated systems: one for space heating and DHW; and another for space heating, cooling, and DHW.

The objective of the present paper is to show an integrated energy efficient building services system supporting a NZEB using available off-the-shelf technologies. The proposed system provides space heating and cooling as well as DHW via a solar-powered heat pump drawing less than 1 kW, while utilizing both electrical and thermal batteries. The key parameters and concepts are identified to size the system for the case study: a single-storey three-bedroom home in Melbourne, Australia, built in 1993. The poorly insulated construction is typical for houses built in that era: single-glazing, brick veneer walls and a tile roof above an unsealed ceiling space. The walls and roof space are insulated with just a single layer of fiberglass insulation, with an estimated R-factor of 3.5 $m^2 \cdot {}^\circ C/W$. Utility and solar PV generation data are available for this home, as it is owned by one of the authors.

The climate of the case study location can be described as mild temperate. During the warmest month of January, the mean maximum and minimum temperatures are 26 and 13 °C, respectively, with the temperature sometimes reaching as high as 45 °C. During the coldest month of July, the mean maximum and minimum temperatures are 13 and 6 °C, respectively, with the overnight low sometimes dropping to 3 °C.

This study is essentially a thermodynamic feasibility study that lays the groundwork for later and more comprehensive transient analyses.

Usually, building mechanical services are sized according to the design (peak) heating and cooling loads. However, this is not always the best way to constrain the capacities and costs of the systems, and it does not encourage changes in behavior and expectations to reduce energy use and greenhouse gas emissions. Typically, this approach leads to unnecessarily large systems requiring more energy. Therefore, new approaches are required such as the one proposed here.

In summary, of the literature reviewed, many papers deal with the self-consumption of solar PV energy. Most of these seek to optimise the periods of self-consumption with appliances and services in a house. However, several issues remain opaque to the reader of these articles. First, the design of the system and selection of appliances and/or size according to existing loads is often not considered. Second, the algorithms and/or actual optimization control of the system itself are often very complex, unobtainable, and therefore not repeatable. Third, for the size and design of a particular service system, the actual variation and seasonal generation of power is generally not discussed, explained or accounted for.

The present paper aims to put together a practical method whereby off-the-shelf products can be used to demonstrate a design retrofit to raise self-consumption of solar energy. It is based on working within the constraints of a heat pump running on less than 1 kW of power and a 6.5 kW solar PV system. More sophisticated methods could involve studying daily energy usage or even half hourly interval data from Smart Meters and solar PV systems. The purpose here is to explore a possible design through a first principle-based solution involving a renewable energy system to establish the extent to which self-consumption can be increased. In the Australian context, this is of vital importance, as many electricity service providers are seeking to limit, control or exclude export of electrical energy to the grid.

This study presents a paradigm shift in how to size a mechanical services system for a home: staying within constraints rather than designing for peak loads. It begins with the question: what can be accomplished by limiting the size of one part of the system? In this case, the project is built around a heat pump operating on about 1 kW of input power. The reasoning for this small size is that, through present DC inverter technology, a compressor can operate directly from solar PV easily and more frequently. This constraint defines all other considerations: how much energy must be generated, stored and used. These considerations are methodically analysed in this study to meet the heating and cooling needs for the case study home.

2. Description of the Mechanical System

2.1. System Components

This section outlines the whole system and introduces the individual components, noting the science, thermodynamics and specifications that will be relied on in the next section to properly size and cost the integrated package.

The reference solar PV system for this study is modelled by that at the Melbourne home of one of the authors. This 6.5 kW microinverter system consists of 20 panels rated at 325 W each. Eight of these panels face northeast and twelve face northwest. The average energy productions for the coldest and warmest months (July and January [15]) are 13 kWh and 31 kWh, respectively. Typical daily productions on completely overcast days in July and January are 7 kWh and 10 kWh, respectively. This system cost AUD 7700, after government rebates.

A vapor compression refrigeration cycle moves energy between thermal reservoirs. It is implemented as a heat pump (HP) consisting of an evaporator, compressor (pump), condenser and expansion valve. A HP provides the mechanism to transfer outdoor heat from the air, also known as the 'source', into a 'sink', which in this case for this study is conditioned hot water that is ultimately used to heat a building. When the objective is to cool a space, transferring heat from a building to the outdoors, this is called a refrigeration cycle. If the heating or cooling is not required immediately, it can be stored in or extracted from a thermal battery (TB).

For the proposed heating system shown in Figure 1, an air-to-water HP transfers heat from the outdoor air to water in its buffer tank. This water is then circulated to a bank of TBs to store enough thermal energy to support space heating in winter (e.g., by hydronic heating and radiant panels). This energy is extracted as required by another water loop.

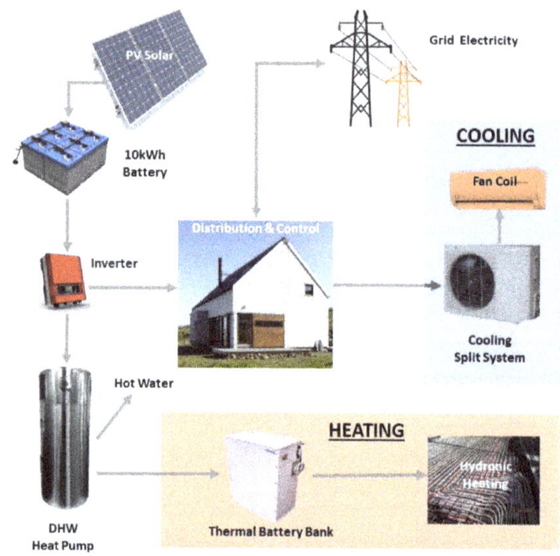

Figure 1. Energy flows in the electrical, heating and cooling systems.

For the cooling system, heat flows in the opposite direction: from the building space to the outdoor air. This could be accomplished by using the same HP system to extract heat from a different TB (storage tank), providing a cold reservoir with which to cool the building via a hydronic loop connected to chilled beams, for example. However, as will be explained later, there is no need for TBs in the case of cooling studied here. It is more straightforward to cool the space directly via the evaporator of an air-to-air HP split system, as shown in Figure 1.

It is acknowledged that, ideally, the cooling side system could not only cool the room air but also heat the DHW with the room heat rejected by the HP. Such systems are not commonly available in Australia for low-capacity applications, such as the case considered in this paper. Furthermore, the company that manufactures the high COP HP selected for the cooling (noted below) does not offer this option of also providing DHW. The authors thought it preferable to rely on such a highly performing HP even if it could not also provide DHW. Instead, the heating side HP is used to meet the DHW needs during the cooling season as well as the heating season. Regardless, Melbourne is a heating-dominated environment, and the cooling system is really only used intensely for two months of the year. Therefore, the missed opportunity to use the cooling side HP for DHW is not significant. This arrangement is in the interest of simplicity and to ensure the use of available high-performance off-the-shelf components. For the much warmer Brisbane, Australia, a cooling system that also provides DHW would be the preferred choice because Brisbane is in a cooling dominated climate.

The heating and cooling systems operate on solar PV energy passing through a bank of electrical batteries (EBs). Domestic demand can draw more solar energy for loads (e.g., appliances) or it can be exported to the grid, if so configured. Otherwise, no additional energy is collected.

The thermal performance of an HP system is given by the coefficient of performance (COP). When operating in the forward or heating mode:

$$COP_{heating} = Q'_{delivered} / W'_{compressor}, \tag{1}$$

where:

$Q'_{delivered}$ = the rate of heat delivered to the high temperature reservoir (W);

$W'_{compressor}$ = the power input to the compressor (W).

When operating in the reverse or cooling mode:

$$COP_{cooling} = Q'_{removed} / W'_{compressor}, \quad (2)$$

where:

$Q'_{removed}$ = the rate of heat removal from the low temperature reservoir (W);

$W'_{compressor}$ = the power input to the compressor (W).

The coefficients of performance for heating and cooling are usually greater than 1. The higher the coefficient, the more effective the HP.

The thermal battery (TB) considered in this study consists of a core of phase change material (PCM) surrounded by water. Table 1 presents the properties of TBs from Sunamp Ltd. (Tranent, UK) suitable for the heating and cooling processes modelled here, although, as will be explained later in this paper, cooling TBs will not be needed [16–18]. The cost per heating TB is approximately AUD 4600 [19]. A similar cost is assumed for the cooling TB (as of July 2021, 1 AUD~0.76 USD).

Table 1. Characteristics of Sunamp TBs [16–18].

Item	Heating Process	Cooling Process
Name	SU58-UniQ9	SU11-UniQ9
Volume of PCM (L)	78	78 (assumed)
$T_{minimum}$ (°C)	45	6
$T_{melting}$ (°C)	58	11
$T_{maximum}$ (°C)	85	16
Energy stored (kW·h)	10.5	3.8 (calculated)

In order to provide space heating at a later time, hot water from a HP flows into a TB, heating it from the minimum temperature of 45 °C until it reaches the melting temperature of 58 °C. Subsequent heat addition just melts the PCM at this constant temperature. Once fully melted, the temperature of the PCM rises. Most of the energy stored in the PCM is from the phase change. This leads to a denser energy storage than, for example, heating a tank of water that remains liquid. When space heating is required later, cool water from the house is passed over the PCM to extract the stored energy, eventually returning the PCM to a solid.

If TBs are used for space cooling, cold water from an HP flows over liquid PCM, cooling it until it reaches the freezing temperature of 11 °C, at which point it begins to solidify. Further heat extraction leads to more solidification without a decrease in temperature but with a significant storage of energy. Once the PCM is fully solidified, the temperature continues to drop. When space cooling is required, warm water from the house is circulated over the PCM to cool the water, eventually melting the PCM.

The TB for heating requires a minimum water supply temperature of 65 °C [17]. This is beyond the range of most conventional HPs using standard refrigerants. An HP with carbon dioxide (CO_2) as the refrigerant allows for higher temperature ranges to be achieved and, conveniently, higher COPs.

A commercial HP was selected to provide realistic specifications. The Sanden Eco® plus system, which uses carbon dioxide as the refrigerant, can supply water for a DHW system at a maximum temperature of 65 °C, with a COP_{HP} of 5.6 while operating at 0.9 kW [20]. For the heating system studied, heated water is passed to the TB and into a DHW tank.

Ideally, the heating and cooling systems should share a reversible HP. However, a commercial reversible CO_2 heat pump with air-to-water heating and a high COP could not be found. The most practical solution is to use a proven conventional HP for the entire cooling process.

The HP selected for the cooling process is the 5.1 kW-rated cooling-only (air-to-air) split system from the Pioneer (Aust.) line of air conditioners [21]. This system has a maximum capacity of 6 kW and uses M20 hydrocarbon refrigerant. The $COP_{cooling}$ for this system is 6.1, with an average power consumption of 0.83 kW.

Table 2 presents the published COPs for the HPs of the heating and cooling systems.

Table 2. COPs for the HPs selected for this study [20,21].

System	COP	At Test Temperatures	Adjusted for Design Operating Temperatures
Sanden Eco® plus	$COP_{heating}$	5.6	4.9
Pioneer WB-18 5.1 kW	$COP_{cooling}$	6.1	6

It is known from the ideal Carnot HP and refrigerator cycles that COPs vary with the temperatures of the hot and cold reservoirs (T_H and T_C) [22]. For this study, the reservoir temperatures approximately correspond to the indoor set points and outdoor conditions. (There is a small difference across each heat exchanger). Since these temperatures differ from those at which the COPs given in Table 2 were determined, the COPs should be adjusted according to the actual temperatures.

Table 3 presents T_H and T_C for the test conditions used to measure the COPs of Table 2, with the test parameters set to the standard conditions to evaluate the performance of HPs [23]. The operating temperatures for this study that were used to calculate the COPs are also tabulated in the last column of Table 2

Table 3. Test and operating temperatures (°C) [15,20–24].

System	Test Temperatures		Operating Temperatures	
	T_H	T_C	T_H	T_C
Heating: Sanden Eco® plus	65	24	65	6
Cooling: Pioneer WB-18	35	27	32	25

In Table 3, the operational outdoor temperatures for the heating and cooling processes (6 °C and 32 °C) are the annual heating and cooling design temperatures for Melbourne [24], each based on a 1% frequency of occurrence. This means that, for 1% of the hours over the year, the air temperature is equal to or less than the heating design temperature—or it is equal to or greater than the cooling design temperature. Relying on the design temperatures helps with sizing the proposed system appropriately. As for the other operating temperatures in Table 3: the T_H for the heating process is the aforementioned maximum water temperature; and the T_C for the cooling process of 25 °C is within the usual range for the indoor temperature under cooling (though TBs, which would require a colder temperature, will not be used in the proposed system).

For the heating system, the COP is adjusted based on the temperature dependence of the COP for a Sanden CO_2 HP system [25], and the conditions presented in Table 3. This reduces the $COP_{heating}$ by 13% from 5.6 to 4.9. The adjusted coefficient is presented in Table 2. A similar adjustment of the quoted $COP_{cooling}$ [21] based on the test and operating temperatures in Table 3 changed this coefficient from 6.1 to 6.0, as also noted in Table 2.

The TB energy storage capacities presented in Table 1 are based on heating and cooling to the maximum and minimum temperatures of the PCM, respectively (85 °C and 6 °C). While the lower temperature can be met with the chosen HP system (if cooling TBs are used), the water supply temperature to the heating TB is limited to 65 °C. Due to the thermal resistances between the supply water and PCM, it is assumed the PCM will only be heated to 63 °C. A recalculation of the energy storage yields an effective capacity of 8.1 kWh for the heating TB under the operating conditions for this study.

The manufacturer of the TB, Sunamp Ltd., is working with Trina Solar to integrate an HP with its TB system [26]. Presumably, this could involve a refrigerant-to-water heat exchanger on the condenser side of the HP, with the water circulated directly to the TB.

However, for the heating system proposed here, it is necessary to place a buffer water tank between the HP and TB. The HP heats the water in the buffer tank, which is then circulated to the TB. Sanden, which makes the HP considered here, markets its product with accompanying hot water tanks [19]. The 160 L size was chosen to serve as the buffer tank for this study. The combined HP and buffer tank system costs about AUD 3000 (the retail price of AUD 4200 less the estimated federal government rebate of AUD 1200).

Another water tank is required for the DHW system, otherwise the heat in the buffer tank could be quickly and completely consumed, for example, by several hot showers. The DHW tank was simulated by another Sanden 160 L water tank, for which the estimated cost is AUD 400.

Electrical batteries are required so that the heating TB can be thermally charged overnight, to run the HP in the early morning and late evening hours and to provide backup electricity on overcast days. The EB unit chosen for this study is the Enphase lithium-ion battery, which can store 1.2 kWh of energy and costs AUD 2000. The battery losses are 4% [27], i.e., the efficiency of the electrical batteries is 96%. Several of these battery units are combined into an EB bank to meet the electrical needs identified by the analysis in Section 3.

2.2. Performance and Equations

For the heating season (winter), the total electrical energy (E_{total}) required to meet all of the heating needs of the house is given by:

$$E_{total,winter} = [(DHL + TB_{losses} + Q_{DHW} + DHW_{losses} + BT_{losses} + E_{pumps})/COP_{heating} + E_{pumps}]/\varepsilon_{EB} \quad (3)$$

where:

DHL = design heating load (the maximum load the system must meet);
TB_{losses} = heat loss from TBs (kWh);
Q_{DHW} = heat required for DHW (kWh)
DHW_{losses} = heat loss from DHW tank (kWh);
BT_{losses} = heat loss from buffer tank (kWh);
$COP_{heating}$ = coefficient of performance for the heat pump that delivers thermal energy to system;
E_{pumps} = energy required to run pumps (kWh);
ε_{EB} = EB efficiency.

The heat required for the DHW is given by:

$$Q_{DHW} = m_{water}\, C_p\, (T_{DHW} - T_{mains}) \quad (4)$$

where:

m_{water} = mass of water in the DHW tank (kg);
C_p = specific heat capacity of water (4.18 kJ/kg·°C for average of temperature range);
T_{DHW} = water temperature maintained in DHW tank (°C);
T_{mains} = water temperature at mains pipe going to DHW tank (°C).

For the cooling season(summer), the electrical energy required to provide the cooling, $E_{cooling}$, is given by: the design cooling load, DCL, in kWh over the COP of the HP for the cooling system ($COP_{cooling}$), while taking into account the efficiency of the EBs:

$$E_{cooling} = (DCL/COP_{cooling})/\varepsilon_{EB} \quad (5)$$

The overall energy required in the cooling season is given by:

$$E_{total,summer} = [(DCL/COP_{cooling}) + (Q_{DHW} + Loss_{DHW})/COP_{heating}]/\varepsilon_{EB} \quad (6)$$

3. Results and Analysis

The detailed calculations for the sizing results given in this section are presented in Appendix A.

3.1. Sizing of Heating System

During the colder months in Melbourne, typically only a few rooms of a house are heated—and usually just during the morning and evening. Sufficient heating for the actual case study, (a modest-sized house in Melbourne) is provided by a 6 kW-rated split-system heat pump servicing a closed off kitchen, lounge and dining area during the day, and the bedrooms in the late evening. On the coldest days in winter, an average of 5 kW of heating is supplied between 6 and 8 am and between 6 and 10 p.m.—a total of six hours, leading to a requirement of 30 kWh of heating. This is the design heating load (DHL) for the system.

Two questions at this point are: can this DHL be met within the constraints of a heat pump operating with less than 1 kW of power (and here a 0.9 kW Sanden HP has been selected) and a 6.5 kW solar PV system; and, if so, what other components of the system are required?

Based on the effective capacity of 8.1 kWh per TB, the DHL of 30 kWh could be met with four TBs (a capacity of 32 kWh). These would be stacked in parallel so that, in warmer weather (spring and autumn), some TBs could be isolated.

Sunamp [17] quotes a heat loss per TB of 0.74 kWh/day (though without specified temperatures), so this battery bank would have a total daily loss of about 3 kWh. Two kilowatt-hours of this can be accommodated by the extra capacity in the battery bank (beyond the required 30 kWh), but an additional 1 kWh must be supplied. Therefore, 33 kWh must be supplied to the TB bank.

The HP would also supply heat for the DHW system. The DHW system consists of a 160 L tank (approximately 160 kg of water), which is assumed to be fully changed out once a day and maintained at 60 °C. The water is heated from the average temperature in June for mains water in Melbourne: 13 °C [28]. The required energy by Equation (4) is 31 MJ or 8.7 kWh. The daily energy loss from the tank is estimated to be 5% (0.4 kWh).

For the buffer tank between the HP and TB, the daily energy loss is also estimated to be 5% (0.4 kWh).

The sum of the heat storage and losses in the TB, DHW and buffer tanks is 43 kWh. This is the thermal energy that must be provided by the HP. For an HP with a COP of 4.9, this would require an electrical input of 8.8 kWh.

Electricity is also required to run circulation pumps in the following water loops: between the buffer tank and TB (assumed to be for 18 h); the hydronic heating loop (for 6 h); and the loop between the HP and the two water tanks. The total daily electrical energy requirement for these pumps, E_{pumps}, is estimated to be 1.1 kWh.

The sum of the above energy requirements is 9.9 kWh. This is what must be supplied by the EBs. However, since these batteries have an efficiency of $\varepsilon_{EB} = 96\%$, a total electrical energy of E_{total} = 10.3 kWh must be supplied by the solar PV system to charge the EBs.

The benchmark 6.5 kW solar PV system will more than satisfy this requirement (average production of 13 kWh/day in the worst-performing month of July). Therefore, this system could be scaled down to meet the 10.3 kWh/day required by the proposed heating system: a 5.2 kW system with 16 panels would suffice. However, since extra solar capacity is required for domestic electricity consumption, the reference solar PV system of 6.5 kW on the case study house will be used here. It is known from the case study that this extra capacity would be insufficient to meet the addition of domestic consumption on most days in July. Deficits could be met from the grid (in which case this will not be a stand-alone system) or by adding more solar panels to the solar PV system.

It was decided to limit the capacity of the EB bank to 10 kWh to reduce the cost of the system. Therefore, the electrical storage consists of 8 EBs providing an electrical storage capacity of 9.6 kWh. Conveniently, this will store just under a day's worth of required electricity for an average day in July.

With the reference solar PV system delivering 7 kWh on a fully overcast day in July (just over half of the daily average for the month), the heating system would be able to meet just over half of the design load. Starting with full TBs, the half-load scenario would occur on the second consecutive fully overcast day. This could be addressed in two ways: the occupants could tolerate colder indoor temperatures; or the shortage could be made up for with power from the grid, in which case this will not be a stand-alone off-grid system.

A summary of the requirements for the heating system is presented in Table 4 below, which also includes the needs of the cooling system, which will now be addressed.

Table 4. System requirements for design heating and cooling loads. Energies are per day.

Item	Heating	Cooling
Space thermal energy load (kWh/day)	30	66
DHW energy required (kWh/day)	8.7	7.2
Thermal losses (kWh/day)	3.8	0.4
Thermal energy required (kWh/day)	43	74
Electrical energy required (kWh/day)	10	13
TB quantity	4	None
EB quantity	8	8
Solar PV system size (kW)	6.5	6.5
Sanden Eco® plus heating system	Yes	No
Pioneer cooling system (6 kW)	No	Yes

3.2. Sizing of Cooling System

The cooling system is sized in a similar manner. In Melbourne, typically only a few rooms of a house are cooled. During the day, an area consisting of the kitchen, lounge and dining area is closed off and cooled. In the late evening, just the bedrooms are cooled. For the case study house, it is known that a split-system HP providing an average of 6 kW of cooling over noon to 11 p.m. suffices for the hottest day. This corresponds to a design cooling load (DCL) of 66 kWh (6 kW times 11 h). This is the scenario used in the present study.

Again, the questions are: can the design load be satisfied within the constraints of a heat pump operating with less than 1 kW of power (the Pioneer 0.83 kW HP in this case) and a 6.5 kW; and, if so, what supporting components will be needed?

The analysis below is based on the same solar PV and EB system required for the heating system, because it is the heating load that determines the capacities for the electrical side which, as will be shown, in this case leaves more than enough capacity to meet the required cooling load.

The storage requirement for cooling differs from that for heating. Most of the cooling load—and indeed the peak cooling—occurs when the sun is in the sky, leading to daytime solar gains through the building envelope. In contrast, the heating load usually occurs when the sun is below the horizon or low in the sky: overnight, morning or evening. Therefore, much of the cooling load can be provided directly by the solar-powered HP during the day (i.e., storage is not needed). However, storage might be required in the evening hours or under cloudy conditions. But, as will be shown, only electrical and not thermal storage is needed for the scenario considered in this paper.

By late January, after 7 p.m., the benchmark solar PV system generates insufficient power to run the HP. Therefore, from 7 to 11 p.m., the cooling system must run mostly on stored electricity; although the sun does not set until about 8:45 p.m., there is a much lower output from the PV system in the two hours leading up to sunset. Therefore, there are two periods to consider: noon to 7 p.m. (requiring 42 kWh of cooling), and 7 to 11 p.m. (requiring 24 kWh of cooling). Whether TBs are required depends on whether the solar PV system and EBs can deliver enough energy to meet not only the daytime needs, but the evening ones as well.

In January, the benchmark solar PV system produces an average of 31 kWh/day. Based on the cooling system COP of 6 and the battery loss of 4% (EB efficiency of 96%), the daytime load of 42 kWh would require only 7 kWh of the 31 kWh of generated solar power. The evening load of 24 kWh would require 4 kWh of the day's solar power—less than the 9.6 kWh capacity of the EBs. This leaves a surplus of 20 kWh of solar power left over (almost two thirds of the daily production) for domestic consumption, export to the grid, extending the cooling overnight if needed, or even running additional cooling systems to cool more of the house. The situation is even better because, with the sun setting at 8:42 p.m. in mid-January, the PV system will still be delivering some electricity to the cooling system over the first half of the evening, without the need to rely exclusively on the electrical batteries, while leaving more energy available for domestic consumption.

Therefore, thermal storage for cooling is not required. Having sized the solar PV system to handle the heating load, when one turns to look at the cooling load, the much larger solar power generation in the summer months ensures there is more than enough electrical power to meet this load.

The situation of a fully overcast day does not present a problem. The benchmark PV solar system typically produces 10 kWh on an overcast day in January. Leveraged through the HP and considering the battery efficiency, this corresponds to 58 kWh of cooling—almost 90% of the design cooling load. Furthermore, under cloudy conditions, the outdoor temperature is usually lower and there will be much less solar gain through the building envelope, ensuring that the actual load that day will be much less.

If there was a compelling reason to rely on thermal batteries for some of the cooling, cooling TBs of Table 1 could be incorporated into the system. Determining the exact number of TBs would require transient modelling over the summer months. However, if it is assumed that for the hottest days a fifth of the design cooling load required TB storage, four TBs would be needed at a cost of AUD 8400.

The summertime DHW requirement is 7.2 kWh, which is slightly less than that for the winter because of the higher mains water in Melbourne of 21 °C [28]. It is recalled from Section 2.1 that, for the sake of simplicity and equipment availability, the heat for the DHW is provided by the heating side HP, even though ideally it could be the heat rejected from the cooling system.

The design requirements, including the daily energy demands, for the heating and cooling seasons are summarized in Table 4. The loads from the first three rows are presented in Figures 2 and 3. The only thermal loss in the cooling season is associated with the DHW tank.

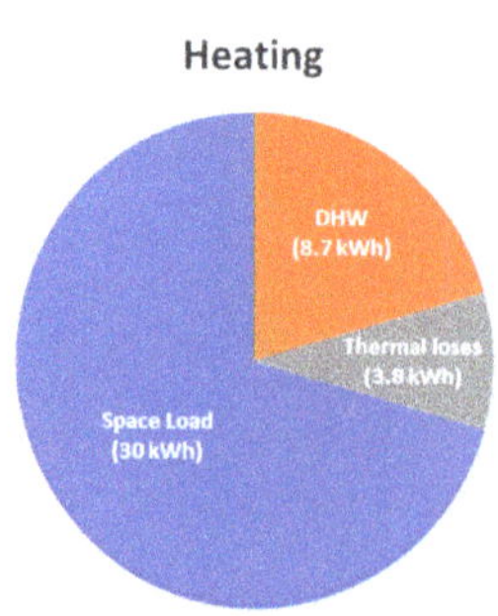

Figure 2. The daily thermal energy demands for the building services system in the heating season.

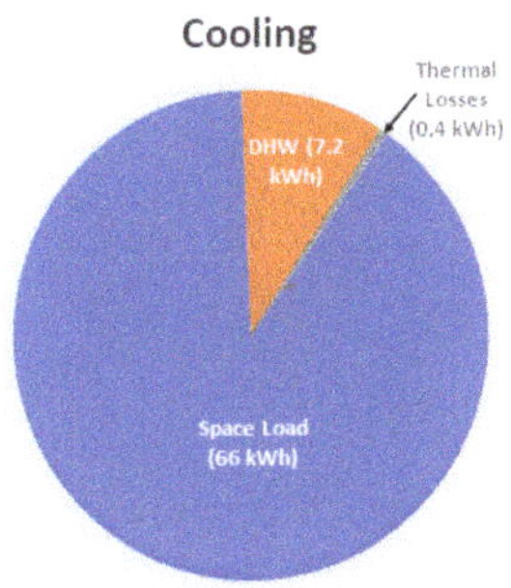

Figure 3. The daily thermal energy demands for the building services system in the cooling season.

3.3. Economics

The costs for the heating and cooling systems are itemised in Table 5. The systems share the solar PV and EB components. The solar system price is after federal and state government solar PV rebates, and the electricity price is after the maximum state battery rebate of AUD 4200. The total cost is AUD 30,100 (USD 22,900).

Table 5. System costs.

Item	Cost (AUD)
CO_2 Heat pump & buffer tank (160 L)	3000
Thermal batteries for heating (4)	8400
Electrical batteries (8)	11,800
Domestic hot water tank (160 L)	400
Solar PV system (6.5 kW)	4500
Split system for cooling (5 kW)	2000
Total	30,100

An exact calculation of the annual cost savings with this solar-powered system would require a transient analysis based on the weather and solar radiation throughout a typical year. However, an estimate can be made on the basis of the energy requirements for the actual house served by the reference solar PV system for this paper. The house is poorly insulated (single glazed windows) and poorly sealed (drafts around doors). This construction is typical for homes in Melbourne of this vintage (about 30 years) or older.

The actual annual energy consumption for this house is 5700 kWh of electricity and 28 GJ of gas. Space heating and cooling for about a quarter of the house (usually closed off from the rest of the house for this) are provided by a split-system reversible heat pump. Occasionally, whole-house heating is supplied by a gas-ducted system, and whole-house cooling by an evaporative cooler. Gas is used to heat DHW, and for the stove.

Some parameters and assumptions were required to interpret the monthly utility bills of the actual house in order to estimate the electricity that must be supplied by the PV system to run the building services system proposed in this paper. Space heating was only required over April to September, and space cooling for October to March. The COP rating for the current cooling system of the house is 3.3, and that for the heating system is 3.7. All of the gas consumption is for space heating, DHW and cooking. Based on the summertime gas consumption (no gas-fired heating), the monthly base load for gas was determined to be 100 kWh equivalent of energy, of which 90% is assumed to be for DHW and the rest for cooking. Based on the electricity demand in the winter (during which most of the heating is by gas), the monthly base load electricity consumption was determined to be 370 kWh. All of the rest of the electricity goes to space cooling, most of which is accomplished through a HP system. It is estimated that a tenth of the space cooling is via the electrically driven evaporative cooling system but, since the efficiency of that system is unknown and the electricity consumption for that is small in any case, it is assumed that all of the space conditioning is through the HP system.

These assumptions were applied to the actual monthly utility bills to determine the equivalent electrical energy (in kWh) that must be generated by the solar PV system to meet the energy needs of the proposed system to provide the required space heating (via the Sanden HP with a COP of 4.9), space cooling (via the Pioneer HP of 6), DHW, cooking (now by electricity) and other domestic electricity consumption. The actual monthly electricity production by the solar PV system was also known.

The proposed system is designed only to condition the closed off spaces noted above, rather than the whole house, and it is assumed this is acceptable to the occupants (though there is sufficient solar power for expanded cooling). Current utility prices for the actual house are: AUD 0.042/MJ of gas; and, for electricity, AUD 0.42/kWh to purchase, and AUD 0.20/kWh to export to the grid (the feed-in tariff).

Based on monthly averages for the solar PV generation and the equivalent electricity needs based on the actual monthly utility bills, the proposed system would require some grid power over May to September. However, over the course of a year, there would be a net electricity surplus of 1500 kWh exported to the grid (2600 kWh exported and 1100 kWh imported). Due to the difference between the purchase and feed-in tariff prices, the net annual electricity bill would be a credit of AUD 58—rounded off here to AUD 100. Given that the annual bill without the proposed system is AUD 2700, the annual savings would be AUD 2800.

Therefore, the proposed system is cost-neutral (and actually slightly profitable). It is also grid-neutral in that there is a net export of electricity over the year. However, it is not grid-independent due to the need to import some electricity during part of the heating season (i.e., the winter).

There are two approaches to calculating the payback period, N. The simplest approach is:

$$N = C/S \tag{7}$$

where:

C = Initial Capital Cost (AUD);

S = Annual Savings (AUD/a).

It is usually more realistic to factor in percentage increases or decreases that affect the savings. In the present case, this is the annual increase in the price of electricity (i). The project will "pay off" when the initial capital cost is equal to the sum of the ever-increasing annual savings:

$$N = \ln(1 + iC/S)/\ln(1 + i) \tag{8}$$

The payback period should properly only consider the cost that is in addition to what a homeowner would usually spend on a conventional system, i.e., the difference between the services proposed in this paper and a conventional system. A conventional system would typically consist of a DHW heater and tank (AUD 2000) and a split system or ducted gas heating (AUD 2000). Additionally, rooftop solar PV (AUD 4500 for this analysis) is so common now in Australia that it is arguably becoming a "standard" new-house feature. Thus, a conventional system costs about AUD 8500. The extra cost, i.e., the initial capital cost for this example, of the proposed system is the total cost in Table 5 (AUD 30,100) less this cost of a conventional system (AUD 8500), or AUD 21,600.

The simple payback period (Equation (7)) based on this cost of AUD 21,600 and the annual savings on the utility bill of AUD 2800 is 7.7 years for no annual change in the electricity price. If one assumes a 3% annual increase in the electricity prices (i = 0.03) (Equation (8)), this payback drops to 7.0 years.

In order to ensure there is no need to import any electricity from the grid over the course of a year (grid independence), the solar PV system size would have to be increased by 90% to 12.5 kW, leading to annual savings of AUD 4700. In this case, the simple payback period for the whole system (including the cost of the expanded solar PV system) would be 5.5 years. With a 3% annual increase in the electricity prices, the payback period is 5.1 years. However, there might not be enough suitably oriented and unshaded roof space for a 12.5 kW system (there is not on the case study home). Furthermore, some utility companies in Melbourne limit the size of grid-exporting rooftop solar systems to 10 kW.

There should be no significant additional maintenance costs compared to a conventional system given that most Australian homes have a heat pump for space conditioning, and a quarter of homes have solar PV—which does not require maintenance beyond occasionally hosing off accumulated dirt. The sealed TBs do not require maintenance over their lifetime. The only extra elements not found in a standard space conditioning system would be the low-power water circulation pumps, for which the maintenance cost (if any) would be expected to be small.

4. Discussion

It is acknowledged that these payback periods are sensitive to the values selected for the electricity and feed-in tariff rates as well as the assumed annual increase in the price of electricity. However, other choices within reasonable limits will not change the fact that the payback period for the system described in this paper will be of the same order of magnitude as the accepted periods for residential rooftop solar PV in Australia.

The typical payback periods for just a 5 kW solar PV system in Australia is 2 to 6 years, depending on: the location, fraction of generated power exported to the grid, feed-in tariffs and government subsidies [29]. Homeowners have already shown this range to be acceptable, given that rooftop solar has been installed on about a quarter of residences in Australia.

In order to achieve a comparable payback for the integrated system presented in this paper, government subsidies or rebates would be required. Since rebates for the solar PV and electrical batteries have already been included in the prices of Table 5, what is needed is either a subsidy for the system as a whole or, more realistically, one for the thermal batteries, which account for a quarter of the total system cost.

Following the success of government subsidies in promoting the installation of residential solar PV systems, some Australian states have in recent years introduced rebates for home electrical battery systems [30,31]. However, there is no known subsidy in Australia for thermal batteries. The justifications for subsidizing EBs are that they help reduce the reliance on the grid and non-renewable power, and they also protect the grid from feed-in overload by facilitating on-site storage at homes.

This paper shows that, as an integral part of a NZEB to facilitate the heating of a home, TBs provide the same benefits. They are just another way of storing energy from the solar PV array—but in a perhaps less well-understood or promoted fashion: via heat transfer rather than electrical charging. The authors believe that government rebates that apply to EBs should be extended to TBs.

It is a fair question to ask why one should invest in the relatively more expensive thermal batteries (based on the costs in Table 5) rather than just adding more electrical batteries to the proposed system. It is important to consider the cost per kWh of energy storage. Based on the effective capacity of 8.1 kWh for each thermal battery used in this case study, the cost is AUD 260/kWh. Assuming that energy stored in an electrical battery would be leveraged through an HP with a COP of 4, the cost of an electrical battery used to support a heating or cooling system is AUD 420/kWh (not including subsidies). Therefore, thermal batteries have an economic advantage (by about 40%) over electrical batteries.

Thermal batteries also have a longer lifespan. Sunamp claims its thermal batteries will experience no noticeable degradation to 10,000 cycles, which is equivalent to about 13 years of daily use of two full cycles per day [17]. The system proposed in this paper relies on one full cycle per day, meaning the TBs would be expected to last about 26 years. In contrast, lithium-ion electrical batteries are commonly reported to have a ten-year lifespan. Specifically, the capacity of such EBs is expected to degrade to 80% after 10 years or 3650 cycles [27].

Although this study proposed that cooling be delivered directly via a conventional HP split-system (a practical and easy option), a case could be made for thermal storage for cooling, with another water loop circulated partly through the piping used for the hydronic heating system. Although more complex, this would facilitate widespread and well-targeted distribution of cooling throughout the house. It could be enhanced with ideally located chilled beams.

5. Limitations

The main limitation of this study is that it is based on peak heating and cooling loads for the house. While this is useful for sizing the system, which is the main focus of this paper, it does not provide detailed information about the annual energy use and potential savings over the course of the year. That would require a dynamic hour-by-hour analysis over the course of a year. In this work, this was estimated with the assistance of utility bills for the case study house.

The house selected for the reference case is another limitation. The construction is typical for Australian houses built in the 1990s and over four decades before then, and the poorly insulated building envelope leads to significant thermal losses and gains over the course of the year. This makes the proposed building services system viable in terms of the payback period. With better built new stock housing, the payback periods would be expected to be longer. Regardless of when the reference house was built, this study is tied to a set of parameters for wall insulation, window types and window-to-wall ratios. It would be worthwhile exploring how the outcomes vary with these parameters.

This study is also limited by location: only Melbourne is considered. Of the eight capital cities in Australia, Melbourne is the only one in climate zone 6 (mild temperate) of the 9 climate zones defined by the Australian Building Codes Board [32]. Different outcomes for the design and payback period could be expected for the other cities. For example, in sunny Darwin, which lies only 12° below the equator, typically there is not a heating load requirement for houses, so the heating part of the host system would not be needed. Also, there would be much more sunshine available for electricity generation through the solar PV system.

The economic justification for this system, i.e., the payback period is strongly dependent on the prices to purchase electricity from the grid and the price to sell it to the grid (the latter being the feed-in tariff), is another potential limitation. The sensitivity of the payback to these variables is worthy of further investigation, set in the context of current and expected trends in electricity prices. This is particularly important given that, as noted earlier in this paper, some utility companies are reducing the feed-in tariffs or eliminating them altogether, effectively no longer allowing rooftop solar to be connected to the grid.

6. Conclusions

An integrated building services system has been investigated as a pathway to designing a residential NZEB. Several aspects are provided in this paper that demonstrate the merits of a unique service system design and its sizing:

- A low input energy (less than 1.0 kW) service system design that can run off a solar PV, batteries and thermal storage.
- A design from readily available off-the-shelf components and appliances.
- A simplified method which takes into account the utility bills of peak daily consumption for winter and summer periods.
- A design constrained by the solar PV system size and the heat pump input power as distinct from designing for peak hourly loads to obtain a system size.
- An investigation of a real case medium sized average insulated house in Melbourne Australia demonstrates that it is possible for the service system to meet the target of above-mentioned constraints.

The system provides space conditioning (heating and cooling) and DHW via heat pumps powered by a solar PV system. Electrical batteries are used to store the solar energy so that it can be delivered to the HPs in the early morning and in the evening hours. Thermal batteries consisting of PCMs are employed to store heat which is delivered to the building via a hydronic heating system. No such thermal storage is required for space cooling, because there should be more than enough solar energy or stored electricity to meet the cooling requirements. The system was sized for the case study of an actual single-storey detached home of modest size in Melbourne, Australia, for which the energy requirements were determined from utility bills.

As for the sizing of the proposed system, the design heating and cooling loads and DHW were met by the following:

- Heating and DHW provided by a highly efficient HP that operates on 0.9 kW of power and uses CO_2 a as the refrigerant;
- Cooling provided by a conventional split-system HP that operates on 0.8 kW and delivers 6 kW of cooling;
- A 6.5 kW solar PV system;
- Four thermal batteries (each of 10.5 kWh rated capacity); and
- Eight electrical batteries (each of 1.2 kWh capacity).

The total cost is AUD 21,600 more than a conventional system, with an approximate payback of 7.0 years based on the annual utility bill savings and the expected annual increase in the electricity price. Although the system would be revenue-neutral (and in fact slightly profitable), it would require importing electricity from the grid during part of the heating season (but exports to the grid would exceed the imports). Grid independence would be achieved by expanding to a 12.5 kW solar PV system.

This paper has proven the thermodynamic feasibility of relying on renewable energy to provide mechanical services to a home of modest size and construction in Melbourne, Australia. The key design parameters have been identified. The next step is to perform a transient analysis (e.g., hour-by-hour) on the same case study and to consider more typical modern homes, e.g., current designs that meet the minimum energy rating of 6 stars (the national requirement) set in Melbourne and elsewhere in Australia. The transient analysis will consider the hourly impact of weather, as well as the effect of changing temperatures on the COP of the heat pump and the efficiency of the TBs. The goal of this further study will be to determine more exactly the energy savings and greenhouse gas emission reductions that could be attained by making such an integrated system part of the National Construction Code for all new builds, and by encouraging retrofits of existing houses with such a system.

Author Contributions: Conceptualization, M.B.L.; methodology, T.S. and M.B.L.; writing—original draft preparation, T.S.; validation, T.S. and M.B.L.; formal analysis, T.S., M.B.L., H.X.L. and P.H.; investigation, T.S., M.B.L., H.X.L. and P.H.; supervision, M.B.L.; writing—review and editing, T.S., M.B.L., H.X.L. and P.H. All authors have read and agreed to the published version of the manuscript.

Funding: This research received no external funding.

Data Availability Statement: The data presented in this study are available in this paper, with the exception of the supporting utility bills and PV generation reports for the case study house. Those bills and reports are not published here for reasons of privacy, but they are available on request from the corresponding author.

Conflicts of Interest: The authors declare no conflict of interest.

Nomenclature

AUD	Australian dollars
BT	buffer tank
C	capital cost
CO_2	carbon dioxide
COP	coefficient of performance
C_p	specific heat capacity
DCL	design cooling load
DHL	design heating load
DHW	domestic hot water
E	Energy
ε	Efficiency
EB	electrical battery
HP	heat pump
HVAC	heating ventilating and air conditioning
i	price inflation
kWh	kilowatt-hour
m	Mass
N	payback period
NZEB	net zero energy building
PCM	phase change material
PEB	positive energy building
PV	Photovoltaic
Q	thermal energy
Q'	rate of heat transfer
S	Savings
T	Temperature
TB	thermal battery
USD	United States dollars
W'	Power

Appendix A. Calculation Method

A. Solar System Size Settings Rated output: 6.5 kW Average output in July: 13 kWh/day Average output in January: 31 kWh/day Battery storage: 10 kWh

B. Heating Analysis (Winter)

1. Estimation of design heating load Average hourly heat loss, Q'_{loss} = 5 kW Hours per day space heating operated, t_{heat} = 6 h $DHL = (Q'_{loss})(t_{heat})$ = 30 kWh/day
2. Heat pump settings COP of heat pump operating in heating mode (heat house), COP_{heat} = 4.9 COP of heat pump operating in cooling mode (cool house), COP_{cool} = 6.0 Maximum hot water temperature, $T_{HW,max}$ = 65 °C
3. Effective Thermal capacity of TBs for heating
    a. PCM component Maximum temperature, $T_{max,PCM} = T_{HW,max} - 2$ °C = 63 °C (accounts for thermal resistances) Minimum temperature, $T_{min,PCM}$ = 45 °C Volume, V_{PCM} = 78 L Heat capacity of solid phase, $C_{p,s}$ = 4 kJ/L·K Latent heat of fusion, h_{sl} = 289 kJ/L Heat capacity of liquid phase, $C_{p,l}$ = 4.5 kJ/L·K Melting temperature, T_{melt} = 58 °C Capacity of PCM = $(V_{PCM})[(C_{p,s})(T_{melt} - T_{min,PCM}) + (h_{sl}) + (C_{p,l})(T_{max,PCM} - T_{melt})]$
    b. Water component Maximum temperature, $T_{max,water} = T_{HW,max}$ = 65 °C Minimum temperature, $T_{min,water} = T_{min,PCM} + 2$ °C = 47 °C (accounts for thermal resistances) Volume, V_{water} = 0.01 m^3 Density at mean temperature, ρ_{water} = 986 kg/m^3 Specific heat at mean temperature, $C_{p,water}$ = 4.18 kJ/kg·K Capacity of Water = $\rho_{water}(V_{water})C_{p,water}(T_{max,water} - T_{min,water})$ Capacity of battery (in kWh), Q_{TB} = (Capacity of PCM + Capacity of water)/3600 = 8.1 kWh Number of TBs required, $N_{TB} = DHL_t/Q_{TB}$ = 4

TBs (rounded up) Extra capacity in TB bank, $TB_{extra\ capacity} = Q_{TB}N_{TB} - Q_{heat}$ = 2 kWh

The need to round up for the number of batteries builds in an overcapacity of 2 kWh. This will offset thermal losses from the TBs in the total thermal requirement calculation below.

4. DHW Tank Volume of water in tank, V_{DHW} = 0.16 m^3 Changes per day, B = 1 Maximum temperature, $T_{max,DHW}$ = 60 °C Mains water temperature in June, T_{mains} = 13 °C Density at mean temperature, ρ_{DHW} = 991 kg/m^3 Specific heat at mean temperature, $C_{p,DHW}$ = 4.18 kJ/kg·K Heat required for DHW (in kWh), $Q_{DHW} = \rho_{DHW}(V_{DHW})C_{p,water}B(T_{max,DHW} - T_{mains})/3600$ = 8.7 kWh
5. Losses and Efficiency TB loss, $Loss_{TB}$ = 0.7 kWh/day per TB DHW tank loss, $Loss_{DHW}$ = 5% or 0.4 kWh/day Buffer tank loss, $Loss_{buffer}$ = 5% or 0.4 kWh/day Battery Efficiency, $\varepsilon_{batteries}$ = 96%
6. Pumps TB Supply Loop Pump (between DHW & TB) 0.012 kW for 18 h/day consumes P_{supply}= 0.216 kWh/day DHW Loop Pump (between TB & DHW) 0.012 kW for 18 h/day consumes P_{DHW}= 0.216 kWh/day Hydronic Heating Loop Pump (between TB & space) 0.12 kW for 6 h/day consumes $P_{hydronic}$ = 0.72 kWh/day
7. Total heat required $Q_{total,winter} = (Q_{TB}\ N_{TB}) + (Loss_{TB}\ N_{TB} - TB_{extra\ capacity}) + Q_{DHW} + Loss_{DHW} + Loss_{buffer}$ = 43 kWh/day
8. Total electricity required (winter peak) $E_{total,winter} = (Q_{total,winter}/COP_{heat} + P_{supply} + P_{DHW} + P_{hydronic})(\varepsilon_{batteries} \times 0.01)$ = 10.3 kWh/day Solar PV system sufficient to meet this need as average output in June is 13 kWh/day 10 kWh electrical battery sufficient as that 13 kWh/day from the solar PV is cycled through it

C. Cooling Analysis (Summer)

1. Estimation of design cooling load Average rate of heat gain, Q'_{gain} = 6 kW Hours per day space cooling operated, t_{cool} = 11 h $DCL = (Q'_{gain})(t_{cool})$ = 66 kWh/day
2. DHW Tank Same procedure as above for heating but: Mains water temperature in January, T_{mains} = 21 °C Density at mean temperature, ρ_{DHW} = 992 kg/m^3 Heat required for DHW (in kWh), $Q_{DHW} = \rho_{DHW}(V_{DHW})C_{p,water}B(T_{max,DHW} - T_{mains})/3600$ = 7.2 kWh DHW tank loss, $Loss_{DHW}$ = 5% or 0.4 kWh/day In the summer: TBs not used so no losses from there; and fluid circuit pumps not operated.
3. Total electricity required (peak summer) $E_{total,summer} = [(DCL/COP_{cool}) + (Q_{DHW} + Loss_{DHW})/COP_{heat}](\varepsilon_{batteries})$ = 13 kWh/day Solar PV system sufficient to meet this need as average output in January is 31 kWh/day 10 kWh electrical battery sufficient to meet needs in evening.

References

1. Stauffer, Y.; Koch, N.; Hutter, A.; Pflugradt, N.D. Quantifying the Potential of Smart Heat-Pump Control to Increase the Self-Consumption of Photovoltaic Electricity in Buildings. 2018. Available online: http://proceedings.ises.org/paper/eurosun2018/eurosun2018-0158-Stauffer.pdf (accessed on 27 August 2021).
2. Sánchez, C.; Bloch, L.; Holweger, J.; Ballif, C.; Wyrsch, N. Optimised Heat Pump Management for Increasing Photovoltaic Penetration into the Electricity Grid. *Energies* **2019**, *12*, 1571. [CrossRef]
3. Beck, T.; Kondziella, H.; Huard, G.; Bruckner, T. Assessing the influence of the temporal resolution of electrical load and PV generation profiles on self-consumption and sizing of PV-battery systems. *Appl. Energy* **2016**, *173*, 331–342. [CrossRef]
4. Beck, T.; Kondziella, H.; Huard, G.; Bruckner, T. Optimal operation, configuration and sizing of generation and storage technologies for residential heat pump systems in the spotlight of self-consumption of photovoltaic electricity. *Appl. Energy* **2017**, *188*, 604–619. [CrossRef]
5. Luthander, R.; Widén, J.; Nilsson, D.; Palm, J. Photovoltaic self-consumption in buildings: A review. *Appl. Energy* **2015**, *142*, 80–94. [CrossRef]

6. Torregrosa-Jaime, B.; Gonzalez, B.; Martínez, P.J.; Payá-Ballester, G. Analysis of the Operation of an Aerothermal Heat Pump in a Residential Building Using Building Information Modelling. *Energies* **2018**, *11*, 1642. [CrossRef]
7. Xiao, B.; He, L.; Zhang, S.; Kong, T.; Hu, B.; Wang, R. Comparison and analysis on air-to-air and air-to-water heat pump heating systems. *Renew. Energy* **2020**, *146*, 1888–1896. [CrossRef]
8. Maleki, A.; Rosen, M.A.; Pourfayaz, F. Optimal Operation of a Grid-Connected Hybrid Renewable Energy System for Residential Applications. *Sustainability* **2017**, *9*, 1314. [CrossRef]
9. Heat Pump Association, UK, 2019 Delivering Net Zero: A Roadmap for The Role of Heat Pumps. Available online: https://www.heatpumps.org.uk/resources/ (accessed on 25 June 2021).
10. Guertler, P. *Getting on Track to Net Zero Ten Million Heat Pumps for Homes by 2030*; E3G: London, UK, 2020.
11. Vaishak, S.; Bhale, P.V. Photovoltaic/thermal-solar assisted heat pump system: Current status and future prospects. *Sol. Energy* **2019**, *189*, 268–284. [CrossRef]
12. Energy Rating—Space Heating and Cooling. Available online: http://www.energyrating.gov.au/products/space-heating-and-cooling (accessed on 2 December 2020).
13. Kolokotsa, D.; Rovas, D.; Kosmatopoulos, E.; Kalaitzakis, K. A roadmap towards intelligent net zero- and positive-energy buildings. *Sol. Energy* **2011**, *85*, 3067–3084. [CrossRef]
14. Fabrizio, E.; Seguro, F.; Filippi, M. Integrated HVAC and DHW production systems for Zero Energy Buildings. *Renew. Sustain. Energy Rev.* **2014**, *40*, 515–541. [CrossRef]
15. Bureau of Meteorology Climate Statistics for Australian Locations. Available online: http://www.bom.gov.au/climate/averages/tables/cw_086071.shtml (accessed on 2 December 2020).
16. Sunamp—UNIQ Heat Batteries Reference Manual V2.3. Available online: http://www.sunamp.com/wp-content/uploads/2019/01/UniQ-Heat-batteries-reference-manual-ver_20180719_v2.3.pdf (accessed on 2 December 2020).
17. Sunamp—Specialists in Thermal Energy Storage. Available online: https://www.sunamp.com/wp-content/uploads/2021/06/Sunamp-Brochure-May-2021-Artwork-LO-RES.pdf (accessed on 28 August 2021).
18. Phase Change Products Pty Ltd.; West Perth, Australia. Personal communication with M.B. Luther, 2019.
19. Bubleshop Sunamp UniQ E—Dual (Heating & Hot Water Storage). Available online: http://www.bublshop.co.uk/sunamp-uniq-e-dual-heating-hot-water-storage/p2040 (accessed on 3 June 2019).
20. Sanden—Sanden Leading Hot Water Technology. Available online: http://www.newgensolar.com.au/heat-pump-hot-water/sanden-eco-heat-pump-hot-water-system.html (accessed on 2 December 2020).
21. *Pioneer_Aircon.pdf*; Pioneer International Pty Ltd.: Minto, Australia, 2021.
22. Moran, M.J.; Howard, N.; Shapiro, D.D.; Boettner, M.B.B. *Fundamentals of Engineering Thermodynamics*, 9th ed.; Wiley: Hoboken, NJ, USA, 2018.
23. Standards Australia. *AS/NZS 3823.2:2013 Performance of Electrical Appliances—Air Conditioners and Heat Pumps, Part 2 Energy Labelling and Minimum Energy Performance Standards (MEPS)*; Standards Australia: Sydney, Australia, 2013.
24. Design Conditions for Melbourne, Australia. Available online: http://cms.ashrae.biz/weatherdata/STATIONS/948680_s.pdf (accessed on 2 December 2020).
25. Sanden—Sanden SANCO2 Heat Pump Water Heater Technical Information. Available online: https://www.sandenwaterheater.com/sanden/assets/File/Sanden_sanc02_technical-info_10-2017_4.pdf (accessed on 3 June 2019).
26. Sunamp—Sunamp and Trina Solar Sign MoU to Collaborate on Energy Saving Solution for Global Market. Available online: https://www.sunamp.com/sunamp-and-trina-solar-sign-mou/ (accessed on 2 December 2020).
27. Enphase—Enphase AC Battery. Available online: https://www4.enphase.com/sites/default/files/downloads/support/ACBattery-DS-EN-UK.pdf (accessed on 28 August 2021).
28. Bors, J.; Kenway, S. Water Temperature in Melbourne and Implications for Household Energy Use. Available online: https://www.waterportal.com.au/swf/images/swf-files/9tr1--001-water-energy-carbon-links_bors_2014_water_temperature_and_implications_for_household_energy_use.pdf (accessed on 29 August 2021).
29. Solar Panel Payback Times. Available online: https://www.choice.com.au/home-improvement/energy-saving/solar/articles/solar-panel-payback-times (accessed on 2 December 2020).
30. Solar Victoria—Solar Battery Rebate. Available online: https://www.solar.vic.gov.au/solar-battery-rebate (accessed on 2 December 2020).
31. South Australia Government—Solar Photovoltaic Systems and Battery Storage. Available online: https://www.sa.gov.au/topics/energy-and-environment/energy-efficient-home-design/solar-photovoltaic-systems (accessed on 2 December 2020).
32. Australian Building Codes Board—Climate Zone Map Australia. Available online: https://www.abcb.gov.au/resource/map/climate-zone-map-australia (accessed on 18 August 2021).

energies

MDPI

Article

Automatically Creating HVAC Control Strategies Based on Building Information Modeling (BIM): Heat Provisioning and Distribution

Andreas Sporr [1,*], Gerhard Zucker [1] and René Hofmann [1,2]

[1] Sustainable Thermal Energy Systems, Austrian Institute of Technology, Giefinggasse 2, A-1210 Vienna, Austria; gerhard.zucker@ait.ac.at (G.Z.); rene.hofmann@ait.ac.at (R.H.)

[2] Institute of Energy Systems and Thermodynamics, Technical University, Getreidemarkt 9/302, A-1060 Vienna, Austria

* Correspondence: andreas.sporr@ait.ac.at; Tel.: +43-664-88256146

Received: 26 May 2020; Accepted: 20 August 2020; Published: 26 August 2020

Abstract: Building Information Modeling (BIM) data are typically exchanged using the Industrial Foundation Classes (IFC) standard. An IFC-based BIM model is a container for data that is created during the design and planning phase and is therefore a rich source of information for the commissioning phase, in which building services are brought to operation. This paper examines the use of BIM data for automated generation of control strategies for energy systems, thus simplifying and accelerating the commissioning phase. We present a methodology to create control strategies of a building heating system with several variations of renewable energy systems and include both heat provisioning and a distribution system. The control goals include favoring the use of non-fossil energy, which is provided by a combination of photovoltaic system (PV), heat pump (HP) and industrial excess-heat source. Thermal energy storages are integrated for load shifting purposes and the control of the heat distribution system is designed towards the requirements of building physics, occupancy and outside climate conditions. A validation of the approach is presented in a combined SIMULINK and TRNSYS simulation environment.

Keywords: automatically generated heating ventilating and air conditioning (HVAC) controllers; Building Information Modelling (BIM); Industry Foundation Classes (IFC); advanced building controls

1. Introduction

Automatic creation of control strategies is an approach to derive the operation of building systems using the information about the building's spatial information, its physical information (e.g., thermal properties) and the energy systems. These data are typically created in the planning phase of the building and can be stored in a BIM model, where all involved stakeholders can access and modify it. Well-known BIM planning tools such as Revit [1], Allplan [2] or ArchiCAD [3] use different proprietary data formats for internal storage, which makes accessing the data unfeasible. However, all of the above tools are capable of exporting data using the Industrial Foundation Classes (IFC) data format, standardized as ISO 16739 [4]. Therefore, this work uses IFC-based BIM data as its main data source.

Buildings are controlled on two levels: process control and building automation. While process control is mainly implemented by means of continuous, linear controllers (such as proportional-integral-derivative (PID) controllers), building automation implements discrete control strategies using textual or graphical programming tools to define system states and the conditions of when to enter and leave states, including definitions of setpoints for PID controllers and enabling conditions for sub-systems. Both building automation and control (BAC) systems are implemented in the commissioning phase and require information about the building and its system. The development process today foresees a control engineer to implement the BAC system based on the planning state that is handed over from the heating, ventilation and air conditioning (HVAC) planner. BIM has the potential to make this process far less tedious and erroneous. At the same time, a well-maintained BIM model contains the information that is needed to automate the implementation of a BAC system.

This work derives the BAC creation process using IFC data as a foundation. It is based on the results of a previous paper [5] and extends the control strategy development beyond mere room automation to a more sophisticated and flexible system. Here, a thermal provisioning and distribution system for buildings is taken into account, where control strategies are developed for variable combinations of renewable and fossil energy sources. First, the IFC data are processed, including room properties such as volume, window positions, floor heating installations, etcetera. This information is used to create a control strategy, which consists of two parts: the energy provisioning system control that contains controllers for various energy provisioning components, and the distribution system control.

2. State of the Art

2.1. Building Automation and Control Strategy Creation

Building automation originates from industrial automation and had its earliest building implementation in heating, ventilation and air conditioning (HVAC) systems—the American ASHRAE Society is still the prevalent organization with regard to standards, methods and protocols in the area of building systems, indoor comfort and energy efficiency of buildings (see e.g., [6]). Building automation operates on linear controllers, as process controllers, with an automation layer to implement supervisory control by defining system states and the transitions between them by means of logic conditions. This combination of traditional logic and continuous control is at the center of this work. In recent decades, approaches were made to develop self-adjusting controllers, such as adapting two-point controllers [7], though this method neither considers the great impact of solar radiation nor can it be used for testing the controller's performance during planning phase. That may lead to a too high temperature yield and thus higher energy consumption [8].

Developing and parametrizing control strategies for HVAC systems of larger buildings make for a complex topic. Katipamula et al. [9] investigated the influence of different control strategies, including varying indoor temperature setpoints and pre-cooling, which were based on a state-space model of a single-family house. However, they focused on another type of building, and did not inspect the provisioning side of a heating system and their strategy as well as its parametrization needed to be made by hand. Song et al. [10] discussed different control systems for improving indoor air quality, but did not take care of the overall control strategy to reduce the buildings' energy consumption and the interaction between different sub-systems within the overall HVAC system. Yang et al. [11] described a method to improve the energy efficiency of a ventilating system by using swarm intelligence. However, within this method the temperature and the indoor air quality of a single room is controlled by the ventilation system, which is uncommon for modern buildings. Further, the results of this method are non-classical control strategies, much more a single room controller and based on a handmade room model. Erickson et al. [12] implemented a control strategy based on a Markov Chain Occupancy Prediction Model. Using this method, two main problems may occur: if the prediction fails, the control strategy will not be able to perform as wished and the desired results will not be reached. Furthermore, the model was handmade too, which is not useful for the task within the present work.

While the existing methods for improving control strategies of HVAC systems provide good results, there is still a need for further research to address the following challenges that the combined provisioning and distribution sides face:

- Handmade models and optimizations, as well as manually parametrized controllers, are not feasible for broad diffusion into the wide variability of existing automation systems due to cost reasons.

- An automation methodology needs to be able to handle different HVAC variants. This calls for a generalized approach to efficiently implement this automation.
- Fallback solutions need to be implemented in order to provide a stable control strategy; otherwise, the development will not be appropriate for usage within existing real-world buildings.
- The current state of research does not discuss the automated development of control strategies for the provisioning side with different variants of renewable and non-renewable energy systems.
- The distribution side influences the provisioning side directly. Therefore, a control strategy needs to regard this physical feedback in its design.

2.2. BIM Data Availability and Organization

Standardization of BIM data has advanced in recent years and covers architectural design information as well as building physics data. Architectural design is well-developed, focusing on geometric properties to create the building envelope as well as the floor-plan layout of a building. A key factor for extending the use of BIM beyond mere 3D-modeling is the use of the meta-data that is defined for each IFC element. Basic elements such as an *IFCWall* have geometric properties including *NominalLength*, *NominalWidth*, *NominalHeight* or *GrossVolume* [13]; but they also have, for example, thermal properties, which are required when running thermal simulations on a BIM model: in an attached *IfcThermalMaterialProperties* element, an *IFCWall* can define thermal transmittance, specific heat capacity and other parameters. Similarly, lighting simulation requires the definition of additional properties for windows and artificial light sources.

For energy provisioning systems, the geometric properties of components are negligible compared to the meta-data that define the functionality of a component. Since BIM-based modeling has its origin in 3D-drawings, this aspect is not yet as developed as geometric properties are. However, development and standardization are continuing. The core standard is IFC [4], which is extended by libraries such as the buildingSmart Data Dictionary (bsDD) [14] on ISO level or national dictionaries such as the Austrian ON A 6241-1 [15]. The latest standard is the ISO 23386 standard, which was published in 2020 [16] and supports the definition of data templates, which contain relevant information for components and cover a broad range of building service components. For the time being, this shortcoming is addressed by augmenting the IFC data with additional auxiliary data (Section 3).

Still, this collection of standards does not fully specify the information of a BIM model; it merely defines the structure of data, but not the content. Material properties, for example, are defined as opaque strings, leaving the content open to the defined by the person or the authoring tool. Planners and architects address this issue by defining modeling guidelines on how to set parameter values and how to name components. These guidelines are proprietary to one architectural office or one developer and therefore cause problems when cooperating over company borders in an open BIM planning and construction process. At the moment, this problem remains unaddressed and would require much more elaborated standardization. An upcoming development is the Employer's Information Requirements (EIR), which shall define standard methods regarding how information is created and named. A first coverage of the EIR is available in ISO 19650 [17], an ISO standard that was derived from the British publicly available specification (PAS) 1192 that provides a general description on how to manage information in BIM over the lifecycle of a building.

2.3. Potentials for Optimization

As elaborated in Section 2.1, current HVAC systems are usually developed based on the expertise of HVAC planners and control engineers and do not have a high degree of automation. Cost planning often follows a standardized costing scheme, such as the German Deutsche Industrie Norm (DIN) 276 [18] or the statement of work for building services by the Austrian Ministry of Digital and Economic Affairs [19]. In this cost schema the distribution side is described in view of room automation, regarding costs, e.g., for the commissioning of data points and definition of setpoints, but not for the programming of control strategies. On the provisioning side there is a need for freely programmed control strategies for the energy system, which is reflected as a separate cost item. This is where the control engineer needs to analyze the schematics (hydraulics, sensors, actuators, etc.) and derive a feasible control strategy.

The costs for building automation depend on the type of project, the involved companies and the level of automation. According to [20], costs in the commissioning phase amount to 2–3% of the total costs of the mechanical systems. Based on the information gained from expert interviews, the costs for building automation range between 2% and 9% of the overall project volume. As an example: a typical office building with 2000 m^2 office space, 600 data points and a total volume of 6 million EUR requires approximately EUR 450,000 for automation hardware and software, with the costs being roughly equally split between hardware and engineering (including 60% budget for programming, 30% for commissioning and 30% for integration). The methodology presented in this paper automates the control strategy programming by using building model data and thus reduces programming efforts. In addition, it is possible to quickly react to changes in the specifications by triggering a rerun of the algorithm based on the updated model data.

3. Model Development

The provided IFC model is the base for the subsequent control generation. Using this data, different heating zones are identified as well as the heat provisioning systems therein. As shown in Section 2.2, the energy distribution side is quite well described within the current IFC standards, however, the provisioning side is still barely defined. Hence, auxiliary data are needed for filling these information gaps. For the scope of this work, the necessary information about energy provisioning systems, such as a heat pump or gas burner, are therefore defined in an auxiliary data file. This will collaborate well with future workflows, since the standard ISO 23386 [16] provides an excellent template for building service components once manufacturers have adapted their product data to the new standard. For the scope of this work, we have defined auxiliary data to provide this information for the following components: heat pumps, thermal energy storages, natural gas-fired boilers, photovoltaic systems, valves, pumps and their interconnection as well as their integration into the overall heating system. It is expected to use the current workflow and abandon the auxiliary data file with only minor adaptations once the ISO 23386 standard is fully adopted. Section 3.2 describes the provisioning system components and their necessary parameters in more detail.

Aside from this component-related information, the creation of control strategies needs additional auxiliary data:

- Setpoints for room automation and energy provisioning systems: These data are typically created during detailed planning phase and implemented in the commissioning phase. Each room is assigned setpoints for temperature, humidity, airflow, etcetera. Additionally, the supply and return temperatures for the heating and cooling system, as well as the ventilation system, are defined.
- Occupancy profiles: Offices, meeting rooms, common areas and other zones have specific usages that have to be defined in terms of "persons per room". This also includes the typical occupancy periods reflected as time schedules for the different zones.
- Outside climate data: Temperature and humidity profiles for the given location are required for defining the default operation of the room controllers and the controllers of the energy provisioning system.

The data provided here reflect only the design guidelines for a building, i.e., the heating system has to be dimensioned so that it can provide indoor comfort on the coldest day at the building location. In daily operation, it is the task of the building automation system to react to the actual conditions in terms of outside temperature or occupancy.

3.1. Building Physics and Distribution System

An earlier publication deals with filtering the IFC data to obtain the necessary information about the floorplan, the separation into rooms and thermal zones, and the information on how the distribution system is linked to the floorplan [5]. Since the BIM models that were examined have proven to be incomplete and inconsistent, an algorithm was developed to create a consistent floorplan layout and distribution system. Within this algorithm, data gaps were closed and the walls (*ifcWall*) were aligned to rooms (*ifcSpace*). Using the material information in the IFC elements, it was possible to create a thermal model of the building.

Based on this previous work, the heating distributing system was identified by extending the algorithm to extract also the heating system components, taking the following steps:

1. Separating data by floor (e.g., *ifcSpaces*, *ifcPipeSegments*, …)
2. Creating a 3D net based on the geometric data of the specific floors for the whole heating distributing system. A parental net is defined by the first pipe element found on this floor. Two levels are spanned, defined by the highest and the lowest points of the net including a determined hysteresis. All elements located between these two points are assigned to the respective net
3. Subdividing the net based on the geometric data of *ifcSpace* elements. This division is named children nets
4. Assigning pipes and relating elements as actuators (e.g., valves), sensors (e.g., temperature), fittings, etc., to rooms by using parental nets
5. Fitting the identified elements into the children nets to check for gaps
6. Integrating all children nets into the parental net
7. Checking for vertical pipes, which are connecting different floors with each other
8. Subdividing into heating zones by using the origin of the energy provisioning system
9. Assigning setpoints and occupancy profiles to heating zones based on *ifcPropertySet* elements and corresponding entities.

The result of these steps is a net describing the heat distribution system, which contains all relevant components; it also identifies which heating zones belong to which heat supply components. Secondly, all heating zones are detected by the position of the actuators and the sensors: all rooms that are supplied after the last flow-regulating actuator are combined into one heating zone, since these rooms cannot be controlled separately afterwards. This results in a set of data for all existing heating zones, which contains all available sensors, actuators, setpoints and occupancy profiles. In combination with the heat provisioning system, this set is used to create control strategies fully automatically, as shown in Section 4.

3.2. Energy Provisioning System

Building energy systems consist of a combination of energy provisioning and storage components which are physically connected by a transport medium (typically water, with a link to the air-based ventilation system). The components can be classified into different, common types, which are found in many buildings. While the components are rather standardized, the hydraulic connections depend on the design intent of the planner and can vary in their realization. Figure 1 shows a superstructure which covers all feasible variants of heat provisioning systems. For the sake of brevity, a few assumptions had to be made in the scope of this work; for once, only the heating case without hot tap water supply is regarded. The cooling case is intentionally left out, since it would require too many additional components and hydraulic connections to be covered here. With regard to distribution systems, only a low-temperature floor heating system is available, leaving out high temperature radiators. In addition, only the most common energy components are included in order not to inflate system complexity. That being said, the superstructure in Figure 1 and its variations cover a broad set of real-world energy system configurations: different components can be removed to achieve the desired configuration, i.e., the energy system at hand is created by removing components that are not needed.

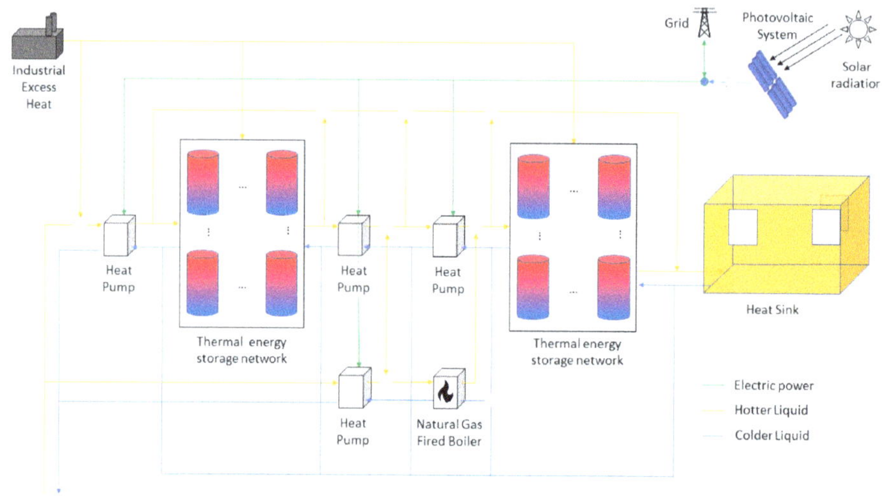

Figure 1. Superstructure of energy provisioning system.

Decisive for the control is the flow of the supply (orange) and the return (blue) as well as the power supply for heat pumps (green). Different configurations are possible: heat pumps that feed a thermal energy storage network of parallel and/or serial storage tanks and use air, brine or an alternative energy source such as industrial waste heat as a source, as well as gas-fired boilers that use the raised temperature of a heat pump as a return source instead of the return flow of the thermal energy storage. Sensors and actuators are not shown since they may differ depending on the configuration. Section 4 then explains how a control strategy is created based on a derivative of this superstructure.

While IFC exports contain most of the required data of the distribution system, the energy provisioning system requires additional data. Basically, three different types of information are needed:

1. Equipment types (heat pumps, thermal energy storages, natural gas-fired boilers, photovoltaic system, actuators, sensors, etc.).
2. Equipment data
    a. Design points of heat pumps (electrical as well as thermal source and sink power at specific input and output temperatures)
    b. Basic information of heat pumps (maximum thermal power, used liquid on both sides including their thermal data)
    c. Basic information of thermal energy storages (heat loss coefficients, conductivity, volume, diameter)
    d. Basic information of natural gas-fired boilers (efficiency curve, maximum thermal power)
    e. Basic information of photovoltaic systems (open circuit and short circuit voltages and currents, maximum power data, number of cells, number of modules)
    f. Positions and types of actuators as well as sensors and their parameters (maximum volume flow, etc.).
3. Equipment connections (both physical flows and information flows).

If this information is available in IFC and is in accordance to standardized naming, it is fetched from the IFC file. Otherwise, the elements and their interconnections need to be defined separately in auxiliary data.

The general structure in Figure 1 allows for various permutations of energy components, where only a subset is viable regarding economic and technical feasibility (a supply that relies only on industrial excess heat is, for example, not regarded, since it is only implemented together with a backup heat pump). These combinations are shown in Table 1. Overall, the presented algorithm can be used to automatically develop control strategies for a total of 26 different configurations.

Table 1. List of feasible combinations of heat provisioning elements based on the superstructure.

Variant	Heat Pump	Natural Gas-Fired Boiler	Industrial Excess Heat	Photovoltaic System	Thermal Energy Storage in Series/Parallel	
Variant A	X			(X)		X
Variant B		X		(X)		X
Variant C	X		X	(X)		X
Variant D	X	X		(X)	X	X
Variant E		X	X	(X)		X
Variant F	X	X	X	(X)	X	X

Elements marked with X are required for a variant. For thermal energy storages, different hydraulic combinations are possible; in general, multiple heat storages can be connected in parallel or serial in both positions in Figure 1 with high and low temperature setpoints. A combination of serial and parallel storages is only possible for variant D and F because of the high temperature, which needs to be provided by a gas-fired boiler for filling multiple storages. Although heat pumps can also provide temperatures above 50 °C, most heat pumps in buildings are designed for lower temperatures. Hence, to reach a higher temperature level, cheaper gas boilers are used. To improve temporal flexibility, any number of stratified heat tanks can be specified, and the control strategy algorithm will combine the given heat producing elements with the storages.

The photovoltaic system (PV) is optional for all variants, since the renewable electricity of the photovoltaic system (PV) can always be integrated. The system is controlled with a simplified energy-balance approach: if electricity is needed and the PV can provide it, it is used for covering parts of the load. If the PV provides more energy than can be used on site, it is fed into the electric grid. For that, different regulations need to be fulfilled: the electricity price is higher than the feed-in price, whereby increasing the share of on-site usage of renewable energy is an important action to improve cost saving.

Despite the advantage of heat pumps of producing an output of about four units heat energy per one unit electricity (coefficient of performance (COP) of brine-water heat pumps equals to about 4 [21], their maximum achievable temperature is at about 80 °C). Even though high-temperature heat pumps exist, their prices are relatively high and their COP is much lower compared to lower temperature heat pumps, [22]. Thus, if higher temperatures are needed or if independency from electrical power sources is desired, the possibility of adding gas-fired boilers is included.

Control of oil-fired boilers is not discussed within this paper, since heat production using this type of technology was banned in Denmark in 2013, [23], and partially in Austria in 2019 [24], and bans will be necessary for reaching the climatic targets of the European Union (EU) [25].

4. Controller Block Library

Once the data of the building energy system has been collected, the control strategy can be created. As a first step, a library of control blocks for the energy components has to be defined. These control blocks are then interconnected and parameterized based on the interconnection of the energy provisioning system. The distribution system and the heating zones that are supplied by this provisioning system are also relevant for the parameterization. This information is used both to estimate any missing parameters (e.g., temperature setpoints of the thermal energy storage networks) and to ensure efficient control of the heating zones.

For the energy components that are listed in Section 3, a controller block and, if necessary, a switching block is created, as shown in Figure 2. These blocks are stored in a library and are automatically imported by the generating algorithm, depending on the energy system structure and the hydraulic circuitry. Depending on the connections between different elements, such as serially connected heat storages, the algorithm changes the controller's behavior. This ensures efficient operation.

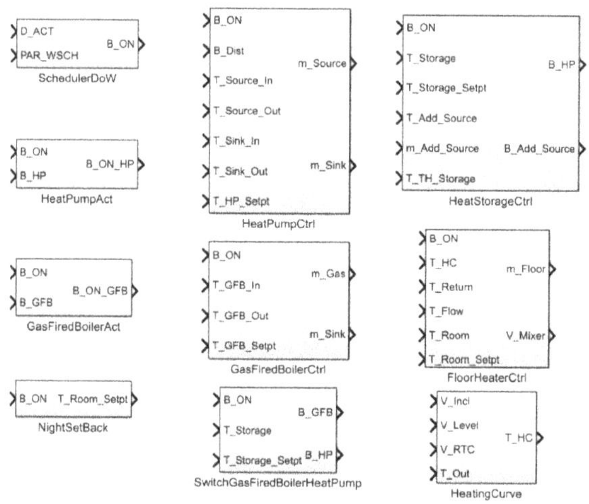

Figure 2. Developed SIMULINK blocks for different types of heating.

The control blocks have three types of variables: design parameters, process variables and dynamic outputs. Design parameters are unchangeable during the runtime of the control strategy and include, for example, the orientation and positioning of the PV system, as well as its inclination and size, window location and sizes, etcetera. Process variables and dynamic outputs may change any time and affect the overall system dynamically.

SchedulerDoW represents the time scheduler of the overall system. Based on the current time *D_Act* and a predefined list of on–off times, which are represented as the input *PAR_WSCH*, this block (de)activates the HVAC system. Specific on–off times are given by underlying occupancy profiles or the type of use of the building, i.e., office buildings are operating within normal business hours, hospitals all day and night long. Every block that has an input named *B_ON* can be (de)activated by *SchedulerDoW*.

Some buildings' inner room temperatures are reduced outside operation times. Therefore, the block *NightSetBack* is used for setting this temperature setpoint to a given value while the overall system is deactivated (*B_ON* = 0). If the inner room temperature reduces too sharply, the system is reactivated.

HeatPumpAct and *GasFiredBoilerAct* are used for HVAC systems, where heat pumps, as well as natural gas-fired boilers, can heat up the same heat storage. Their inputs *B_HP* and *B_ON_GFB* are switched by the block *SwitchGasFiredBoilerHeatPump*. Basically, this block defines the need of a natural gas-fired boiler and the usefulness of activating the heat pump based on different conditions:

- Setpoint temperature of the heat storage: If the heat pump is not able to reach the needed temperature, it is only used when the lowest node temperature is lower than the provided one, and energy from the photovoltaic system is available. After reaching the maximum possible temperature, the heat pump is deactivated, and the natural gas-fired boiler is activated.
- Needed energy from the distributing side: If the heat pump is not able to provide this energy, either due to errors during design phase or if the heat pump is only used for partial-load operation, the natural gas-fired boiler is used for provisioning.

HeatPumpCtrl is used to provide a pre-defined flow temperature, which can be the target temperature of the underfloor heating or of the storage tank. The target temperature of the underfloor heating is determined by Figure 3. In the test case shown later in Figure 4, *Pump Brine* and *Pump Storage* are used for controlling the mass flows of the source and the sink side of the heat pump. On both sides, PI-controllers are used. *Pump Brine* changes the mass flow for reaching a specific temperature difference between brine intake and return: $\Delta T_{br} = 4K$. A higher temperature difference than 4 K may lead to environmental problems. Thus, the pumps energy consumption increases. Depending on the power usage on the heat pump's hot side, the extraction of heat from the ground changes and the controller tries to compensate this load change. For that, a PID controller is used, which compares the flow and the return temperatures on the cold side to be able to guarantee the temperature difference: ΔT_{br}. The controller can either be used for regulating the mass flow by using a valve or a pump directly.

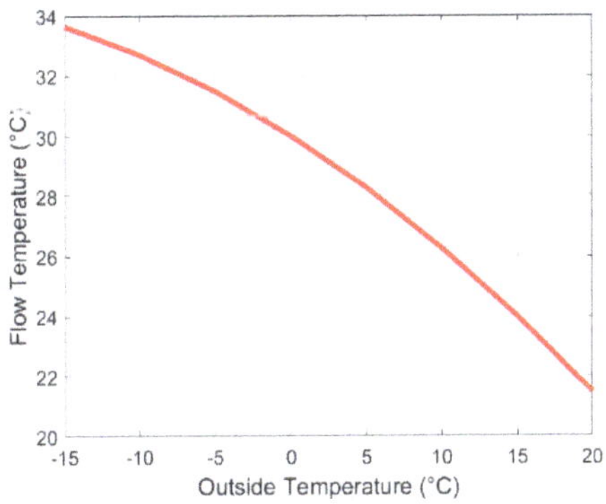

Figure 3. Heating curve based on pre-defined support points.

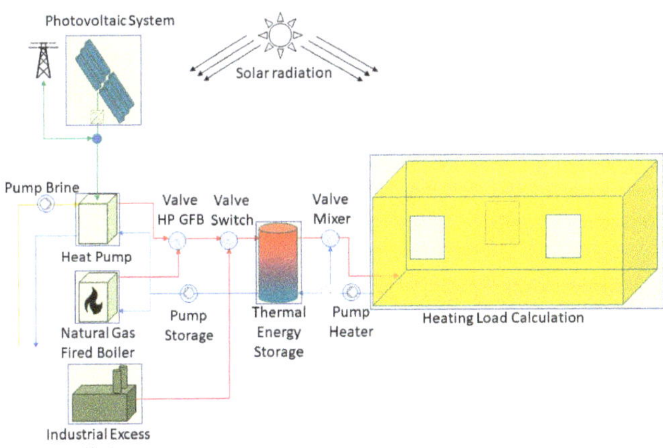

Figure 4. Heating system derived from the superstructure, including actuators.

For designing the hot side controller, two different situations are compared: if industrial excess heat is available, this energy is used to fill the thermal energy storage. Though, if this temperature is too low, the heat pump needs to be activated—switching between these two variants carried out by the valve *Valve Switch* (Figure 4). Since the industrial excess heat does not need an additional pump for passing water through the thermal energy storage, the hot sides' water needs to get pumped through the storage and the heat pump. Pump Storage is controlled using a PI-controller, which compares the return temperature with the flow temperature and tries to reach the given flow temperature setpoint. Depending on the heat pump's nominal output, this maximum temperature difference is calculated using Equation (1).

$$\dot{Q}_{nom} = \dot{m}\ c_p \left(T_{hp,flow} - T_{hp,ret}\right) \tag{1}$$

Thus, the heat pump's performance is limited, and possible miscalculations are avoided. Finally, the heat pump's activation period depends on the thermal energy storage temperature and its se-point. Attention should be paid to the heat pump's duty cycle, because of the high start-up current and the high stress during the start-up phase.

The input variable *B_Dist* decides, if industrial excess heat is currently used; *T_Source_In* and *T_Source_Out* gives information about the temperature on the brine side, which is necessary for controlling the mass flow of the heat-pump's cold side; *T_Sink_In* and *T_Sink_Out* are necessary for controlling the mass flow on the hot-side to set the correct temperature difference. The output variables *m_Source* and *m_Sink* are the mass flow control inputs for the installed pumps or valves, which are set by PI-controllers. The setpoint of the heat pump depends on the following components, which are connected, and is set by using the input *T_HP_Setpt*.

If a natural gas-fired boiler is part of the system, the block *GasFiredBoilerCtrl* is used for controlling the boiler. This controller is simple, since neither energy from the PV system needs to be considered, nor two hydraulic sides must be controlled. Hence, the mass flow *m_Gas* for heating up the heat storage depends on the difference between return temperature from the heat storage *T_GFB_In* and flow temperature into the heat storage *T_GFB_Out* and the temperature setpoint of the heat storage *T_HS_SP* as well as the current temperature *T_Storage*. The circulation pump is regulated by a PI controller, using *m_Sink* as the control variable.

HeatStorageCtrl switches between using industrial excess heat *B_Add_Source* and using another heat providing component *B_HP* as a heat-pump or a gas-fired boiler. Depending on the temperature *T_Add_Source* and mass flow *m_Add_Source* of the industrial excess heat and the other heat providing component, free energy from industrial excess heat can be used. If the storages' temperature *T_Storage* is higher, or a higher setpoint *T_Storage_Setpt* must be reached, industrial excess heat may be used till the transferred energy is non-economic.

The blocks *FloorHeaterCtrl* and *HeatingCurve* are used for controlling the temperature within a single room, which can be affected by changing the flow temperature and the mass flow from the floor heating system. The first parameter is outside temperature-regulated and the second is constant. In real systems, a constant mass flow is reached by using a valve at the heating manifolds return flow. The valves position will normally not be changed within the building's lifetime. The flow temperature can be calculated using a heating curve (example shown in Figure 3).

The configuration of the heating curve is provided by an expert and depends on different parameters, which can be changed for improving the energy efficiency and the comfort level, as:

- Inclination (*V_Incl*)
- Level (*V_Level*)
- Room temperature change (*V_RTC*)

By changing these parameters, the curves shape and the heating systems behavior change. Usually, they are adjusted once depending on experience and are constant during the heating systems life cycle. Partially, these values are time-dependent, e.g., if night setback is activated.

The output *T_HC* relates to the same named input from *FloorHeaterCtrl*. Within this block, the valve position *V_Mixer* of the mixing valve is set depending on the storage, the return *T_Return* and the heating curve *T_HC* temperature. Depending on the current temperature of the storage, the valve may mix liquid from the return flow. The room's temperature is controlled based on the room's temperature setpoint *T_Room_Setpt* and the current room temperature *T_Room* by changing the mass flow using a PI-controller.

5. Topology-Aware Algorithm for Control Strategy Assembly

The workflow so far has collected all relevant information from an IFC-based BIM model and has augmented it with auxiliary data. A library of controller blocks has been designed to support the control strategy creation for a superstructure of feasible energy system variants. This chapter now covers the creation of the control strategy for a given system. The algorithm that has been developed is capable of creating valid control strategies for all variants of the superstructure, including the management of different combinations of serial and parallel storages. For the scope of this paper, one test case variant of the energy system is selected, which contains PV, heat pump, natural gas-fired boiler and industrial excess heat as sources, one storage tank, and a simplified distribution system with only one thermal zone (Figures 4 and 5). The resulting control strategy at the end of the workflow is shown in Figure 6.

The system shown in Figure 4 also corresponds to the simulation setup used for the plausibility check. The models for the supply side are based on already validated ones, such as a heat pump model, whose thermal output power is calculated based on the thermal input and the electrical power as well as the temperature spread between source input and sink output temperature. In this test scenario, the thermal storage corresponds to a decoupling of the supply and distribution side. The calculations of the interior temperature are obtained by the *Heating Load Calculation*. The heating system in use, in this case underfloor heating, is simulated and the heat losses and gains are taken into account. A certain inertia results from the storage mass of the installed screed and the heat transfer of the heating pipes to the screed. This inertia is considered in the design of the controllers, since a more inert system requires a different control system than a highly dynamic one.

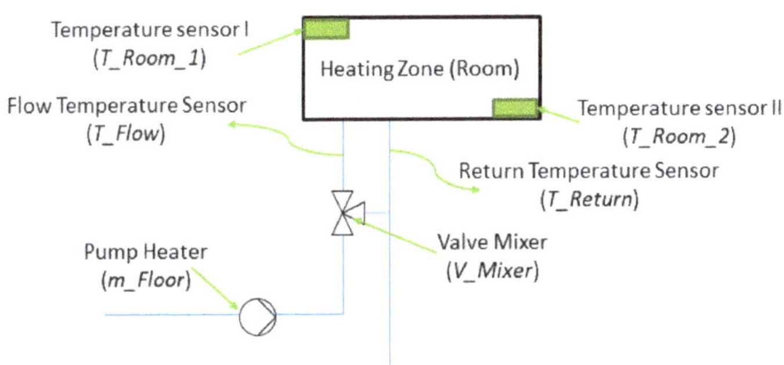

Figure 5. Example of typical sensors and actuators within a heating zone or room.

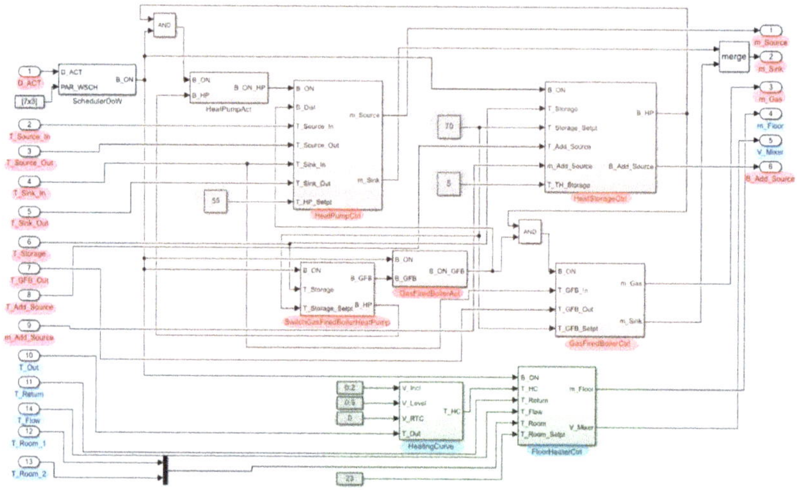

Figure 6. Automatically created baseline rule strategy for the heating system of a single room.

Based on the test case information, a topology map shown in Table 2 is created; the inputs and outputs of the provisioning components are derived based on the definitions in Section 4 and are matched with the component names in the topology map in order to reflect the hydraulic connections. The provision side is represented from *Pump Storage Out* to *Photovoltaic System* and the distribution side from *Thermal Energy Storage Discharge Out* to *Pump Heater Out*. A distinction is made between inputs and outputs, which also define the flow direction of the liquid used.

Table 2. Topology map.

Source	Sink
Pump Storage Out	*Heat Pump Hot_In*
Pump Storage Out	*Natural Gas Fired Boiler In*
Heat Pump Hot_Out	*Valve HP GFB In*
Natural Gas Fired Boiler Out	*Valve HP GFB In*
Valve HP GFB Out	*Valve Switch In*
Industrial Excess Heat Out	*Valve Switch In*
Valve Switch Out	*Thermal Energy Storage Charge In*
Thermal Energy Storage Charge Out	*Pump Storage In*
Photovoltaic System	*Heat Pump Power In*
Thermal Energy Storage Discharge Out	*Valve Mixer In*
Heating Zone Out	*Pump Heater In*
Valve Mixer Out	*Heating Zone In*
Pump Heater Out	*Valve Mixer In*
Pump Heater Out	*Thermal Energy Storage Discharge In*

5.1. Provisioning Side

Figure 4 shows components highlighted in green, which require an according control block, while blue components are directly controllable actuators, such as valves and pumps. In the next step, their interconnections need to be implemented using the topology map in Table 2. The algorithm terminates when all inputs and outputs are connected.

The algorithm starts at the source side; in this test case, this is the brine side of the heat pump connecting the *Pump Brine* pump with the heat pump. Thus, the control blocks *HeatPumpCtrl* and *HeatPumpAct* are used. The hot side of the heat pump (on the right) leads into the first valve, *Valve HP*

GFB. Its outlet leads to the next valve, *Valve Switch*, and then to the thermal energy storage tank *Thermal Energy Storage*. Thus, the *HeatStorageCtrl* control block is required. A further inlet of the valve *Valve HP GFB* corresponds to the *Natural Gas Fired Boiler*, whose flow is fed by the Pump Storage and is connected to the return flow of the storage tank. Thus, the blocks *GasFiredBoilerCtrl* and *GasFiredBoilerAct* are used. The Pump Storage also feeds the heat pump, which means there is a direct link between the heat pump and the gas boiler. Therefore, the block *SwitchGasFiredBoilerHeatPump* is required. The discharge side of the storage tank is connected to valve *Valve mixer*, which leads directly into the heating zone. Since this is part of controlling the distributing side, no further control of the provisioning side is required.

5.2. Distribution Side

The distributing side is simplified to only one thermal zone. Figure 5 shows the placement of the sensors, valves and pumps that are necessary for maintaining indoor comfort in the thermal zone.

The algorithm starts by identifying all rooms within a building. They are represented by *ifcSpace*. Based on this information, all sensors, which are represented as *IfcDistributingControlElement* and *IfcSensorType* "Temperature Sensor", within the rooms' boundaries are identified. Either they are directly linked to this room within IFC, or they are assigned to a specific room by knowing its three-dimensional coordinates. Then, all room-depending sensors need to be assigned to one specific room. Referring to the example from Figure 5, Temperature sensor I and Temperature sensor II are assigned to the Heating Zone (Room).

Second, all components, which are used for heating the specific heating zone are identified. Initially, heating pipes, represented as *ifcPipeSegment*, are identified. Using their location, they can be assigned to a specific room, too. Then, all connected pipe segments are traced, which leads to a net of the heating system and an easier assignment of sensors, valves and pumps. Using this net, all further elements are assigned to specific rooms by using the nearest ones to the first room-exiting pipe. Finally, the sensors *Flow Temperature Sensor* and *Return Temperature Sensor*, as well as the *Mixing Valve* and the *Heating Pump*, are assigned to the same *Heating Zone (Room)*. This procedure is performed for every room. The piping map has, of course, only be created once.

5.3. Control Strategy Synthesis

Depending on the identified components, the according control blocks and their sensors as well as actuators are added to the control strategy; they are highlighted in blue in Figure 6.

The connection of the provisioning side with the distribution side is controlled by a Pump *Pump Heater* and the mixing valve *Valve Mixer*. Due to this, the control block *FloorHeaterCtrl* is added and both outputs (*m_Floor, V_Mixer*) are connected. If there is no back mixing of the return flow, the output *V_Mixer* of the control block *FloorHeaterCtrl* would not be connected. This would affect the entire control strategy, since the flow temperature of the heating system can no longer be controlled by the flow temperature sensor, which is located at the entry to the room, but by the setpoint of the thermal energy storage. Changing this would force a change in the heat pump and gas boiler setpoints. Thus, the interconnection of the individual components is influenced by both the source and the sink side.

In Figure 6, all predefined parameters are highlighted in grey (boxes with constant values). The storage's setpoint *T_Storage_Setpt* is calculated for the known nominal temperature of the heat pump and the known flow temperature of the heat distributing side. Using the storage and the zone sizes, which need to be supplied with hot water, the temperature threshold *T_TH_Storage* is calculated. So, the storage will always be able to provide enough thermal energy and is refilled early enough without stressing the heat pump more than needed.

Other configuration parameters, such as *V_Incl, V_Level* and *V_RTC*, are either given or calculated by provided measured values for the outside and the associated flow temperature. Scheduler values need to be defined or are assumed depending on the building's usage, e. g., an office building's operation hours for the energy systems between 7:00 and 20:00 during workdays are assumed as default values.

As referenced earlier, Figure 6 shows the complete control strategy for the system in Figure 4, including both provisioning and distribution side. This structure is created fully automated by the topology-aware algorithm.

The controllers for the provisioning system are in the upper area (light red) and the distribution side controllers are highlighted with a light green background. *D_ACT* is an external input providing the current time. In the present case, the additional source refers to industrial excess heat. Hence, its inputs (temperature and mass flow) are connected to the corresponding inputs of the heat storage control. This free energy is preferably used, if it meets the corresponding setpoints. The PV system serves as a simple energy source, whose output is used in addition to that from the electrical grid. Subsequently, these data can be used to change the setpoints and switch-on times, but this is not discussed in detail in this paper.

The parameters determined were based on the interconnection and configuration of the individual components. The heat pump was designed with a flow temperature of 55 °C (*T_HP_Setpt*), but the thermal storage tank has a setpoint value of 70 °C (*T_Storage_Setpt*). This configuration would give a warning in the plausibility check, which is scheduled after the control strategy development. However, since a gas boiler and industrial waste heat are available as additional sources, this setpoint is valid. The switchover via *SwitchGasFiredBoilerHeatPump* is performed via these two temperature levels.

The indoor temperature setpoint is either defined in the room model of the BIM model, or is defined manually. The parameters for the heating curve are identified by the specified support points.

All these parameters can be adjusted in subsequent optimizations. These can be carried out automatically or can be changed by the respective users.

6. Plausibility Check Based on Simulation Results

Control strategies should already be checked during the planning process. This allows users to test different energy provisions and distribution systems as well as their dimensioning and to identify the optimal system. Simulations, which can be integrated in different simulation environments, support this process. The resulting control strategies can be tested with more complex models for fine tuning or can be modified during the commissioning of the finalized building. The resulting control strategy is tested in a simulation environment, which is coupled in the co-simulation environment PTOLEMY [26]. Simulink is used for the control strategy implementation that operates a thermal simulation of the building, the provisioning and the distribution system. In this test case, the system shown in Figure 4 is checked and the entire process described in the previous chapters is performed. If there are changes in the interconnection or in the components, these are recognized and are already considered in the control concept shown in Figure 6 as an example. Currently, the corresponding simulation environment of the components is set up manually. This can be automated in subsequent work by identifying relevant parameters in BIM models, but is not part of the present work. The simulation is run for a period of 48 h, the results for the provisioning side are shown in Figures 7 and 8.

Plot (a) in Figure 7 shows the usage of industrial excess heat (blue), the activation of the heat pump (green) and the gas-fired boiler (red). The storage tank temperature at the outlet and the upper storage tank temperature (plot (b)) have an initial temperature of 20 °C and should be raised to 70 °C (*T_Storage_Setpt*). Since the control strategy prioritizes renewable energy, the temperature is first raised to 55 °C only by the heat pump. As the average temperature of all layers of the storage tank is relevant for the controller, the loading by the heat pump continues until approximately 10:30. Then the heat pump is deactivated, and the gas boiler takes over to raise the temperature from 55 °C to the final temperature of 70 °C. Once this temperature has been reached, the boiler is deactivated and is regularly activated for reheating. Between simulation time 24:00 and 35:00, the overall control was overruled by a free heat source (*Industrial Excess Heat*). The energy is fed directly into the thermal energy storage and the standard control takes over again as soon as the heat source has faded. The frequency of activation can be changed by modifying the storage tank temperature hysteresis. If the target temperature of the storage tank was ≤55 °C, the gas boiler would not be activated, and only the heat pump would be responsible for charging the storage tank.

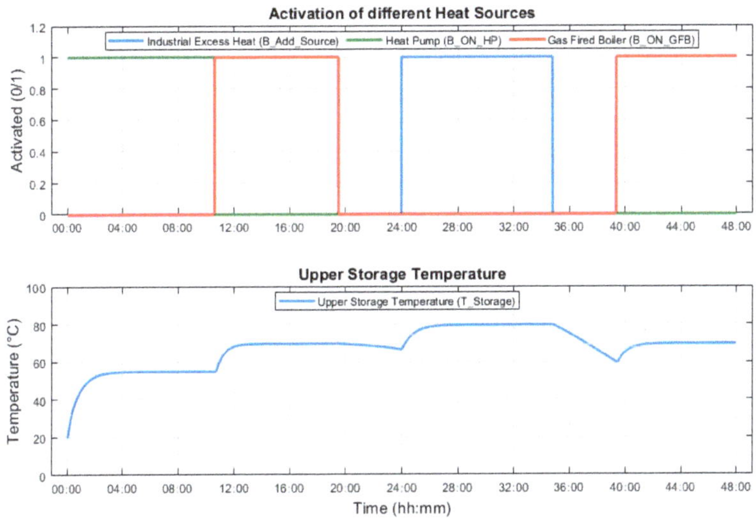

Figure 7. Switching between heat pump, natural gas-fired boiler and industrial excess heat.

(a)

(b)

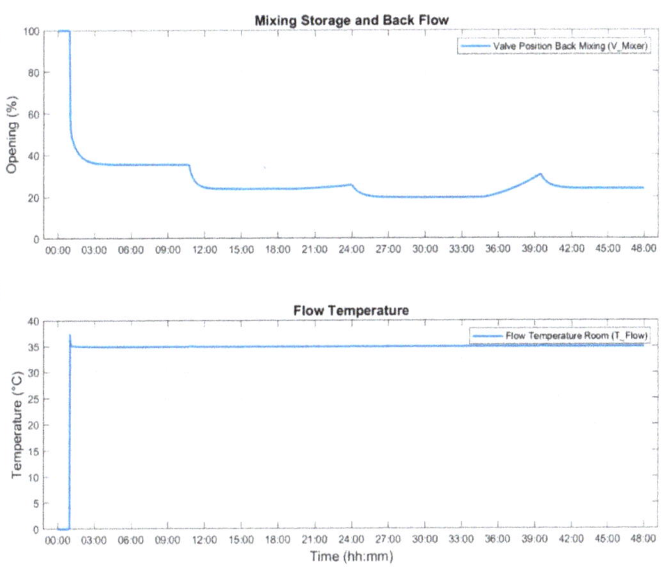

Figure 8. Supply temperature and back-mixing valve position for room temperature controlling. (**a**) the opening of the mixing valve to achieve the desired flow temperature; (**b**) the mixed flow temperature.

Figure 8a shows the opening of the mixing valve to achieve the desired flow temperature. In Figure 8b, the mixed flow temperature can be seen. The mixing valve is responsible for providing the desired flow temperature. The room temperature is then controlled via the mass flow rate (*m_Floor*), which is provided by a separate pump.

Figure 9 shows the indoor temperature of the controlled heating circuit as well as the floor temperature. Similar to the discharged storage tank, the indoor temperature starts at 20 °C and is controlled and kept at a constant level of 22 °C. The flow temperature corresponds to 35 °C (Figure 9) which is controlled by the mixing valve and corresponds to the heating curve for the existing outdoor temperature.

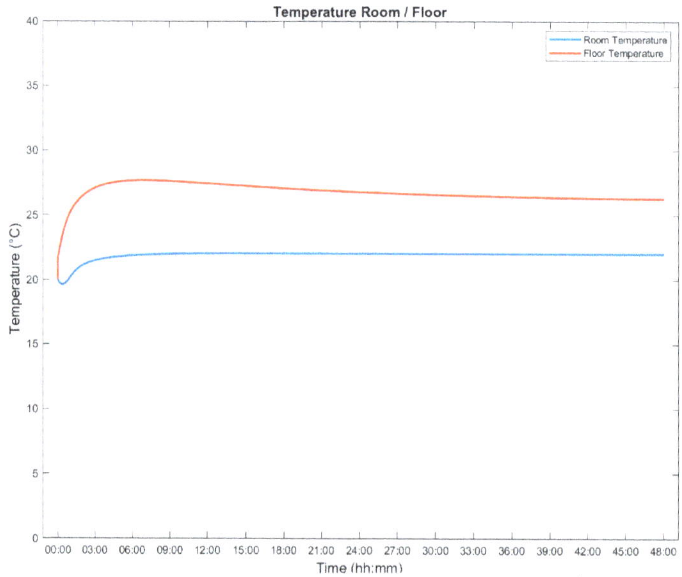

Figure 9. Room and floor temperature control.

7. Discussion

As shown in the figures, the provisioning as well as the distribution side are controlled accordingly. Switching between the heat pump and the gas boiler works based on pre-defined temperatures and the temperature of the storage tank is also constantly within the desired limits. The indoor temperature is constantly controlled to the predefined value and has been quasi-decoupled from the storage tank temperature by using a mixing valve.

The present simulation results were generated by coupling the automatically generated control strategy with manually created, classical provisioning and distribution systems. These results can be used for comparison with other systems or for comparison with changed building or system parameters. These include, for example, different thermal energy storage tank sizes, heat pumps with different performance levels, other interconnection variants or fundamental changes to the building in terms of thermal insulation or windows. The benefit of this approach is that provision and distribution systems based on empirical values can be tested using dynamic simulations. The results will change in each planning stage, as the first BIM models will have a lower quality than those in the later planning phase.

The strategy creation process is expected to reduce development costs in different areas; we estimate that the highest impact on cost and resource savings will be the following:

- A priori testing and optimizing of control strategies before a building is commissioned: this increases the overall quality of the building operation and will reduce operation costs.
- Automatic creation of control strategies: our BIM-based approach is expected to cut the costs for programming in half. In the example building shown in Section 2.3, this would address the 60% programming part of the building automation share of EUR 450,000.

- Evaluation of different HVAC configurations in early design stage: while this is possible today by thermal simulation, our approach refines the accuracy of simulation by automatically adding a control strategy to an HVAC system, thus providing high indication on dynamic behavior in high time resolution.
- Fast reaction to changes during planning by triggering reruns of the control strategy creation process: upon significant changes in the system, a manual reprogramming of control strategies can be replaced by our automated approach; we expect that each of these events is halved in its resource demand.

In the presented method, simulations are performed with widely used models. These include, for example, the Carnot model of the heat pump or room temperature calculations based on heat flows through walls and windows. However, an exact representation of reality is not possible with current models, so a validation based on real data would provide a further benefit. The control strategy development described above was tested based on different BIM models of real planned buildings; this led to promising results. However, no building could yet be accompanied from the planning to the construction phase. There are various reasons for this: Building data have to be provided in early planning phases. However, the BIM models have to be of a certain quality in order to obtain satisfactory results. Since BIM is still in its infancy in the Mechanical, Engineering and Plumbing (MEP) industry, projects have yet to be acquired in which not only BIM models can be obtained, but also the entire planning and construction process can be monitored. Since these processes can take years to complete, validations based on real data will only be able to be submitted later.

8. Outlook

The new methodology presented in this paper allows the automated development of a general-purpose building energy provisioning and distribution system; it uses both IFC-based BIM data and auxiliary data to extract the hydraulic structure of the energy system and derive a control strategy from it. While this new approach is capable of solving a broad variety of energy component combinations, it is limited to a subset of applications, namely, only addressing the heating case, not the cooling case. Future research will extend in this direction, so that it is possible to apply the methodology to the areas of cooling and, consequently, control of air humidity.

Furthermore, the created controller strategy can be used to perform online optimization, i.e., tuning the controller parameters while the system is in operation. This would allow users to further increase energy efficiency of the building energy system.

In further work, the developed methods will be tested on buildings designed and constructed using BIM. Since BIM has yet to gain a foothold in most of the construction industry and the model quality has to meet certain standards, validations must be carried out based on direct and intensive cooperation with building contractors. This also increases the likelihood of acceptance to use the developed methods and validates the results by comparing them with older buildings. However, the developed methods offer a solid basis for acquiring partners for the validation, as they can already be tested on BIM models and show feasible results.

Author Contributions: Conceptualization, A.S., G.Z. and R.H.; methodology, A.S. and G.Z.; software, A.S.; validation, A.S. and G.Z.; formal analysis, A.S. and G.Z.; investigation, A.S.; resources, G.Z. and R.H.; data curation, A.S. and G.Z.; writing—original draft preparation, A.S. and G.Z.; writing—review and editing, A.S., G.Z. and R.H.; visualization, A.S.; supervision, G.Z. and R.H.; project administration, G.Z.; funding acquisition, G.Z. All authors have read and agreed to the published version of the manuscript.

Funding: This research was funded by Austrian Research Promotion Agency (FFG), grant number 861710 and The Article Processing Charge (APC) was funded by AIT Austrian Institute of Technology GmbH.

Conflicts of Interest: The authors declare no conflict of interest.

Abbreviations

The following abbreviations are used in this manuscript:

B	Binary signal (-)
c_p	Heat capacity (kJ/kg·K)
$\dot{m}$	Mass flow (kg/s)
PAR	Parameter (-)
$\dot{Q}$	Heat flow (W)
ΔT	Temperature difference (K)
T	Temperature (°C)
ACT	Activation
Add_Source	Additional Source
br	Brine
$Dist$	Disturbance
GFB	Gas-Fired Boiler
hp	Heat pump
In	Incoming
$Incl$	Inclination
Out	Outcoming
RTC	Room Temperature Coefficient
ret	Return
$Setpt$	Setpoint
TH	Threshold
$WSCH$	Weekly Scheduler
BIM	Building Information Modelling
bsDD	buildingSmart Data Dictionary
COP	Coefficient of Performance
EIR	Employer's Information Requirements
HC	Heating Curve
HP	Heat Pump
HVAC	Heating, Ventilation and Air Conditioning
IFC	Industrial Foundation Classes
PV	Photovoltaic

References

1. Multidisciplinary BIM Software for Higher-Quality, Coordinated Designs. Available online: https://www.autodesk.com/products/revit/overview (accessed on 28 August 2018).
2. Allplan. Available online: https://www.allplan.com/ (accessed on 26 April 2020).
3. Archicad. Available online: https://www.graphisoft.com/archicad/ (accessed on 26 April 2020).
4. *Industry Foundation Classes (IFC) for Data Sharing in the Construction and Facility Management Industries*; EN ISO 16739:2019; ISO International Organization for Standardization: Geneva, Switzerland, 2019.
5. Sporr, A.; Zucker, G.; Hofmann, R. Automated HVAC Control Creation Based on Building Information Modeling (BIM): Ventilation System. *IEEE Access* **2019**, *7*, 74747–74758. [CrossRef]
6. American Society of Heating. *ASHRAE Handbook: Heating, Ventilating, and Air-Conditioning Systems and Equipment*; ASHRAE: Atlanta, GA, USA, 2004.
7. Siemens. Self-Learning Room Temperature Controller REV23. 23 April 2002. Available online: https://assets.new.siemens.com/siemens/assets/api/uuid:2143089670688f5b36b074c018c603c148f70350/version:1502454463/rev23-datasheet.pdf (accessed on 20 June 2020).
8. Zhang, L.; Huang, X.; Liang, L.; Liu, J. Experimental study on heating characteristics and control strategies of ground source heat pump and radiant floor heating system in an office building. *Procedia Eng.* **2017**, *205*, 4060–4066. [CrossRef]

9. Katipamula, S.; Lu, N. Evaluation of Residential HVAC Control Strategies for Demand Response Programs. *ASHRAE Trans.* **2006**, *112*, 535–546. Available online: https://www.ashrae.org/File%20Library/Technical%20Resources/ASHRAE%20Transactions%20and%20Conferences%20Programs/2006-Chicago-TOC.pdf (accessed on 10 May 2020).
10. Song, Y.; Wu, S.; Yan, Y.Y. Control strategies for indoor environment quality and energy efficiency—A review. *Int. J. Low-Carbon Technol.* **2015**, *10*, 305–312. [CrossRef]
11. Yang, R.; Wang, L. Optimal control strategy for HVAC system in building energy management. In Proceedings of the PES T&D 2012, Orlando, FL, USA, 7–10 May 2012; pp. 1–8. [CrossRef]
12. Erickson, V.L.; Cerpa, A.E. Occupancy based demand response HVAC control strategy. In Proceedings of the 2nd ACM Workshop on Embedded Sensing Systems for Energy-Efficiency in Building, Zurich, Switzerland, 2 November 2010; pp. 7–12. [CrossRef]
13. buildingSMART-IFC4 Add2 Release. Available online: http://www.buildingsmart-tech.org/specifications/ifc-releases/ifc4-add2 (accessed on 27 March 2017).
14. buildingSMART Data Dictionary (bsDD). Available online: http://bsdd.buildingsmart.org/ (accessed on 9 March 2019).
15. Austrian Standards International Committee 011. Available online: https://shop.austrian-standards.at/action/en/public/details/545938/OENORM_A_6241-1_2015_07_01 (accessed on 10 May 2020).
16. ISO International Organization for Standardization. *Building Information Modelling and Other Digital Processes Used in Construction—Methodology to Describe, Author and Maintain Properties in Interconnected Data Dictionaries*; ISO/TC 59/SC 13, ISO 23386:2020; ISO International Organization for Standardization: Geneva, Switzerland, 2020.
17. ISO International Organization for Standardization. *Organization and Digitization of Information about Buildings and Civil Engineering Works, Including Building Information Modelling (BIM)—Information Management Using Building Information Modelling*; DIN EN ISO 19650:2018; ISO International Organization for Standardization: Geneva, Switzerland, 2018. [CrossRef]
18. DIN. DIN 276. December 2018. Available online: https://www.din.de/de/mitwirken/normenausschuesse/nabau/normen/wdc-beuth:din21:293154016. (accessed on 4 August 2020).
19. Haustechnik, L.B. Available online: https://www.bmdw.gv.at/Services/Bauservice/Haustechnik.html (accessed on 4 August 2020).
20. Wilkinson, R.J. Establishing Commissioning Fees. *ASHRAE J.* **2000**, *42*, 41–47.
21. Wärmepumpen Testzentrum. Test Results of Brine to Water Heat Pumps Based on EN 14511:2013/EN 14511:2018 and EN 14825:2013/EN 14825:201. 30 August 2018. Available online: https://www.ntb.ch/fileadmin/NTB_Institute/IES/pdf/Projekte_WPZ/PruefResSW180830.pdf (accessed on 11 March 2019).
22. Arpagaus, C.; Bless, F.; Uhlmann, M.; Schiffmann, J.; Bertsch, S.S. High temperature heat pumps: Market overview, state of the art, research status, refrigerants, and application potentials. *Energy* **2018**, *152*, 985–1010. [CrossRef]
23. International Energy Agency. IEA—Danish Energy Agreement for 2012–2020, June 2012. Available online: https://www.iea.org/policiesandmeasures/pams/denmark/name-42441-en.php (accessed on 18 February 2019).
24. Land Niederösterreich. Mit 1. Jänner tritt in Niederösterreich das Ölheizungsverbot in Kraft, 28 Dectomber 2018. Available online: http://www.noe.gv.at/noe/Mit_1._Jaenner_tritt_in_Niederoesterreich_das_Oelheizungs.html (accessed on 18 February 2019).
25. European Union. A Clean Planet for All a European Strategic Long-Term Vision for a Prosperous, Modern, Competitive and Climate Neutral Economy, Brussels, November 2018. Available online: https://eur-lex.europa.eu/legal-content/EN/TXT/PDF/?uri=CELEX:52018DC0773&from=EN (accessed on 19 February 2019).
26. Buck, J.T.; Ha, S.; Lee, E.A.; Messerschmitt, D.G. Ptolemy: A Framework for Simulating and Prototyping Heterogeneous Systems. 1994. Available online: http://s-space.snu.ac.kr/handle/10371/7554 (accessed on 22 October 2015).

energies

MDPI

Article

IFC BIM Model Enrichment with Space Function Information Using Graph Neural Networks

Adam Buruzs *, Miloš Šipetić, Brigitte Blank-Landeshammer and Gerhard Zucker

AIT Austrian Institute of Technology, 1210 Vienna, Austria; milos.sipetic@ait.ac.at (M.Š.); brigitte.blank-landeshammer@ait.ac.at (B.B.-L.); gerhard.zucker@ait.ac.at (G.Z.)
* Correspondence: adam.buruzs@ait.ac.at; Tel.: +43-664-8890-4316

Abstract: The definition of room functions in Building Information Modeling (BIM) using IfcSpace entities is an important quality requirement that is often not fulfilled. This paper presents a three-step method for enriching open BIM representations based on Industry Foundation Classes (IFC) with room function information (e.g., kitchen, living room, foyer). In the first step, the geometric algorithm for detecting and defining IfcSpace entities and injecting them into IFC models is presented. After deriving the IfcSpaces, a geometric method for calculating the graph of connections between spaces based on accessibility is described; this information is not explicitly stored in IFC models. In the final step, a graph convolution-based neural network using the accessibility graph to classify the IfcSpace entities is described. Local node features are automatically extracted from the geometry and neighboring elements. With the help of a Graph Convolutional Network (GCN), the connection and spatial context information is utilized by the neural network for the classification decision, in addition to the local features of the spaces which are more commonly used. To evaluate the classification accuracy, the model was tested on a set of residential building IFC models. A weighted version of the common GCN was implemented and tested, resulting in a slight improvement in the classification accuracy.

Keywords: BIM; IFC; architecture model enrichment; machine learning; IfcSpace

Citation: Buruzs, A.; Šipetić, M.; Blank-Landeshammer, B.; Zucker, G. IFC BIM Model Enrichment with Space Function Information Using Graph Neural Networks. *Energies* **2022**, *15*, 2937. https://doi.org/10.3390/en15082937

Academic Editor: Francesco Nocera

Received: 8 February 2022
Accepted: 14 April 2022
Published: 16 April 2022

1. Introduction

BIM modelling is the de facto standard in architecture, engineering, and the construction (AEC) industry. Various proprietary closed software packages are used to create BIM models. For the sake of interoperability, the buildingSMART consortium created the IFC (Industry Foundation Classes) standard [1], which is the established open data format for the BIM industry. IFC provides a rich vocabulary for the specification and classification of building components, as well as the properties needed for alphanumeric information. Unfortunately, in practice, the IFC files are most often incomplete. There exist solutions for the automatic checking of building code compliance, such as the Solibri Model Checker, but these tools employ hard-coded rules that require exact and complete BIM model specifications. In practice, incorrect design, lack of information provided by the user in the design phase, or the use of various local languages for labelling elements strongly limits the applicability of these automatic model checkers. For the validation of models and for further calculations, planning, or for cost estimations, the semantic enrichment of the IFC models is needed to inject this missing information into the IFC model.

In our work, we focus on the semantic enrichment of spaces (IfcSpace entities) with the function of the corresponding room. The exact spatial definition of the space is an important piece of information for stakeholders, for example, to quickly create a list of apartments with exact size information in residential buildings. Intended usage and occupancy of spaces is an important piece of information for facility managers: corridors, halls, and basement spaces require less ventilation and heating than office rooms or apartments.

Construction companies benefit from correctly labelled spaces by being able to calculate various cost factors and procurement plans more precisely (for example, calculating the required amount of tiling from the aggregated surface area of toilets and bathrooms).

IfcSpace entities are graphically not visible when rendering the BIM model; our analysis showed that the IfcSpace entity for a room is often omitted in the modeling process. However, even in higher quality models—where IfcSpaces are defined and thus PropertySets could be defined—the PropertySets are often left blank, leaving IfcSpace entities with only the space boundaries and without the room function defined.

IfcSpace entities are important elements that are required for HVAC simulations and control applications [2], and are needed for thermal zoning in thermal dynamic simulations [3]. The intended room function influences the different parameters that are used in the simulation, such as temperature, humidity, or air exchange setpoints and ranges (e.g., lower temperature setpoints in bedrooms). Similarly, storage rooms, corridors outside of apartments, and garages have lower or no heating requirements. Fine-grained control of related parameters of the building model in the design phase results in lower estimated energy needs, and, if those controls are correctly implemented in the commissioned building, also in a lower final energy consumption. Additionally, having the spaces automatically derived and their functions determined saves time and reduces the need for manual, error-prone work. Exact thermal simulations also require second-level space boundaries that can be derived from the defined spaces, as described in [4]. Room functionality is also an important information for facility managers throughout the lifecycle of the building.

Our paper structure is as follows: In Section 2, the state of the art of machine learning-based automatic enrichment of BIM models is reviewed, especially with regards to room-function classification. In Section 3, a geometric method is presented to automatically detect spaces in BIM models, if they are missing. We present a method for extracting connections between spaces and for building a graph of accessibility. Furthermore, a Graph Neural Network (GNN)-based classification algorithm is described to classify these spaces. This classification is based on this accessibility graph and the calculated geometric features that are extracted from the raw BIM Models. We also present a modification of the Graph Convolutional Network to strive for higher accuracy. In Section 4, the results of the classification are presented as an accuracy analysis and a confusion matrix. The paper is closed with Section 5.

2. Related Work

Machine learning methods have gained increasing popularity in recent years in the field of automated BIM processing. Krijnen and Tamke [5] used neural networks to classify buildings into residential and non-residential categories based on floor plans.

The semantic enrichment of IFC BIM Models has been addressed in multiple studies; Sacks et. al. developed a rule-based system that evaluates IF–THEN types of rules based on prepared features [6,7]. To extract geometric relations that are not explicitly contained in IFC files, they used Query Language for 4D Building Information Models (QL4BIM [8]).

Bloch and Sacks [9] examined the use of semantic enrichment in BIM models; they compared neural network-based rules against hand-crafted rules for IfcSpace classification (building rooms) and concluded that the neural networks based on geometrical features and spatial connections strongly outperformed the rule-based method in terms of accuracy.

Koo et al. [10] applied support vector machines (SVM) to classify building elements into eight of the most frequent IFC categories (Column, Beam, Slab, Wall, Covering, Door, Window, Railing). They used simple geometric features (bounding box size, volume, gyration, etc.) that were calculated using the opencascade geometry engine [11], as well as features derived from IfcRelations (number of related Walls/Windows). They trained their SVM classifier with six building IFC models and reached a classification accuracy of over 90%. The same team later used multi-view convolutional neural networks (MVCNN) to classify wall and door elements in BIM models [12]. The MVCNN [13] generates 2D renderings of objects, and practically transforms the 3D object recognition problem into the well-studied 2D recognition problem, which can be solved efficiently using convolutional neural networks [14].

Lumio [15] et al. used deep learning methods to classify images generated from BIM model external views into three categories (residential, industrial, and other buildings). Besides classical SVM classifiers, they also tried deep neural networks (ResNet, MobileNet), which are very successful in machine vision tasks. They found that these deep neural nets considerably outperformed SVMs for this task and reached over 90% accuracy in predicting the building type based on the generated 2D images.

Edmunds et al. [16] created the IFCNet benchmark dataset for IFC element classification. They also tested the MVCNN architecture for classifying these elements; it showed very good results (over 85% accuracy). However, in their method, the context information of the elements is not used, and the elements are classified based on their geometry alone.

Many scientific and engineering datasets have an inherent structure that can be represented with a graph. Graph applications range from social networks, citation databases, molecular chemical datasets, recommendations, predicting infectious diseases, etc. Advanced neural network models for prediction and analysis tasks over graph data have attracted considerable research attention in recent years. Applications of deep learning methods on graph datasets include node classification, graph classification, link prediction, and graph generation tasks [17]. The methods applied on graph datasets include graph convolutional networks, recurrent networks, autoencoders, adversarial methods, and reinforcement learning. A common advantage of applying these methods to node classification is that they inherently incorporate graph network information into the node classification besides the single node information represented with the node features. A general mathematical framework for the application of deep learning on graphs is given by Bronstein et al. [18], where the analogy between graphs and manifolds is shown, and the extension of the successful Convolutional Networks to graphs is derived.

Kipf and Welling [19] proposed a graph based neural network model with a simple layer-wise propagation rule:

$$H^{l+1} = \sigma(\widetilde{A}H^l W^l) \quad (1)$$

where H^l is the input of the l-th layer, W^l are the weights of the layer, and the $\widetilde{A}$ matrix is calculated from the adjacency matrix, thereby establishing connections between adjacent nodes in one layer. It was shown how this formula can be interpreted as an approximation of a spectral graph convolution; therefore, the layer was named graph convolution. Additionally, the application of a graph convolutional network was demonstrated on a node classification task.

IFC BIM models contain numerous relations among their elements; thus, a graph of elements can be extracted with the IfcRelationships acting as edges. Additional relationships can be derived by applying different geometric reasoning methods. For example, using the calculated geometric distance between the extracted element shapes, a space adjacency graph structure can be derived by connecting spaces that are geometrically close to one another. After such graphs are derived, it becomes possible to apply graph neural networks to train models for inductive reasoning, such as element classification.

Several recent studies have dealt with building elements classification based on floor plans: Su et al. classifies structural elements (walls, windows, doors, stairs) based on floor plans with graph neural networks [20]. A recent paper by Paudel et al. [21] solved room-type classification based on the floor plan data of single apartments using various graph neural networks. Hu et al. [22] presented room-type classification based on the floor plans of university buildings using random forest and graph convolutional networks. All these studies report high (around 80%) classification accuracy. However, one has to note two things: 1. By selecting the model training and test set, the information concerning the parent building of each apartment is not preserved; it is likely that one floor of a building is included into the training set, while another floor of the same building belongs to the test set. One can assume that there are considerable similarities between the floors of the same building. 2. The number of common room types is relatively low.

3. Materials and Methods

We present a method where two types of information are extracted from IFC files to serve as input for a neural network-based supervised classification trained on labelled data: First, numerical features calculated from the geometric shapes of IFC elements are extracted. Second, the accessibility between spaces is calculated. The spatial proximity and traversability of spaces can be best described by a graph. The spaces are represented as nodes and connected with an edge if a person can walk from one space to the other. This condition is satisfied if spaces touch without a boundary between them or are connected via a door or staircase. The proposed graph convolutional network merges both numerical features and graph structure to train the weights of a neural network to classify the nodes in this graph. This method is not only promising for spaces, but also for other kinds of classifications and automatic IFC model quality assessments.

3.1. Space Extraction from IFC Elements

The IFC standard provided the IfcSpace (https://standards.buildingsmart.org/IFC/RELEASE/IFC4/ADD2_TC1/HTML/link/ifcspace.htm, accessed on 15 January 2022) entity for defining spaces or rooms in BIM models. IfcSpace elements can be defined and exported from closed BIM software packages. The correct definition of the spaces is important for subsequent analyses such as area calculation or apartment size determination. Further quality checks can also leverage the properties of the IfcSpace entities.

In our approach we started from the walls, which we extracted separately from each floor. First, the *Opencascade* TopoDS [23] shape was extracted from each wall with the help of *ifcOpenShell* [24]. Then, the oriented bounding box for each wall and the projection of that bounding box onto a horizontal plane was calculated. With this projection, two-dimensional rectangles were obtained for each wall. Using the *shapely* python package, these two-dimensional shapes were processed; the convex hull was calculated from the polygons and then the geometric difference of the convex hull and the rectangles corresponding to the walls was calculated. This difference resulted in a list of polygons that were closed polygons encompassed by the wall-rectangles. Polygons that were touching the convex hull had to be removed from this list, because these polygons were not contained within the building bounding walls. The polygons were then extruded along the z-axis (resulting in IfcExtrudedAreaSolid objects), assuming vertical walls and using the height of the storey that we calculated as the median of the wall heights belonging to the given IfcStorey. The obtained IfcSpace entities could then be written back into the original IFC file with the *ifcopenshell write* function, thus augmenting the initial BIM model with IfcSpaces. This procedure is visualized on Figure 1. This method assumes vertical walls and horizontal slabs. The processing of slanted slabs and roofs was out of scope of this paper. A further limitation of the method is that incorrectly modelled walls can also lead to problems with space detection (for example, when small gaps exist between the walls).

3.2. Accesssibility Graph Calculation

An accessibility graph is used to encode relations among spaces in the building. Adjacent graphs are commonly used in different kinds of analysis of spatial layouts, e.g., using the space syntax theory [25]. In architecture, this is often done to analyze the layout, connectedness, and relations between the spaces of the building. Another application of accessibility graphs is the automated calculation of emergency exit pathways; the maximum distance from each room to a fire-safe room is regulated by the Austrian building code (https://www.oib.or.at/de/oib-richtlinien/richtlinien/2019/oib-richtlinie-2, accessed on 15 January 2022). Using the accessibility graph, the shortest route to reach a fire-safe place can be easily calculated utilizing standard graph algorithms (for example, implemented in the *networkx* python package [26]). Adding length measures to the space-graph edges (distance between rooms), the path length can be obtained in meters.

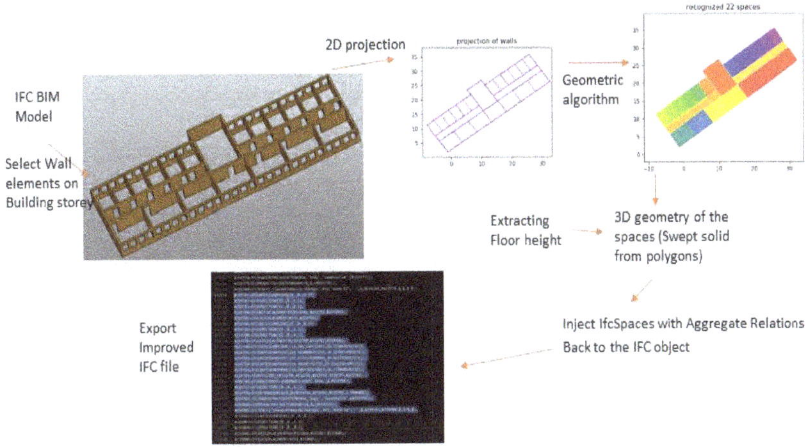

Figure 1. The visualization of the IfcSpace detection workflow.

To classify spaces, we filtered the information that is available in the IFC BIM Model; only Spaces, Doors, and Stairs were extracted. In order to place the Space objects into context, we calculated which rooms were connected, i.e., directly traversable from one to another by humans. To create a graph of accessibility, we calculated which spaces were directly reachable from other spaces through doors, or were in direct spatial contact (spaces are touching). Here, we calculated the spatial proximity with the help of the open source *Opencascade* geometry library (*pythonocc* interface [23])—see Figure 2. There are four types of space–space connections that we checked:

- two spaces are touching (zero distance);
- two spaces are connected through a door (IfcDoor);
- two spaces are connected through an opening (unpopulated IfcOpeningElement);
- two spaces are connected through stairs (IfcStair).

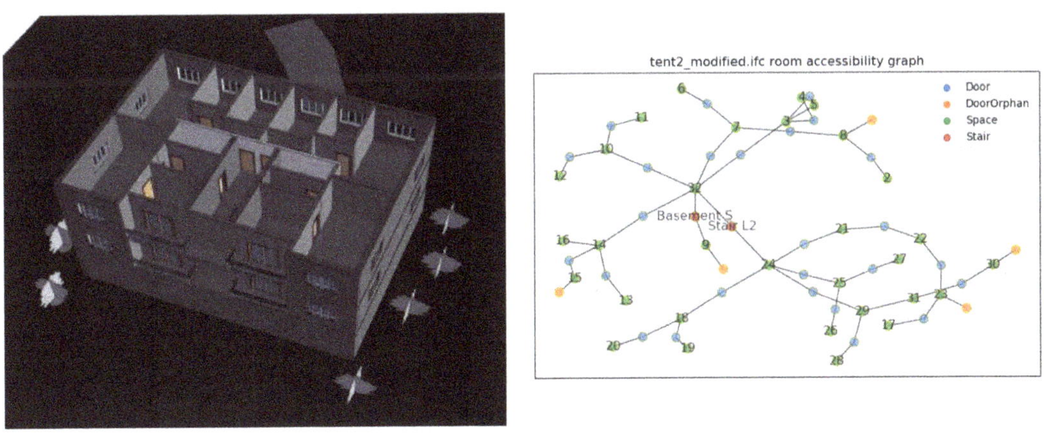

Figure 2. A building model and the corresponding calculated accessibility graph. Orphan doors were the doors with only one connected space (external or terrace doors).

If any of these four connections existed between two spaces, we created an edge between the two spaces in the graph created from the spaces as nodes.

3.3. Extracting Features

To perform the classification, certain geometric features were extracted from the shapes of the spaces and doors. These geometric properties were then used as features for supervised learning. Raw geometric features for the shapes of spaces were calculated using the the *Opencascade* G_prop package (https://dev.opencascade.org/doc/refman/html/class_g_prop___g_props.html, accessed on 15 January 2022). These raw features included:

- Total Volume;
- Principal moments of inertia;
- Gyration radius in the 3 axes in the principal coordinate system;
- Total Volume;
- Static moments in x, y, z directions;
- Height of the shape (in vertical z direction);
- Dimensions of the oriented bounding box of the space shape.

Additionally, the number of windows, doors, openings, and stairs belonging to a space was calculated using geometric proximity. In order to speed up the relatively slow *Opencascade* distance calculation between shapes, we first calculated a distance lower limit with the help of bounding box corners—with much lower computation time and complexity—and only if the distance between the bounding boxes lied below a certain limit were the time-consuming exact geometric distances calculated between the *Opencascade* shapes. The shape representation type of the space (Brep, Surface Model, or Swept Solid) was added as a categorical feature.

As a next step, the normalized features were calculated from these raw features. These included volume/height as the area of the space (in case of horizontal upper slabs), the ratio of the maximum and minimum bounding box dimension, or the ratio of the volume of the bounding box divided by the volume of the space—giving only one for cuboid shaped spaces.

Using these features, a classifier was trained to label each room with one of the following 15 labels: Bathroom, Toilet, Living Room, Kitchen, Bedroom, Corridor, Staircase, Foyer, Storage, Terrace, Office, Kitchen/LivingRoom, Living Room/Bedroom, Other, or Technical Room. Additionally, if the space definition was incorrect in the original file (for example whole floor was marked as one space), the space was labelled as "DELETE". We also let the network predict such incorrect spaces.

3.4. Neural Networks on Graphs

A multi-layer preceptor model (MLP) was implemented to serve as a baseline reference. MLP does not use the available graph information; it consists of two hidden layers and accepts all available numerical and categorical features as input, and outputs the predicted probabilities for each of the 16 room types.

Our first graph-based reference neural network model was the Graph Convolutional Network from Kipf [19]. It operates on a matrix of N nodes, each with F number of features. The Activation in hidden layer l can be given as:

$$h_{l+1} = \mathrm{f}(h_l, A) = \sigma(A h_l W_l) \tag{2}$$

where A is the normalized adjacency matrix (with the size of $N \times N$); A_{ij} is nonzero if and only if i and j nodes are connected. W_l is the trainable weight matrix in the hidden layer l. The size of W_l does not depend on the size of the graph, just on the number of features F and on the size of the hidden layers. These W_l weights are shared between the nodes, just like the convolution matrix is shared by convolutional networks. For the first layer, H_0 is initialized with the input features. The activation function σ is usually implemented with the tanh function.

With these layers, the prediction for a two-layer GCN, as an example, can be written as:

$$Z_{i\gamma} = softmax\left(\sum_{\kappa}\sigma\left(\sum_{j\beta}A_{ij}\sigma\left(\sum_{k\alpha}A_{jk}X_{k\alpha}W^{0}_{\alpha\beta}+b_{\beta}\right)W^{1}_{\beta\kappa}+b_{\kappa}\right)W^{D}_{\kappa\gamma}\right) \quad (3)$$

Here, we only get contributions to the output if i and j are indices of neighboring spaces and j and k are also neighbors, so k is second neighbor to i. The Greek indices run over the number of features. The trainable bias vectors are denoted by b. The formula shows that in case of two GCN layers for the prediction of the class of a given element, anything up to the second neighbors of this element contribute (for 1-layer GCN, only the first neighbors show a contribution). $X_{k\alpha}$ is the feature α for node k; this is the first layer input $X = h_0$.

$W^{D}_{\kappa\gamma}$ is the weight matrix of the last dense softmax layer. The softmax function returns a vector $Z_{i\gamma}$ for each space i with the size equaling the number of classes and sums up to 1 (so $\gamma = 1 \ldots 15$ for the 15 room categories). The elements of this vector can be treated as probabilities of classes; the one with the highest "probability" is the predicted class.

The GCN in the stellargraph [27] python package is implemented for transductive inference (classifying the unlabeled nodes in a known graph), which does not naturally generalize to unseen nodes. This problem of inductive inference (predicting labels of the nodes of a previously unseen graph) was solved by saving the learned weight matrices (W_l) and creating a new stellargraph GCN instance from these previously learned weights for an unseen graph without re-training.

In the course of training the GCN model we have realized that it often does not outperform the simple MLP network in the inductive setup. Combining the two predictions was attempted with the hope that an ensemble model would perform better. In order to merge the predictions of the two models, we calculated a joint prediction. For this, we simply summed up the output probability matrix of the two classifiers and determined the joint prediction as the argmax of the single lines of this averaged probability matrix. This method is often referred to as ensemble averaging.

3.5. Weighted Graph Convolutional Network Architecture

As the classification accuracy of the GCN for the inductive setup was not satisfactory, a modification of the GCN architecture was attempted.

For the particular problem of assigning labels to IfcSpaces, the weakness of the GCN is that the importance of the single site vs. neighbor nodes features cannot be adjusted, i.e., they are simply averaged. In the original GCN, the l-th layer output h_j^{l+1} is obtained from the layer input $h^{l}{}_{k\alpha}$ with the formula in Equation (2):

$$h^{l+1}_{j\beta} = \sigma\left(\sum_{k\alpha}A_{jk}h^{l}_{k\alpha}W^{l}_{\alpha\beta}\right) \quad (4)$$

where $j, k = (1 \ldots N)$ are the node indices, the Greek letters are the feature indices, A_{jk} is the adjacency matrix that depends only on the graph structure, and $W^{l}{}_{\alpha\beta}$ are the trainable weights in layer l. We proposed an extended layer, with separate weight matrix for the single site elements (diagonal of the adjacency matrix) and separate weights for the off-diagonal elements, corresponding to the neighbors:

$$h^{l+1}_{j\beta} = \sigma\left(\sum_{k\alpha}A_{jj}h^{l}{}_{j\alpha}W^{diag}_{l\alpha\beta} + A_{jk}h^{l}{}_{k\alpha}W^{off-diag}_{l\alpha\beta}\right) \quad (5)$$

Here, we have two matrices ($W^{diag}_{\alpha\beta}, W^{off-diag}_{\alpha\beta}$) instead of one, which were both separately trained on the data (to optimize the cross-entropy loss function).

With this novel architecture, a higher accuracy for the transductive learning was achieved, and we also obtained higher inductive accuracy, as shown in the last column of Table 1.

Table 1. Inductive Inference evaluation of the Neural Network models: MLP = simple multi-layer preceptor neural network, GCN: Graph convolutional network of Kipf and Welling [11]. GCN_MLP_joint mixture of MLP and GCN classifier, WGCN = novel Weighted Graph Neural Network Model.

File	MLP	GCN	GCN_MLP_Joint	WGCN (Ours)
Duplex_A_20110907_optimized.ifc	20.0%	25.0%	30.0%	35.0%
FJK-Project-Final-gen.ifc	42.9%	50.0%	35.7%	42.9%
IFC_Schependomlaan_most_spaces.ifc	39.8%	28.9%	39.8%	39.8%
KIT_Smiley-West_Arch.ifc	51.4%	34.3%	64.3%	78.6%
SGD_Munkerud_Arch-1.ifc	23.3%	58.3%	46.7%	50.0%
Tent2_modified.ifc	46.6%	30.0%	28.6%	53.3%
Total accuracy	38.3%	37.6%	45.0%	50.0%

4. Results

We created a labelled dataset of spaces for eight IFC models of residential buildings that were freely available on the internet. The source of the IFC files were archives of the Ghent University IFC repository and archives of the *duraark* repository (https://www.re3data.org/repository/r3d100012506, accessed on 15 January 2022). For these BIM models, we used 13 room types.

For training and validation, we used the 320 spaces from the IFC models and randomly selected 80% of these rooms as a training set; the remaining 20% of the nodes we used as validation set. We optimized the neural network weights for 1500 iterations with the Adam optimizer using the *stellargraph/tensorflow* GCN implementation. The training time was under 1 min on an intel i9 CPU.

In Figure 3, the resulting confusion matrix of the classification is shown. We can see that bedrooms, the most common room types, were also the most often misclassified room types. A 62% classification accuracy was reached on the validation dataset.

(a) (b)

Figure 3. Confusion matrix for classification (**a**) of all the training and validation data, for which we obtained an accuracy of 72%; (**b**) For the validation nodes alone (which were 20% of the total nodes), we obtained an accuracy of 62%.

The models were tested in an inductive setup as well; the models were trained on 1500 spaces of nine proprietary IFC files, then they were tested on six other unrelated IFC files that were neither seen nor processed during the training. With this independence of the training and testing dataset we reached the most realistic accuracy assessment. The obtained accuracy for this test is shown in Table 1. With GCN, we reached an average classification accuracy of 50%, where the accuracy ratio strongly varied between the considered building models.

One example space classification result can be seen on Figure 4.

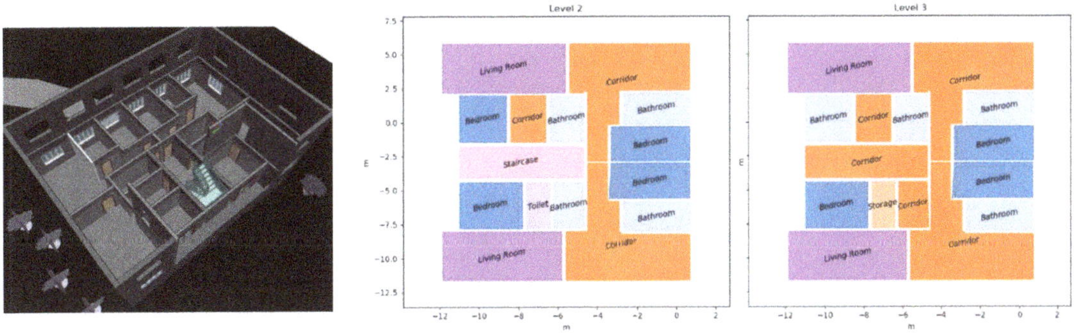

(a) (b)

Figure 4. (**a**) An IFC model visualized with BIMVision software (https://bimvision.eu/, accessed on 15 January 2022); (**b**) The output of our WGCN space classifier for the spaces on the two floors of the building model.

Figure 5 shows the confusion matrix for the inductive inference test. Some prediction errors are understandable; for example, the network often mistakes Corridors for Foyers (14 times). The IfcSanitaryTerminal elements (used to represent elements such as drains, water closets, or sinks) were not used as space features, as around half of the used models did not contain such elements. Without this feature, it was hard for the model to differentiate between bathrooms and storage rooms (21 cases).

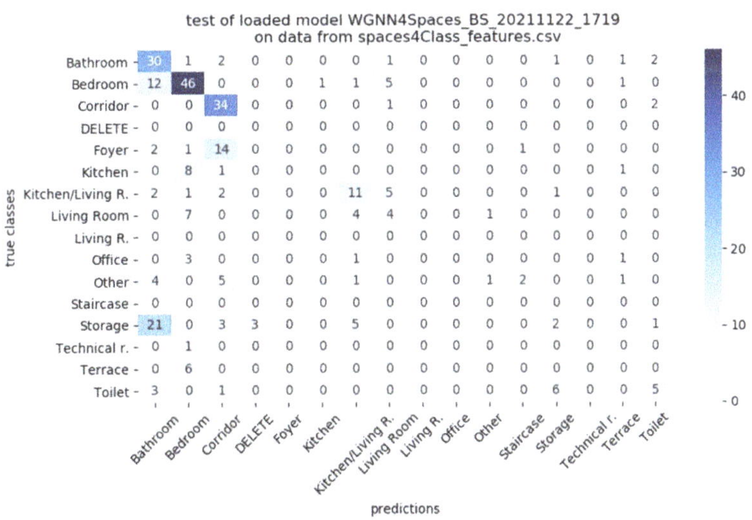

Figure 5. Confusion matrix for the inductive inference test of the WGCN model. The model was trained on a proprietary set of IFC files with 1500 spaces and validated on 283 other spaces from other IFC models as a realistic inference test.

The relatively low classification accuracy stemmed from the high number of room-types—some of them having just a few examples in the training set. Some room-types are hard to differentiate without additional information (like Storage rooms vs. Toilets); we also see that Foyers and Corridors are often confused by the neural model. They have a similar room function, although the difference is only that Foyers are within the apartments. Therefore, it made sense to merge some room types and recalculate both the confusion matrix and accuracy. We grouped the following room types to create the following three aggregate classes:

("Corridor", "Foyer", "Staircase", "Other")
("Kitchen/Living Room", "Living Room", "Living Room/Bedroom")
("Storage", "Toilet", "Technical room")

With this change, we reached an accuracy of 64%; the resulting confusion matrix is shown in Figure 6.

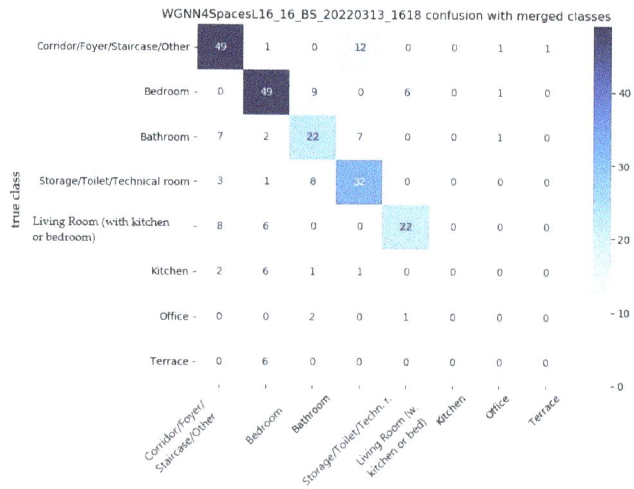

Figure 6. Confusion Matrix of the WGN classification with layer size of two, after merging classes. The accuracy reached 64%.

5. Conclusions and Outlook

A study for classifying IFC spaces with calculated geometrical features, accessibility graphs, and a graph-based machine learning model was presented here. The presented results demonstrate the operation of the workflow; however, there is still room for accuracy improvement before production deployment. For future work, the classification accuracy shall be improved by fine tuning the current workflow—there is an improvement potential that could be achieved by increasing the number of regarded features. Additionally, the inclusion of additional IFC elements such as IfcSanitaryTerminal would improve classification accuracy to differentiate spaces that have those elements (such as toilets, bathrooms, kitchens) from those which usually do not (living rooms, bedrooms, storage). We also did not use the information on whether a space was outside or inside the building, as this would require a time-consuming geometric calculation. However, having such information would be very useful to differentiate between, e.g., a terrace and a bedroom. Another huge quality improvement is expected by using higher quality labelled input data for training the neural networks. For this study, only a couple of IFC files were available, where the spaces within one IFC model were highly correlated (i.e., spaces in different floors or within one floor had very similar geometry). Even a data stock of several thousand spaces would yield only a limited number of independent samples compared to the variety in existing buildings. It also has to be noted that in low-detailed IFC models (e.g., with the duct system missing from the model), sometimes the labelling of the space type was not obvious for human experts either. For example, toilets and storage rooms have very similar features; they have similar geometry, are windowless, and are typically located towards the core of the building. Additionally, the present analysis does not involve parent apartment-related information (e.g., small apartments usually have only one bathroom, or living rooms are usually the biggest rooms of an apartment). To incorporate such knowledge rules in the classification decision, first the individual apartments would need to be detected. Based on the segmentation of the spaces into apartments and public areas, new features could be calculated, which could be very useful for the space classification. This apartment information inference and processing should be a subject of a follow-up research project. Still, the presented study shows that it is possible to achieve an automated classification of an incomplete BIM model (i.e., one that is missing the classification of rooms), and that the method has the potential to automate the room classification process. It is expected that a similar approach can be used whenever BIM models are handed over from one stakeholder to another, e.g., in the case of issuing a building permit based on the BIM model data.

Author Contributions: Conceptualization, A.B. and G.Z.; methodology, A.B. and M.Š.; software development, A.B. and M.Š.; validation and software testing, B.B.-L.; data curation, B.B.-L. and M.Š.; writing—original draft preparation, A.B.; writing—review and editing, M.Š., B.B.-L. and G.Z.; visualization, A.B.; supervision, A.B. and G.Z.; project administration, A.B. and G.Z.; funding acquisition, A.B. and G.Z. All authors have read and agreed to the published version of the manuscript.

Funding: This research was funded by Forschungsförderungsgesellschaft (FFG), Austria, grant number 880874.

Institutional Review Board Statement: Not applicable.

Informed Consent Statement: Not applicable.

Data Availability Statement: Not applicable.

Conflicts of Interest: The authors declare no conflict of interest.

References

1. ISO International Organization for Standardization. Industry Foundation Classes (IFC) for Data Sharing in the Construction and Facility Management Industries. ISO. Available online: https://www.iso.org/standard/70303.html (accessed on 31 January 2022).
2. Sporr, A.; Zucker, G.; Hofmann, R. Automated HVAC Control Creation Based on Building Information Modeling (BIM): Ventilation System. *IEEE Access* **2019**, *7*, 74747–74758. [CrossRef]
3. Hitchcock, R.J.; Wong, J.; Consulting, H. Transforming ifc architectural view bims for energy simulation: 2011. In Proceedings of the Building Simulation 2011: 12th Conference of International Building Performance Simulation Association, Sydney, Australia, 14–16 November 2011; p. 7.
4. Lilis, G.N.; Giannakis, G.I.; Rovas, D.V. Automatic generation of second-level space boundary topology from IFC geometry inputs. *Autom. Constr.* **2017**, *76*, 108–124. [CrossRef]
5. Krijnen, T.; Tamke, M. Assessing Implicit Knowledge in BIM Models with Machine Learning. In *Modelling Behaviour: Design Modelling Symposium 2015*; Thomsen, M.R., Tamke, M., Gengnagel, C., Faircloth, B., Scheurer, F., Eds.; Springer: Cham, Switzerland, 2015; pp. 397–406. [CrossRef]
6. Belsky, M.; Sacks, R.; Brilakis, I. Semantic Enrichment for Building Information Modeling. *Comput.-Aided Civ. Infrastruct. Eng.* **2016**, *31*, 261–274. [CrossRef]
7. Sacks, R.; Ma, L.; Yosef, R.; Borrmann, A.; Daum, S.; Kattel, U. Semantic Enrichment for Building Information Modeling: Procedure for Compiling Inference Rules and Operators for Complex Geometry. *J. Comput. Civ. Eng.* **2017**, *31*, 04017062. [CrossRef]
8. Daum, S. QL4BIM. 2021. Available online: https://github.com/SimonDaum/QL4BIM (accessed on 23 November 2021).
9. Bloch, T.; Sacks, R. Comparing machine learning and rule-based inferencing for semantic enrichment of BIM models. *Autom. Constr.* **2018**, *91*, 256–272. [CrossRef]
10. Koo, B.; La, S.; Cho, N.-W.; Yu, Y. Using support vector machines to classify building elements for checking the semantic integrity of building information models. *Autom. Constr.* **2019**, *98*, 183–194. [CrossRef]
11. Open CASCADE Technology | Collaborative Development Portal. Available online: https://dev.opencascade.org/ (accessed on 1 February 2022).
12. Koo, B.; Jung, R.; Yu, Y. Automatic classification of wall and door BIM element subtypes using 3D geometric deep neural networks. *Adv. Eng. Inform.* **2021**, *47*, 101200. [CrossRef]

13. Su, H.; Maji, S.; Kalogerakis, E.; Learned-Miller, E. Multi-view Convolutional Neural Networks for 3D Shape Recognition. *arXiv* **2015**, arXiv:1505.00880.
14. LeCun, Y.; Bengio, Y.; Hinton, G. Deep learning. *Nature* **2015**, *521*, 436–444. [CrossRef] [PubMed]
15. Lomio, F.; Farinha, R.; Laasonen, M.; Huttunen, H. Classification of Building Information Model (BIM) Structures with Deep Learning. In Proceedings of the 2018 7th European Workshop on Visual Information Processing (EUVIP), Tampere, Finland, 26–28 November 2018; pp. 1–6. [CrossRef]
16. Emunds, C.; Pauen, N.; Richter, V.; Frisch, J.; van Treeck, C. IFCNet: A Benchmark Dataset for IFC Entity Classification. *arXiv* **2021**, arXiv:2106.09712.
17. Zhang, Z.; Cui, P.; Zhu, W. Deep Learning on Graphs: A Survey. *IEEE Trans. Knowl. Data Eng.* **2020**, *34*, 249–270. [CrossRef]
18. Bronstein, M.M.; Bruna, J.; LeCun, Y.; Szlam, A.; Vandergheynst, P. Geometric deep learning: Going beyond Euclidean data. *IEEE Signal Process. Mag.* **2017**, *34*, 18–42. [CrossRef]
19. Kipf, T.N.; Welling, M. Semi-Supervised Classification with Graph Convolutional Networks. *arXiv* **2017**, arXiv:1609.02907.
20. Song, J.; Yu, K. Framework for Indoor Elements Classification via Inductive Learning on Floor Plan Graphs. *ISPRS Int. J. Geo-Inf.* **2021**, *10*, 97. [CrossRef]
21. Paudel, A.; Dhakal, R.; Bhattarai, S. Room Classification on Floor Plan Graphs using Graph Neural Networks. *arXiv* **2021**, arXiv:2108.05947.
22. Hu, X.; Fan, H.; Noskov, A.; Wang, Z.; Zipf, A.; Gu, F.; Shang, J. Room semantics inference using random forest and relational graph convolutional networks: A case study of research building. *Trans. GIS* **2021**, *25*, 71–111. [CrossRef]
23. PythonOCC | Open CASCADE Technology. Available online: https://dev.opencascade.org/project/pythonocc (accessed on 16 November 2021).
24. IfcOpenShell. Available online: http://ifcopenshell.org/ (accessed on 16 November 2021).
25. Kim, H.; Jun, C.; Cho, Y.; Kim, G. Indoor Spatial Analysis Using Space Syntax, The. *Int. Arch. Photogramm. Remote Sens. Spat. Inf. Sci.* **2008**, *37*, 1065–1070.
26. Hagberg, A.A.; Schult, D.A.; Swart, P.J. Exploring Network Structure, Dynamics, and Function using NetworkX. In Proceedings of the 7th Python in Science Conference, Pasadena, CA, USA, 21–22 August 2008; pp. 11–15.
27. StellarGraph Machine Learning Library. CSIRO's Data61. 2022. Available online: https://github.com/stellargraph/stellargraph (accessed on 2 February 2022).

energies

Article

Development and Evaluation of Occupancy-Aware HVAC Control for Residential Building Energy Efficiency and Occupant Comfort

Christina Turley [1], Margarite Jacoby [1], Gregory Pavlak [2] and Gregor Henze [1,3,4,*]

1 Department of Civil, Environmental, and Architectural Engineering, University of Colorado, Boulder, CO 80309, USA; christina.turley@colorado.edu (C.T.); margarite.jacoby@colorado.edu (M.J.)
2 Department of Architectural Engineering, Pennsylvania State University, University Park, PA 16802, USA; gxp93@psu.edu
3 National Renewable Energy Laboratory, Golden, CO 80301, USA
4 Renewable and Sustainable Energy Institute, Boulder, CO 80309, USA
* Correspondence: gregor.henze@colorado.edu; Tel.: +1-303-492-1094

Received: 25 August 2020; Accepted: 2 October 2020; Published: 15 October 2020

Abstract: Occupancy-aware heating, ventilation, and air conditioning (HVAC) control offers the opportunity to reduce energy use without sacrificing thermal comfort. Residential HVAC systems often use manually-adjusted or constant setpoint temperatures, which heat and cool the house regardless of whether it is needed. By incorporating occupancy-awareness into HVAC control, heating and cooling can be used for only those time periods it is needed. Yet, bringing this technology to fruition is dependent on accurately predicting occupancy. Non-probabilistic prediction models offer an opportunity to use collected occupancy data to predict future occupancy profiles. Smart devices, such as a connected thermostat, which already include occupancy sensors, can be used to provide a continually growing collection of data that can then be harnessed for short-term occupancy prediction by compiling and creating a binary occupancy prediction. Real occupancy data from six homes located in Colorado is analyzed and investigated using this occupancy prediction model. Results show that non-probabilistic occupancy models in combination with occupancy sensors can be combined to provide a hybrid HVAC control with savings on average of 5.0% and without degradation of thermal comfort. Model predictive control provides further opportunities, with the ability to adjust the relative importance between thermal comfort and energy savings to achieve savings between 1% and 13.3% depending on the relative weighting between thermal comfort and energy savings. In all cases, occupancy prediction allows the opportunity for a more intelligent and optimized strategy to residential HVAC control.

Keywords: HVAC control; occupancy prediction; energy consumption; thermal comfort

1. Introduction and Background

The finite quantity of fossil fuels and the mounting concern of climate change makes reducing energy use a global necessity. Buildings are major consumers of energy worldwide, and used around 3060 million tons of oil equivalent (Mtoe) in 2018 according to the International Energy Agency (IEA) [1]. In the United States, heating, ventilation, and air conditioning (HVAC) systems account for 50% of all building energy consumption [2], while U.S. homes alone are responsible for the use of approximately 4.7 quadrillion British thermal units (Btu) for space heating and air conditioning per year [3]. Therefore, reducing energy consumption associated with residential heating and cooling has the potential to result in large energy savings when applied across the sector.

Traditionally, heating and cooling in residential buildings has been controlled by a thermostat that has a single setpoint temperature, which keeps the indoor temperature constant whenever the thermostat is in use. Over time, different technologies have been added to HVAC systems to improve temperature control and reduce energy use, one of which is occupancy-based control. This method controls the indoor temperature to provide thermal comfort only when the building is believed to be occupied, and turns the HVAC system off when it is vacant. This typically results in reduced energy use during unoccupied hours. Previous studies have estimated that the potential savings when using these systems is between 5–23%. The magnitude of savings depends on various factors, such as climate, building vintage, and occupant behavior [4]. While occupancy-based HVAC control has potential benefits, questions still remain on how best to detect occupancy and implement control decisions. To provide the best experience for occupants, thermal comfort standards should be met during all occupied hours. Thus, a good control strategy needs to not only know when a building is currently occupied, but also needs to accurately predict occupancy ahead of an occupants arrival. This allows the space to be appropriately conditioned in advance of the arrival.

1.1. Historical Trends in U.S. Housing

An understanding of how buildings are changing is critical to reducing energy use in the built environment. Looking at how buildings and control systems have performed historically can highlight opportunities for improvement and can indicate the ways in which current trends may shape the future. The 2015 Residential Energy Consumption Survey found that the United States residential sector is comprised of 118.2 million homes, totaling 223 billion square feet of floor space [3]. Residential buildings currently use 22% of U.S. annual energy, and in the three decades from 1980 to 2009 residential building site energy use increased by 8.9% [5]. This growth can be attributed to increases in three factors: home size, number of homes, and appliance use. For instance, the number of households in the U.S. increased by 33%, while the average size of a single-family detached home also increased from 2100 square feet to 2688 square feet, as depicted in Figure 1. This led to a 52% increase in total floor space [5]. Additionally, appliance electricity consumption during the same time period increased by 30.6%, with the largest increases being from microwave ovens, personal computers, air-conditioners, and clothes dryers [5].

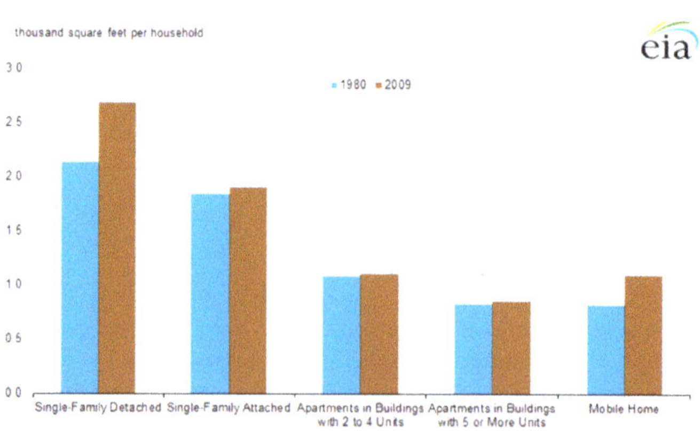

Figure 1. Size of residences by home type for 1980 and 2009 (ft^2) [5].

At the same time that energy use was increasing due to changes in housing characteristics, other factors were leading to decreases in energy use. These decreases are attributable to: (1) population shifts in the U.S., as large numbers of people moved from the Northeast and Midwest to the less heating-intensive regions of the South and West; (2) changes in weather patterns, both heating-degree days and cooling degree-days were lower across the nation in 2009 than in 1980 [5]; and (3) a decline in energy intensity, led by advances in engineering and a promotion of energy efficiency standards for household appliances. For example, the annual fuel utilization efficiency (AFUE) of a standard furnace increased from 78% to 97%, leading to decreased energy consumption for the same heating output. The largest change in consumption occurred from 1990–2001, which coincides with an era when federal efficiency programs, like ENERGY STAR, were enacted. The combined effect of all contributors in the 30-year period was an increase in energy consumption over time, with U.S. homes consuming 9.1 quads per year [5].

1.2. Temperature Control in Buildings

Despite changes in the housing sector, occupants' desires for thermal comfort have remained constant. One method of measuring occupant comfort is the predicted mean vote (PMV), first developed by Povl Ole Fanger [6], which predicts the average comfort level of a hypothetical group of people in a space. ASHRAE Standard 55, first published in 1966, specifies the fraction of occupants that find a space comfortable using the predicted mean vote must be at least 80% [7]. PMV, which ranges from −3 (too cold) to +3 (too hot), is based on the combined effects of air temperature, mean radiant temperature, relative humidity, air speed, metabolic rate (based on activity), and occupants' clothing levels.

In buildings, comfort requirements are met by using a thermostat and control system to maintain a setpoint temperature. The thermostat measures the indoor air temperature and compares it to the setpoint temperature, while the control system manages how the HVAC system tracks the indoor air temperature, attempting to keep it within small deviations of the setpoint. The interaction of these two components, and the programming of the system, determines how well the setpoint is tracked and how effectively the system achieves thermal comfort conditions.

Manual, programmable, and "smart" thermostats are the three main categories of thermostats in-use today. In manual thermostats, the setpoint is a single temperature that the system always tries to maintain when it is on. To change the temperature of the space, you must manually change the setpoint. Programmable thermostats are similar, but with different temperature setpoints for different times of the day or days of the week that can be programmed by users [8]. This allows temperatures to be setback during nighttime hours or during daytime vacancies, and automatically adjusts to more comfortable temperatures when people are frequently home. These thermostats often have modes for different days (weekday and weekend), and modes for different times of day (e.g., morning, day-time, evening, and night). Endorsed by Energy Star at their 1995 release, initial demonstrations showed that programmable thermostats could reduce heating and cooling bills by 10%–30%. However, the U.S. Environmental Protection Agency ultimately suspended Energy Star certification of programmable thermostats in 2009 since a lack of undisputed energy savings materialized [8]. Investigations revealed that 30% of households had failed to set them properly, and over 89% had not set separate weekend and weekday schedules [9]. Due to their complexity of operation, most programmable thermostats were operated manually, negating their energy savings potential.

Connected, or "smart", thermostats have emerged in response to consumer aversion to programmable thermostats. Like programmable thermostats, connected thermostats create a setpoint schedule but the operation is designed to be user-friendly and may change over-time, given occupancy patterns. Products, like the Nest Learning Thermostat, Honeywell, or Ecobee thermostat, are internet or "cloud" connected and can be controlled by phone, web interface, or a touchscreen. The system comes with a preset schedule that modifies itself based on the user's manual adjustments during initial use [10]. First introduced in 2011, the Nest Learning Thermostat catalyzed the market, with over 100% market growth per year in the first three years. In 2015, 40% of the thermostats sold were connected thermostats [10]. The main features of most connected thermostats include extensive data tracking, remote accessibility, local sensors to track occupancy, and web-enabled weather forecast data. While all connected thermostats are designed to enhance temperature control, they also create an opportunity to save energy through the use of setback temperatures that are automatically programmed though the initial use.

1.3. HVAC Control Strategies

The control strategy utilized by an HVAC system determines how the heating and cooling is managed. The two main controller types use in residential buildings are discrete controllers and continuous controllers. A discrete controller, in its simplest form, simply turns devices on and off, allowing only two states of operation. For example, when indoor temperatures are below a certain threshold (often a degree below the setpoint temperature) the heater is actuated. The heater is then turned off once the setpoint temperature is reached. These controllers offer ease of installation and operation, however, they often suffer from overshoot and undershoot, making it difficult to maintain a precise indoor temperature. This occurs because of the large thermal inertia of buildings and their engineered systems, which can result in large deviations from setpoint temperatures [11]. If temperatures are more precisely maintained, then the system will be frequently cycling on and off. Frequent cycling can be damaging to equipment and annoying to occupants.

Continuous controllers, on the other hand, modulate heating and cooling to provide heat transfer at the rate that it is needed to reject disturbances and track the setpoint. Although many continuous controllers only use current indoor temperature as an input, they can provide much more precise setpoint tracking, as the building response dynamics are accounted for in the set-up (i.e., tuning) process. Continuous control is normally provided by proportional-integral (PI) or proportional-integral-derivative (PID) local loop feedback controllers, which attempt to minimize undershoot, overshoot, rise time, settling time, and steady-state error [9].

More advanced control systems can take additional inputs, such as future building occupancy and predicted weather. One such strategy considered in this work, called Model Predictive Control (MPC), predicts the future state of the building by incorporating weather forecasts and current indoor temperature. These inputs are fed into a model to predict how the building will change under a variety of different HVAC actions. An optimization is then performed to determine which action will achieve the required temperature while minimizing energy use. The optimal control action is then sent to and implemented by the HVAC system [12,13], and the cycle will be performed again. By predicting future states and correcting for state prediction errors at every time interval, MPC acts as closed loop, real-time building controller. Known as receding horizon control, a newly optimized control strategy is determined as temperature and weather forecasts are updated [14].

MPC frequently utilizes a reduced order linear dynamic model that represents the building as an equivalent circuit of thermal resistances and capacitances (RC) [13]. This means that the heat transfer in and out of the building is simplified to linear expressions, making the optimization problem convex and easier to solve. Buildings, however, do not always act linearly, which leads to modeling errors and mismatch [13]. The extensive time and effort required for properly calibrating a model for individual buildings has kept MPC from widespread adoption [15]. Yet, MPC continues to show promise and is predicted to gain traction with research showing residential energy savings of 28% on average and cost savings of 16% [15,16].

1.4. Potential of Occupancy-Aware HVAC Systems

When occupancy is included in HVAC control, the system behaves conventionally when the building is believed to be occupied, attempting to meet a pre-programmed temperature setpoint. In contrast, when the space is believed to be vacant, the HVAC system allows the indoor temperature to drift somewhat by using a more liberal setback temperature setpoint, minimizing total energy use. Occupancy-based controls can be either reactive or predictive. In reactive control, the system detects an occupant in the space and then turns on the system. This can lead to uncomfortable temperatures when an occupant first enters a space, as the system may not be able to immediately reach the new setpoint. In predictive control, the arrival time of the occupant is predicted and the system preheats or precools the space so that indoor temperature reaches the setpoint just before the occupant arrives, minimizing energy while maintaining comfortable temperatures during all occupied hours [17]. However, correctly predicting when an occupant will arrive is challenging, as the behavior of individuals is difficult to model.

Industry professionals have been working to ascertain the energy savings potential for occupancy-aware HVAC control. In 2014, Nagele et al. conducted a survey of 30 households in southern Germany over a period of 14 months [9]. They then used the data collected, such as temperature setpoints and house characteristics, to calculate energy use under eight different control strategies for ten simulated households. Using a constant temperature on/off controller as the reference case, they showed that PID controllers, setback temperatures, model predictive control (MPC), and occupancy-based HVAC control all have the ability to reduce energy use, when implemented correctly. See Figure 2 for a comparison of the results.

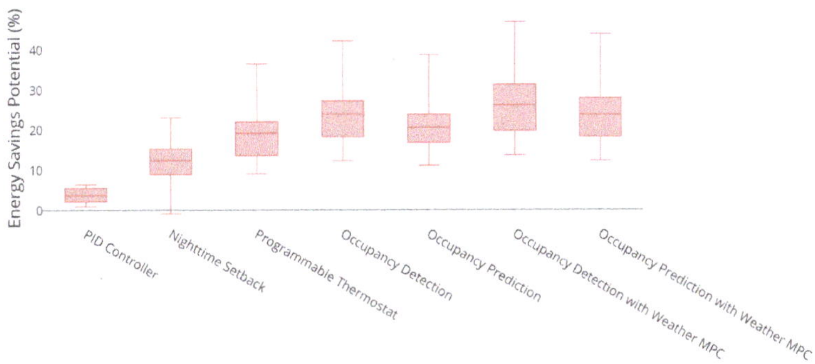

Figure 2. Boxplot of potential savings by control strategy for ten households [%]; adapted from Reference [9].

The reactive strategy of simple occupancy detection offers the largest potential energy savings, but can also increase unmet comfort hours. If a space is routinely uncomfortable when an occupant comes home, then they are likely to turn the detection control off. Thus, to gain consumer adoption, unmet hours need to be low enough that consumers use the functionality. This makes occupancy prediction the preferred control choice. In simulation studies of occupancy-prediction control, savings are estimated to be between 6%–48%, and depend on factors, such as climate, insulation levels, and occupancy schedules [17,18]. Beyond simulations, utilities have measured the energy savings of connected thermostats, which often employ occupancy-aware controls [19]. In reviewing 35 studies from 2007 to 2016, the U.S. Department of Energy reported energy savings that ranged from 1% to 15% [10]. Definitive values for savings are hard to determine due to variances in hardware, software, buildings, occupant behavior, and local weather.

1.5. Occupancy in Buildings

In order to effectively incorporate occupancy information into building controls, occupancy patterns and their impacts on energy use must first be understood. A 2017 international survey of building energy professionals and researchers listed occupant behavior as the largest contributing factor to energy modeling errors [20]. This is because of the varied and stochastic nature of human behavior, which changes dramatically from person-to-person and from day-to-day. This makes correctly predicting occupant behavior, and therefore its effect on buildings, extremely difficult.

In 1978, Robert Socolow published a 5-year observational study on occupant behavior where they tracked gas consumption of 205 identical townhouses in Twin Rivers, New Jersey, finding a 33% variation in consumption [21]. This revealed that seemingly identical buildings can vary due to factors, such as occupants' setpoint temperatures and hot water use. Similarly, a study in Kuwait showed that residents used setpoint temperatures that varied between 19 °C and 25 °C for air-conditioning, with electricity use increasing 21% with a 2 °C change [22]. In Denmark, Rune Andersen collected four years of annual heating data from 290 identical townhouses. Again, a wide variation was found with annual heating consumption ranging from 9.7 kWh/m^2 to 197 kWh/m^2, a ratio of 20 to 1 [23]. These studies, carried out in different climates, across continents and several decades, show the significant variation of occupant behavior and the impact it has on energy consumption.

1.6. Modeling Occupant Presence

Modeling the impact of occupancy behavior on energy consumption is comprised of two steps. First, a researcher must create a reasonably accurate model of occupancy. Second, this model is incorporated into a building performance simulation (BPS). Research in the past decades has investigated the best method for performing each step, both of which are necessary to understand the impact of occupancy-based HVAC control on energy use.

While an accurate occupancy model is critical to understanding the impact of occupant behavior, a 2017 industry survey showed that industry professionals believe current models over-simplify real behaviors, leading to inaccurate predictions [20]. A model can be either overly optimistic, in which case actual energy consumption in a building performs below expectations, or overly conservative, which leads to oversized mechanical equipment.

While many occupant models have been published in scientific papers, an industry consensus on what the best model is has not yet been achieved. Occupancy can be modeled and predicted at two levels: group or individual. In the group level, one model is created for the entire group occupying a building. In this method, which is currently the most widely used, the building is essentially the entity being modeled. At the individual level, a separate model is created for each occupant of a building. Some of the most common occupancy models are described below [24].

- *Schedules* are the current industry standard for modeling occupancy presence. A predetermined fraction of occupancy is multiplied by the space density to determine the number of people during each hour.
- *Deterministic models* use a rule-based approach to represent occupancy behavior. Unlike schedules, deterministic models incorporate environmental triggers that can affect actions.
- *Non-probabilistic models* use historical data to create a model. The aggregated data is averaged to create a probability profile, with each time interval having a probability between 0 and 1. If the probability is above a threshold, the building is predicted to be occupied; below the threshold, vacant. Because the profile is created from a training set, the accuracy of the model depends highly on the data used. The model created does not include a stochastic term.

- *Probabilistic or stochastic models* incorporate the variability of human behavior by using randomization. Like non-probabilistic models, stochastic models use historical data to create a model. A probability profile is created and compared to a randomly generated number to classify the space as occupied or vacant. Because a random number is used, a different profile will result each time the model is generated. Stochastic models require multiple runs to achieve reliable results.
- *Agent-based models* model occupants individually, aggregating multiple prediction models to create a full building model. Because modeling is done on an individual basis, the complexity is extremely high.

1.7. Modeling Building Performance

The second step to incorporating occupancy is loading the model into the building simulation, for which there are many simulation programs available. The International Building Performance Simulation Association (IBPSA) lists sixty-seven whole building energy simulation programs [25]. EnergyPlus, developed and distributed by the U.S. Department of Energy (DOE), is used most commonly in occupancy research [20]. EnergyPlus is a compiled physical model, which means the characteristics of the building, such as insulation values, window size, and orientation, are built into the model itself [17], while the mechanical equipment and schedules, such as occupancy, are included as inputs to the building operation. When executed, the simulation calculates the heat and mass transfer for each time step [4]. Simulations are normally performed per annum to integrate both heating and cooling seasons [17]. ASHRAE occupant schedules are embedded within the example EnergyPlus models but different occupant models can also be incorporated in custom models.

1.8. Review of Commonly Used Occupant Models

Past studies have sought to answer the question of which occupant model works best to predict occupancy [14,18,26–34]. Since the published studies were conducted using individually collected data, such as occupancy and climate, and utilized building-specific performance simulations, it has been difficult to directly compare different occupant models [24]. Individual analyses have sought to solve this by comparing different occupancy models made with the same occupancy data. A review of occupant presence comparison studies is summarized in this section.

A study by Mahdavi and Tahmasebi [35] compared three models: two probabilistic models from literature (Reinhart 2004 and Page 2008) and a non-probabilistic model the authors developed. Using data from eight workspaces, the three models were created using four weeks of training data to predict building occupancy over the next 90 working days. Predicted occupancy was compared to measured ground-truth occupancy to analyze the prediction model's capability. The model was evaluated by comparing the arrival time, departure time, duration of occupancy, fraction of correct occupancy state, and number of transitions to the ground-truth data. Analysis showed that the two stochastic models performed similarly, while the non-probabilistic model performed best. Mahdavi and Tahmasebi conclude that, while probabilistic models are suitable for annual simulations, non-probabilistic models are more effective in providing short-term occupant presence predictions [35].

Following their 2015 study, Tahmasebi and Mahdavi [36] input a variety of occupancy models into a building simulation program to determine the effect of the occupancy prediction on building performance. The first model used the ASHRAE 90.1 office schedule. The second used the average group occupancy data for the year, while the third used the average individual occupancy data for the year. A stochastic model for each of the previous three was created to generate a total of six models. An EnergyPlus performance simulation was executed to calculate energy use under each occupancy model. Stochastic models were executed using 100 Monte Carlo runs to find the average performance. The performance of the models was evaluated using the key performance indicators of annual and peak heating and cooling loads; see Figure 3. It was observed that the ASHRAE schedules performed poorly in all metrics. The stochastic models of individual and grouped occupancy performed better when simulating heating loads, while the non-probabilistic models performed better when simulating cooling loads. Tahmasebi and Mahdavi conclude that known occupancy data is critical for accurate building performance simulation, while stochastic models are not [36].

Duarte et al. [26] performed an occupancy study on a multi-tenant 11 story office building in Boise, Idaho. Using data from 223 private offices over two years, probabilistic models and ASHRAE 90.1 schedules were compared to a non-probabilistic model. Comparing the different occupancy models, the ASHRAE 90.1 schedule overestimated occupancy by as much as 46%. Using data from ten offices for training, the stochastic model matched the training data but not the overall measured occupancy. The authors recommend using a low and high non-probabilistic model because it represents occupancy well without increasing modeling complexity [26].

In all comparison studies, the authors agree that the best model is case specific [24]. Most models are developed using single data and building sets and do not transfer effectively to different building types or occupant behaviors [37]. Despite this, some general conclusions can be drawn:

- ASHRAE occupancy schedules are not reflective of actual behavior.
- Model complexity, such as stochasticity, does not always improve results.
- Models perform best when applied to the case study used to derive them [24].

Since there is no universal occupant prediction model, the IEA Annex 66 consortium recommends choosing a model that matches the complexity levels of the occupant model to the case study. The study presented in this paper aimed to evaluate the possible energy savings on short-term occupancy-based HVAC predictive control. Thus, a non-probabilistic model, which was shown to have the best short term presence prediction, was used [26,35].

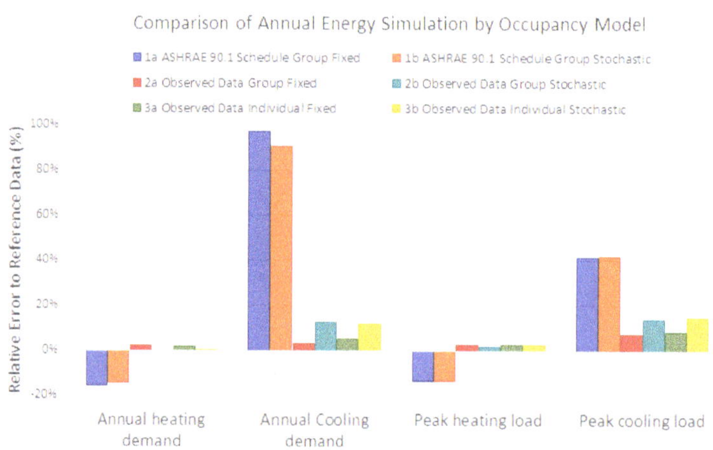

Figure 3. Results of building performance simulation (BPS) model accuracy from 2017 study. Adapted from Reference [38].

2. Occupancy Model Generation and Discussion

The goal of this work was to evaluate the impact of occupancy-based HVAC control on residential building models and associated potential energy savings. To facilitate this, actual residential occupancy had to first be determined. Real occupancy data was collected from multiple homes located in Boulder, Colorado, and used to create non-probabilistic occupancy prediction models. The effectiveness of the model was determined by comparing the predicted occupancy against actual occupancy. The details of this process are explained in following subsections.

2.1. Ground Truth Data Collection

Occupancy and eight different physical modalities (e.g., temperature and CO_2) data was collected from six homes for a period of 4–9 weeks each. Occupancy data was collected using a geo-fencing application installed on the users' cell phones, as well as with a paper sign-in sheet by the front door. The two collection methods were cross-referenced by the researchers to confirm correctness. Through both methods, occupant arrival and departure times were recorded for each person residing in the home. Individual occupant data was combined in order to determine the binary occupancy state of the residence. General occupancy information for each residence is shown in Table 1.

Table 1. Summary of residences measured during study.

House #	Occupant Count	House Type	Days Measured	Avg. Occupancy
1	4	house	64	86%
2	1	apartment	45	56%
3	3	house	71	75%
4	3	apartment	29	82%
5	2	apartment	27	81%
6	1	apartment	63	52%

Residences used for the study were chosen from volunteer participants at a university, and consisted of graduate and undergraduate students, a post-doctoral researcher, and a university professor. Since most of the participants were students of some type, their occupancy patterns may be quite different from those collected in a different segment of the population. For instance, the fact that several of the homes contained multiple graduate students with a variety of class and work schedules meant that someone was nearly always home (yielding occupancy rates of 82% and 86%). Furthermore, none of the homes contained children, which might have led to different occupancy patterns. Additionally, occupancy was impacted by extended absences during spring and fall breaks for a few of the homes studied.

While the average occupancy for the testing period ranged from 52% to 86%, the daily occupancy of each home varied. Distributions of daily occupancy rates for each residence are shown in Figure 4. Home 6 (a young graduate student who lived alone) had the largest distribution in daily occupancy, while home 5 had the lowest (married postdoctoral researchers, one of whom worked from home).

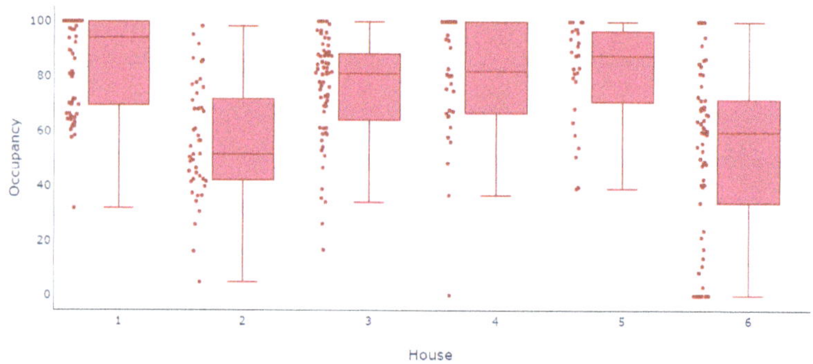

Figure 4. Daily occupancy rate for each measured day by house [%].

Beyond daily occupancy, arrival time, departure time, and number of occupancy state transitions were also analyzed. Arrival is defined as the transition of the residence from an unoccupied state to occupied state; departure is the inverse transition. In residences with more than one occupant, the arrival and departure times identified only indicate cases where the residence transferred to or from vacant. Distributions in daily occupancy, arrivals, departures, and occupancy state transitions demonstrate the stochasticity of human behavior. For all residences measured, the arrival and departure times for weekdays differed significantly from those of weekends.

2.2. Occupancy Model Generation

Based on the literature review, a non-probabilistic model was chosen to model occupancy. Since non-probabilistic models use past data to create an occupancy probability, the model can be optimized by establishing what training data to include. Model optimization was done by splitting the collected occupancy data into separate training and testing data sets. Models were trained using the first set, and performance was evaluated by testing the trained model on the second dataset. The only exception to this is the case of the moving training mode, which used a receding horizon. In this case, the model was continually being trained, then tested on unseen data, and then the model was updated after comparison to ground truth.

The percentage of data used for training versus testing was varied, along with three other parameters. Up to 96 different non-probabilistic models were created for each house, with each model using a different subset of training data to create the occupancy profiles. Table 2 shows the different values for each parameter that was used when creating the models.

Table 2. Variables used for occupancy model creation.

Day Categorization	Training Time	Training Mode	Time Resolution
day of week	1	fixed	1 min
week/end	2	moving	5 min
mfweekend	3		15 min
	4		60 min

The four group parameters are defined as follows:

- Day categorization: This determines how each day of the week is categorized. For example, in *day of week*, only training data that matches the day being predicted is used. In *week/end* categorization, all Mondays, Tuesdays, Wednesdays, Thursdays, and Fridays are used to predict occupancy for weekdays. Finally, in *mfweekend* categorization, Tuesday, Wednesday, and Thursday are used to predict weekdays with Monday and Friday kept as separate individual days.

- Training time: This determines how many weeks are used when training the model, ranging from one week to four weeks. Up to 50% of the collected data was used for training, meaning that in residences where only four weeks of total occupancy data were recorded, only two weeks were available for training. In these cases, only 48 total models were generated because the 3-week and 4-week training time was unavailable.
- Training mode: This determines whether the training time is used in a fixed mode (static training set) or in a moving mode (where a trailing horizon is used). For example, in a 1 week moving mode, only the last seven days are used to predict occupancy for that day.
- Time resolution: This determines how often occupancy is sampled. Time is shown in minutes.

After a training set was generated from the collected data, the average occupancy was determined for each interval of the day, resulting in an occupancy probability between 0 and 1. To create a binary occupancy schedule, a threshold probability was set for each day. This threshold was determined by finding what value produced the same minutes of predicted occupancy as the summed occupancy minutes in the input data. An example of a single day for House 1 is shown in Figure 5. Occupancy prediction models were created for a time period of two to five weeks, depending on the length of total measured data.

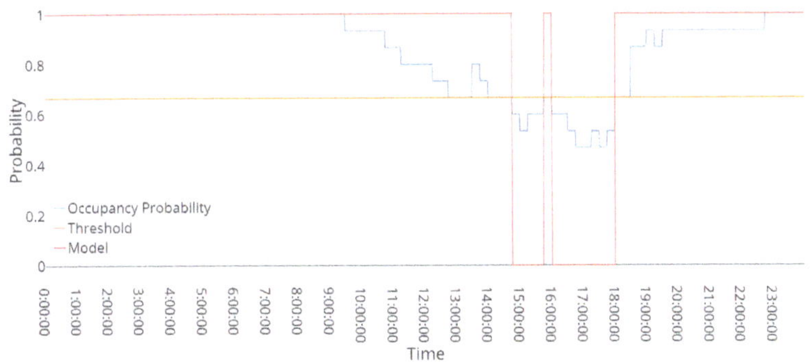

Figure 5. Occupancy probability, threshold, and resulting model for single day, by time.

2.3. Occupancy Model Accuracy

Once the non-probabilistic models were generated for each home, the resulting predicted occupancy state was compared against the actual occupancy state. To evaluate the effectiveness of the occupancy prediction models, three metrics were used:

- **False negative rate:** Percentage of minutes that the model incorrectly predicted the house was vacant when it was occupied.
- **False positive rate:** Percentage of minutes that the model incorrectly predicted the house was occupied when it was vacant.
- **State matching error:** Percentage of minutes that the model incorrectly predicted occupancy. This is the inaccuracy of the model. The state matching error is the sum of the false negative and false positive rate.

All the metrics are error rates, and so should be minimized. The best models had low values for false negative rate, false positive rate, and state-matching error. False negative errors and false positive errors have different impacts. When a false negative occurs, the house is actually occupied when predicted vacant. In this case, indoor temperature may not be in the comfort range because the control has been changed to a setback temperature. When a false positive occurs, the house is actually vacant when predicted occupied. In this case, the HVAC system may be running to maintain an unnecessarily tight temperature setpoint range, resulting in higher energy use.

Training and evaluation times varied by house due to differences in data collection periods. Table 3 lists how the collected data was used in generating the prediction models. **Training period** shows how many weeks were set aside for determining occupancy probability, and **evaluation period** shows how many weeks were used to evaluate the generated model. The training period is the maximum number of weeks available. Training time, used as a variable, determined how many of the weeks were used in training the model. During the evaluation period, the predicted occupancy was compared against the measured occupancy at one-minute intervals.

Table 3. Summary of model training and evaluation for each house.

House #	Models Created (Count)	Training Period (Weeks)	Evaluation Period (Weeks)
1	96	4	5.1
2	96	4	2.4
3	96	4	6.1
4	48	2	2.1
5	48	2	1.9
6	96	4	5.3

The model configuration with the lowest state matching error is shown for each house in Table 4. House 5 had the lowest state matching error with an error rate of 8%, while House 6 had the highest at 35%. Each house, due to its occupancy pattern, had a different optimal occupancy model, which indicates the value in tailoring the model to the specific use case.

Table 4. Best occupancy prediction model for each residence.

House	Day Categori-Zation	Training Time	Training Mode	Time Resolution	False Negative	False Positive	State Matching Error
1	mfweekend	4	fixed	15	12%	4%	16%
2	weekend	3	moving	5	13%	13%	26%
3	weekend	1	moving	1	28%	6%	35%
4	weekend	1	moving	60	8%	2%	10%
5	weekend	1	moving	60	5%	3%	8%
6	day of week	2	moving	15	7%	30%	37%

To understand the effect of each parameter on the resulting prediction model, the performance results were compiled and analyzed. The parameter with the largest effect on state matching error was the occupancy pattern, shown in Figure 6. The results show that each house, with its different occupancy patterns, has a strong influence on the effectiveness of creating a prediction model. As previous studies showed, the behavior of people has a large variance and can drastically affect prediction models.

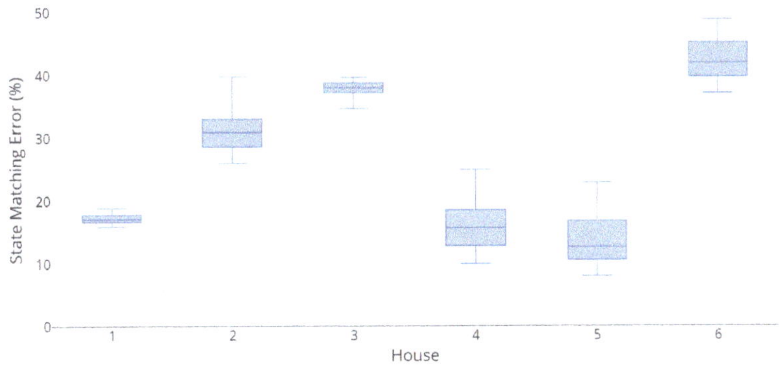

Figure 6. State matching error for each house [%].

In contrast to the differences in house occupancy patterns, the parameters that were used to develop the training set each had a small influence on the prediction error. The results of all four parameters are discussed below.

Day categorization On houses with fewer weeks of collected data (House 2, 4, and 5), using *day of week* categorization resulted in the least accurate prediction models. This is likely due to the extremely limited training data for each day. In a *day of week* model, each weekday is treated individually. Thus, if only two weeks of training data are used, then there would be only two instances of each day. In contrast, in House 6, where the occupant had a part-time job that she attended three days a week, the *day of week* method increased the prediction accuracy.

Training time: As would be expected, using more training data improved the accuracy of models for most of the houses. House 3, the exception, had a shift in occupancy patterns halfway through data collection, when some of the occupants went on vacation and extended visitors arrived. This indicates that when new occupants join a household, the previous training data will not effectively predict the new occupancy pattern. To explore this theory further, Figure 7 shows the resulting state-matching error when the training time is extended to seven weeks. With additional training weeks, the error is reduced, indicating that the longer the training data is accumulated, the more the error can be reduced.

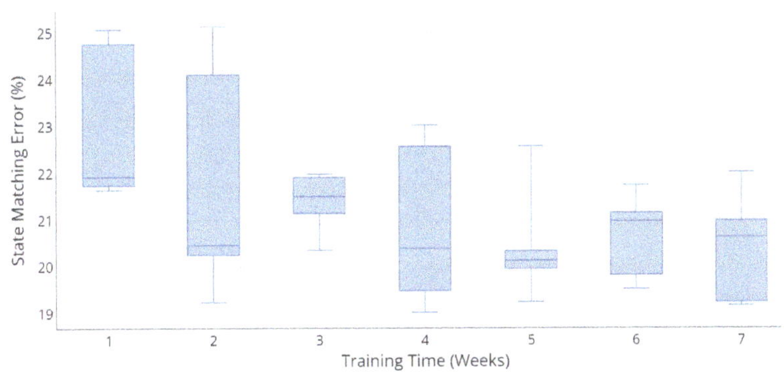

Figure 7. State matching error for House 1, by number of training weeks.

Training mode: The moving training mode had improved prediction accuracy for five of the six houses. By allowing the model to adjust over time, the moving training mode adapts to shifting behavior. House 1 had the best prediction accuracy with the fixed mode. The difference between the highest performing fixed mode and highest performing moving mode for House 1 was 0.3%. The higher performance using fixed mode indicates the initial data reflected the general behavior more than later weeks.

Time resolution: Results are nearly identical for each time resolution variable signaling that sampling time does not play a large role in increasing prediction accuracy.

While the lowest state matching error can be achieved by optimizing a non-probabilistic occupancy prediction model to a specific house, a universally applicable model is desirable. This would allow a single model to be deployed in different houses, without the need for preliminary data gathering to determine the parameters. To find the best occupancy model for all the houses surveyed, the state matching error results were normalized by dividing the results by the lowest state-matching error achieved by that house. The lowest error for each house was used to ensure that all of the houses were considered equally. Figure 8 shows a parallel category plot of the results. Occupancy models that were within 5% of the best model for that house are shown in green, while other models are shown in red.

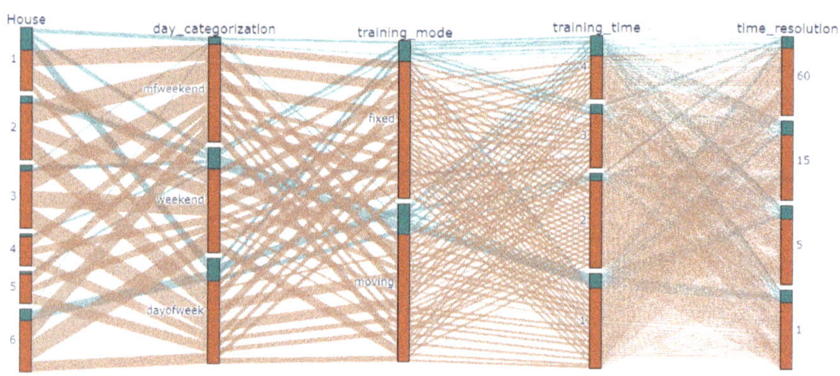

Figure 8. Parallel category plot of occupancy models.

Results show that the *mfweekend* day categorization does not work well for the houses surveyed. *Day of week* and *week/end* both work well, with *week/end* performing better for most homes, suggesting the six houses surveyed do not have distinct Monday or Friday schedules. Four and one week training times were the best, with two and three week training times performing slightly worse. Houses with consistent schedules benefited from the increased data of a longer training time. Houses in which occupants were absent for days benefited from the faster reaction of the shorter training time adjusting to their absence and return. The moving training mode produced the highest number of low state matching errors, which is likely due to its ability to continuously adjust to occupants' behavior changes over time. As seen in the individual evaluations, different time resolutions produced equivalent results, although the 15-min time resolution yielded a slightly higher number of low normalized state matching errors. Based on these results, the optimal universal model across all houses was a *one week* training time *moving* model that uses *week/end* day categorization and a 15-minute time interval. The state matching error for each house with the universal prediction model is shown in Table 5.

Table 5. Results of universal occupancy prediction model for each residence.

House	Day Categori-Zation	Training Time	Training Mode	Time Resolution	False Negative	False Positive	State Matching Error
1	weekend	1	moving	15	15%	2%	17%
2	weekend	1	moving	15	22%	10%	32%
3	weekend	1	moving	15	28%	6%	35%
4	weekend	1	moving	15	8%	2%	10%
5	weekend	1	moving	15	5%	3%	8%
6	weekend	1	moving	15	11%	33%	44%

3. Building Simulation Setup

Building performance simulations were conducted in EnergyPlus (Version 9.1, National Renewable Energy Laboratory, Golden, CO, USA) to understand the impact of residential HVAC control on energy use for a number of representative home scenarios. Multiple home types, climates, seasons, and occupancy patterns were used to more globally represent the breadth of scenarios, as well as to understand the range of possible outcomes.

3.1. Building Performance Simulation Settings

The six previously collected occupancy data sets were used as the possible occupancy scenarios. Data from these homes were assumed to represent occupancy patterns over all climates and seasons for which the simulations were performed. Five prototype home styles were used, as provided by National Renewable Energy Laboratory (NREL) for the building models [39]. Each home had a different climate and building construction that was representative of the national housing stock. The five locations used were: Boston, Phoenix, Atlanta, Seattle, and Houston. House sizes averaged 2000 ft^2 with typical home construction and vintages for each region. Details on the home and construction parameters are shown in Table 6.

The occupancy prediction models developed in Section 2.3 ranged from 13 to 36 days depending on the house. A two-week period with one-minute timestep intervals was used for the building performance simulations. Building simulations were performed for two different seasons using the first two weeks of January and the first two weeks of July. Including both winter and summer runs allowed the impact of HVAC control to be ascertained for both heating and cooling modes.

Boston and Atlanta have cold, near-freezing winters, and hot summers. In both of these locations, the outdoor air temperature was well outside of the comfort range for the majority of the simulations. Houston has moderately cold winters and hot summers, while Phoenix has mild winters and extremely hot summers. Both Houston and Phoenix require significant cooling in the summer. During winter in Phoenix, the outdoor air temperature oscillates within the comfort range. Seattle, in contrast, experiences both cool winters and cool summers, requiring some heating year-round.

Heating and cooling temperature setpoints were established using the method of predicted mean vote (PMV). ISO EN 7730 establishes three comfort categories using operative temperature. These categories are shown in Table 7. Class A and B were used as the defined comfort range to maintain a PMV within 0 ± 0.5.

Table 6. Summary of house model constructions.

	Boston, MA	Phoenix, AZ	Atlanta, GA	Seattle, WA	Houston, TX
Climate	Cold 5A	Hot-Dry 2B	Mixed-Humid 3A	Marine 4C	Hot-Humid 2A
Vintage	<1950s	1970s	1970s	<1950s	1970s
House Size	2589 ft^2	2013 ft^2	2013 ft^2	1938 ft^2	2013 ft^2
Envelope					
Attic	Uninsulated	Ceiling R-13, Vented	Ceiling R-19, Vented	Ceiling R-13, Vented	Ceiling R-13, Vented
Wall Cavity	Uninsulated	Uninsulated	Uninsulated	Uninsulated	Uninsulated
Foundation	Uninsulated	Uninsulated	Uninsulated	Uninsulated	Uninsulated
Windows	Clear, Double, NM, Air	Clear, Double, Metal, Air	Clear, Single, Metal	Clear, Single, Metal	Clear, Double, Metal, Air
Air Leakage	15 ACH50	15 ACH50	15 ACH50	15 ACH50	15 ACH50
HVAC					
Heating	Gas Boiler, 80% AFUE	Gas Furnace, 80% AFUE	Gas Furnace, 80% AFUE	Gas Furnace, 80% AFUE	Gas Furnace, 80% AFUE
Cooling	Room AC, EER 10.7	Central, SEER 13	Central, SEER 13	None	Central, SEER 13

Table 7. Three categories of thermal comfort (ISO EN 7730, 2005).

	Thermal State of the Body as a Whole		Operative Temperature °C	
Category	**PPD %**	**PMV**	**Summer (0.5 clo)**	**Winter (1 clo)**
A	<6	$-0.2 < \text{PMV} < +0.2$	23.5–25.5	21.0–23.0
B	<10	$-0.5 < \text{PMV} < +0.5$	23.0–26.0	20.0–24.0
C	<15	$-0.7 < \text{PMV} < +0.7$	22.0–27.0	19.0–25.0

While operative temperature, as used in the calculation for PMV, defines thermal comfort, HVAC systems are controlled by measuring zone air temperature. The ambient air temperature setpoints that were used to control the heating and cooling systems in the simulations are given in Table 8.

Table 8. Zone air temperature setpoints and setback temperatures used in building simulations.

	Setpoint Temperature	Setback Temperature
Heating	22.0 °C	18.0 °C
Cooling	24.5 °C	28.0 °C

Three HVAC control scenarios were modeled in the building performance software: conventional operation, occupancy-based HVAC control, and occupancy-based MPC. Conventional operation used a constant heating and cooling setpoint, and was used as the baseline. Results for each strategy are discussed in the subsequent sections.

3.2. Conventional Control (Baseline) Results

When all home scenarios were operated under conventional control, Boston winters showed the highest amount of energy use among all scenarios. This is logical due to the cold ambient environment, older vintage house, and larger size.

In addition to energy use, thermal comfort was evaluated by calculating operative temperatures. Temperatures between 20 °C and 26 °C were considered comfortable to accommodate 0.5 to 1.0 clo clothing levels (typical of a person inside their home). By using a constant setpoint temperature that was within the comfort range, it was expected that the percentage of time temperatures were within the comfort range would be high. Comfort was high except for Phoenix and Seattle in summer. Further examination shows Phoenix achieved a constant internal air temperature of 24.5 °C in summer. However, the operative temperatures were higher, which created uncomfortable conditions. This case demonstrates the impact of using air temperature, rather than operative temperature to drive system controls. In Seattle, both the indoor air and operative temperature were too high for comfort. High temperatures were caused by the hot outdoor temperatures which could not be mitigated without an air conditioner.

4. Occupancy-Based HVAC Control Results

Occupancy information can be incorporated into HVAC control in a multitude of ways. In all cases simulated in this study, knowledge of building occupancy, either through prediction or detection methods, was used to establish the setpoint temperature for HVAC control. When the space is believed to be vacant, the temperature is allowed to drift to a more relaxed setback temperature, reducing use of the HVAC system when unoccupied. However, when the space is either believed to be occupied or predicted to soon be occupied, the space is maintained at the setpoint temperature, which ensures that temperatures are within the comfort zone.

4.1. Occupancy Control Schemes

Five total occupancy detection models were considered and simulated, which fall under three main strategies:

1. **Reactive:** Occupancy is detected and setpoint temperatures are adjusted accordingly. In this case, occupancy is sensed and no prediction is used.
2. **Predictive:** Occupancy is predicted using two different non-probabilistic models, as developed in Section 2.3.
    - *Universal model*: This is the prediction model that performed best for all houses and used a one week, 15 min, week/end categorization, moving training set.
    - *Individually tuned model*: This is the prediction model that performed best for the specific house. The models used are listed in Table 4.
3. **Hybrid:** A hybrid of predictive and reactive occupancy models. Occupancy is first predicted using the non-probabilistic models developed in Section 2.3. During operation, if an occupancy change from vacant to occupied is detected that was not predicted, the control will react and reset the temperature control to occupied settings. In order to maintain the predictive aspect of the model, this control method does not react to changes from occupied to vacant states, which would have made for purely reactive control.
    - *Universal hybrid:* This is the prediction model that performed best for all houses, and used the same universal model as described above, but with the reactive component.
    - *Individually tuned hybrid:* This is the prediction model performed best for the specific house, with the reactive component. Models used are listed in Table 4.

4.2. Results

Building simulations were conducted for all five occupancy-aware control methods for all homes, climates, and seasons. The energy savings of each method, relative to the conventional baseline model, are shown in Figure 9, with the average savings for that model shown in Table 9. While all of the control strategies reduced the total energy used during the simulation, the two pure predictive models (universal and individually tuned) had the largest energy savings potentials, with 10.9% and 9.6% savings, respectively. The reactive control model has similar energy savings to the prediction models, with an average of 9% of total energy consumption. This method was particularly helpful in homes where long periods of unpredictable vacancy, such as a vacation, occurred. The hybrid approaches, which used both predictive and reactive occupancy, saved the least energy with an average of 3 %–5% savings.

Table 9. Average energy savings by control method.

Control Method	Energy Savings
Reactive	9.1%
Universal Model	10.9%
Individually Tuned	9.6%
Universal Hybrid	4.3%
Individual Hybrid	5.7%

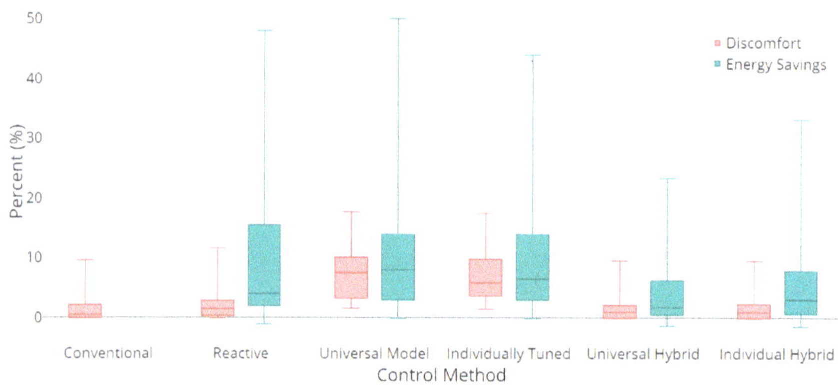

Figure 9. Energy savings and discomfort percent for simulation period by control method [%].

Of equal importance to energy savings are the comfort of occupants under each control method, which was assessed using building operative temperature. Table 10 shows the percentage of time that the occupants were predicted to be uncomfortable, as classified by time spent occupying the residence when it was not within Class A or B comfort temperatures. The hybrid models achieved the highest levels of comfort, with time spent in discomfort similar to that seen with conventional setpoint control. In contrast, the purely predictive models, which did not react to incorrect predictions, led to the largest discomfort percentages. Figure 9 shows the savings over conventional, compared with the percentage of occupied time in the uncomfortable range.

Table 10. Average discomfort by control method.

Control Method	Unmet %
Conventional	2.4%
Reactive	2.6%
Universal Model	7.3%
Individually Tuned	6.9%
Universal Hybrid	2.0%
Individual Hybrid	2.1%

Achieving energy savings without disrupting occupant comfort is the primary goal of effective HVAC control. Since energy savings are achieved by allowing the temperature to drift to uncomfortable conditions when the space is believed to be unoccupied, occupancy-aware controls can only be truly effective if occupancy is accurately predicted and detected. Parameters, such as city, season, and occupancy patterns, all affect comfort, according to the simulations. Phoenix and Seattle have high unmet comfort ratios in summer, as discussed in the conventional control.

The fact that energy savings are dependent on city indicates that either vintage of the home or climate (and likely both) determine relative savings potential. Since home styles vary according to the region, it is difficult to decouple these two effects. In Figure 10, energy savings are shown by city and season simultaneously. Energy savings are highest in Atlanta, Boston, and Houston, especially during summer. All three climates are extremely hot and humid during the summer months, thus reducing unnecessary air-conditioning results in large energy savings.

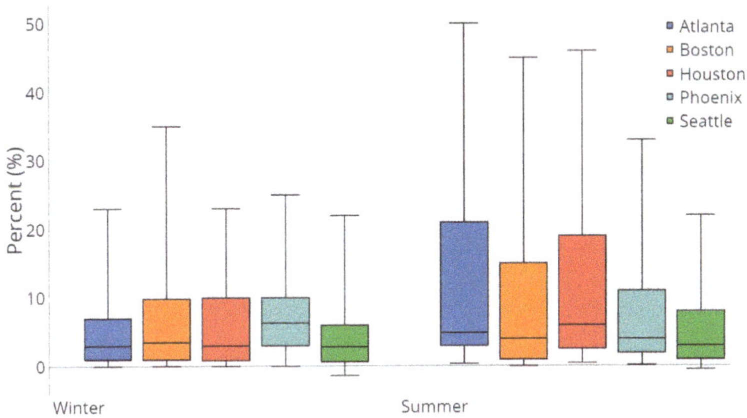

Figure 10. Unmet comfort and energy savings for simulation period, by city and season [%].

Both energy savings potential and discomfort were affected by occupancy patterns. Houses 2 and 6, which had the highest vacancies, also had the highest median energy savings. This indicates that the higher the vacancy rate, the higher the energy savings potential. This is not surprising, as more unoccupied hours means more opportunities for the temperatures to drift outside of comfort. In the cases of predictive models, the energy savings potential is also dependent on how well the vacancy is predicted. The savings can only be realized when the house is both vacant and correctly predicted to be so. Despite low vacancy rates, House 5, which had the most accurate model, also had the highest energy savings in some simulations. This indicates that the better the prediction model is, the higher the possible energy savings can be, as time spent heating or cooling a house that has been incorrectly predicted as being occupied is reduced. In analyzing comfort, House 2 had the highest discomfort portion and a high prediction inaccuracy, signalling the importance of accurate prediction to comfort. Overall, these figures indicate that although large energy savings are possible, the prediction model needs to be accurately calibrated to achieve energy savings and comfort.

Due to the large ratio of discomfort, prediction-only models were not an effective HVAC control strategy. Both hybrid models were able to achieve comfort at the same level of conventional control. Since occupant comfort is not degraded, these methods are more likely to be used by occupants. Conventional, reactive, and hybrid methods all achieve discomfort below 3% on average. Reactive control has the highest average energy savings at 9.1% but at the cost of reduced comfort in comparison to conventional control. Conventional control, which is the baseline, has no energy savings. The two hybrid controls, universal and individual, achieve an average energy savings of 4.3% and 5.7%, respectively. Individual hybrid is able to achieve the highest energy savings while maintaining comfort levels at or below conventional control. Therefore, the individual hybrid control is the recommended occupancy-based HVAC control.

5. Model Predictive HVAC Control Results

Model predictive control (MPC) was the final HVAC control scheme considered. In MPC, an algorithm is used to predict and proactively react to upcoming temperature disturbances or setpoint changes. MPC has been used in the past to optimize a number of parameters in building control, from incorporating weather forecasts for temperature control to shifting peak loads for the power grid [11,13,15,16,40–50]. MPC has the advantage of being a proactive rather than reactive control strategy. For example, by predicting the effect of an increase in outdoor temperature before it occurs and overheats the space, indoor temperature can be gradually reduced, and the amount of energy used by the HVAC system can be minimized.

In this work, MPC was used in conjunction with weather and the occupancy-prediction models from Section 2.3 to optimize the temperature setpoints. An optimization algorithm was executed to find the setpoint temperature that minimizes both energy use and discomfort. In Section 4, occupancy prediction models were used to change the setpoint temperature. In that case, four total temperature setpoints were allowed: the heating setpoint temperature, the cooling setpoint temperature, the heating setback temperature, and the cooling setback temperature (Table 8). MPC optimization considered not only those four temperatures, but also temperatures within those bounds, to find the optimal solution. In order to run MPC, parameters, such as the optimization algorithm, cost function, the prediction horizon, the execution horizon, and building model, all had to be determined.

5.1. Model

Commonly, MPC is performed utilizing reduced-order linear system models. This allows the optimization to be performed more quickly and easily. However, whole building energy simulations, like EnergyPlus, allow the calculation of radiant heat balances and non-linear part-load system performances, which simplified models cannot capture [44]. In cases where thermal comfort is being evaluated, these calculations are essential, and so EnergyPlus was chosen as the modeling engine to perform the task. However, to reduce computation time, the model was reduced as much as possible. These reductions were achieved by hard-sizing the HVAC equipment using TMY3 data for the simulation period, increasing the simulation timestep from one to fifteen minutes, and reducing the numbers of reported variables [44,51].

5.2. Optimization Parameters

Model predictive control requires a prediction horizon to designate how far into the future the model is predicting and optimizing. In this study, a 24-hour prediction horizon was used to account for diurnal temperature swings and internal gains from daily occupancy patterns and equipment use. The execution period, which dictates how often the optimization is conducted, used a one-hour horizon to adjust to actual occupancy and indoor temperature values. Once an optimum temperature setpoint was found, it was implemented in the model and the simulation via EnergyPlus was stepped forward one hour in time. A new optimization process was then started with the current state values and a new 24-hour prediction horizon.

The MPC utilized a Matlab-based particle swarm optimization algorithm to determine the optimal control actions. Particle swarm optimization (PSO) uses a group of candidate solutions as beginning values. Simulations using these seeds are executed, yielding initial results for the cost function. As the particles are evaluated and move throughout the decision space, they *swarm* towards the optimum solution. By using a swarm, the possibility of finding a local minimum, rather than global minimum, is reduced [52].

5.3. Objective Function

The cost function combines all the factors that are to be evaluated and optimized into a single formula. How the cost function is configured determines which values are considered most important in finding the best solution. In this study, the cost function minimized energy use and occupant discomfort. Optimization of this function was constrained by the allowable temperature band. Discomfort was calculated using predicted mean vote (PMV). Since thermal comfort was not always achieved in the baseline model, the PMV from the baseline was used as the maximum allowed PMV in the optimization run, which prevented the optimization algorithm from penalizing solutions that provided comfort performance equivalent to (or better than) the baseline model. By minimizing energy use and occupant discomfort concurrently, MPC could reduce energy use without sacrificing occupant comfort using Equation (1).

$$\min\left(\sum_k E_k + P\right) \tag{1}$$

$$\text{subject to: } T_{lower,k} \leq T_{optimal,k} \leq T_{upper,k},$$

where P is the occupied discomfort, E_k is hourly HVAC consumption at timestep k, T is the temperature setpoints at timestep k, and k is the number of timesteps in the evaluation. Occupant discomfort is calculated as shown in Equation (2).

$$P = C\sum_k(\mid PMV_k \mid - PMV_{max}), \tag{2}$$

where C is the comfort penalty slope, PMV_k is predicted mean vote during timestep k, and PMV_{max} is the PMV comfort threshold. Since the PMV during the baseline run may exceed 0.5, such as in the case of Seattle in summer, the threshold is adjusted to allow the optimized MPC to use an equitable PMV during optimization. This threshold is calculated with Equation (3).

$$PMV_{max} = \max(0.5, \mid PMV_{base,k} \mid), \tag{3}$$

where PMV_{max} takes the higher value between 0.5 and the PMV from the baseline run at timestep k.

The goal of MPC is to optimize for all the factors within the cost function, of which there are two in this case: energy consumption and thermal discomfort. The relative importance of the two factors is controlled by C, the comfort penalty slope, which determines the scaling of discomfort costs. With a smaller C value, setpoint temperatures leading to uncomfortable hours do not increase the cost function as much, allowing some thermal discomfort to occur in favor of energy savings. Thus, the comfort slope allows flexibility in the cost function and can be tuned to meet the individual goals of the occupant, depending on how much comfort they are willing to sacrifice.

To determine an appropriate comfort slope, MPC simulations were completed for two days in House 1. During the modeled time period, the house was vacant for approximately six hours of each day, allowing the comfort slope to be evaluated during both occupied and unoccupied states. Figure 11 shows the summarized results from this experiment. Discomfort was evaluated by classifying hours within ±0.5 PMV as comfortable, ±0.7 PMV as Class C discomfort, and beyond ±0.7 PMV as excessive discomfort. As the comfort slope value increased, hours of discomfort decreased, while energy use increased. A comfort slope was chosen to allow only a few hours of Class C discomfort. For all simulations in this study, a comfort penalty slope of 1000 was used, which allowed some energy savings, while still maintaining comfort most of the time.

5.4. Simulation

MPC was used to simulate three different scenarios, in order to understand the factors that make it more or less effective. The individually tuned hybrid models found in Section 2.3 were used in two different locations (Atlanta and Houston) in the summer, and the Atlanta results were compared to those found from a model that used perfect occupancy forecasting. Atlanta and Houston were chosen due to their relatively average energy use and temperatures in Section 4. By using a climate that was neither mild nor extreme, the MPC results should apply to more regions.

In each scenario, two different homes' occupancy schedules were utilized in the simulations to show the effects of differing occupant profiles. The three scenarios yielded different results and insights, which are discussed in the following subsections.

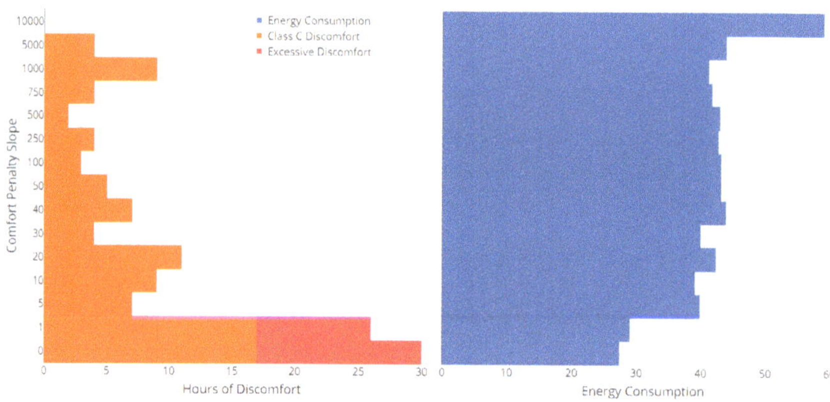

Figure 11. Resulting comfort and energy use, by comfort slope value.

5.4.1. MPC Case 1: Houston with Occupancy Prediction

In this simulation, MPC was used with the individually tuned hybrid models found in Section 2.3 to model the cooling requirements for a home in Houston in summer. Occupancy profiles for Houses 2 and 5 were chosen, as these exhibited very different patterns. A summary of the EnergyPlus model settings, optimization parameters, and objective function is provided in Table 11. The maximum computation time allowed for each optimization was 30 min, which yielded a minimum of 300 function evaluations for each execution horizon. Overall, the MPC simulation for each house took 60 h to complete the five-day run period.

Table 11. Settings used for Model Predictive Control (MPC) optimization in Case 1.

Parameter	Value
City	Houston
Season	Summer
Houses	2 & 5
Prediction model	Individual Hybrid
Run period	5 days
Timestep	15 min
Planning horizon	24 h
Execution horizon	1 h
Occupied allowed temperatures	22 °C was $\leq T_{optimal,k} \leq$ 24.5 °C
Unoccupied allowed temperatures	18 °C $\leq T_{optimal,k} \leq$ 28 °C
Temperature increments	0.5 °C
Comfort penalty slope (C)	1000
Optimization time per execution horizon	30 min

Results for Case 1 are shown in Table 12. Discomfort is measured by exceedance above Class A and B comfort. This value was calculated by summing the operative temperature deviations above 26.0 °C or below 23.0 °C for all occupied hours. Results are measured in Kelvin-hours (Kh). For both houses simulated, the energy saved is very low, with an average savings of 1%. However, little to no thermal discomfort was achieved. The MPC optimization results found that the highest allowed temperatures provided the lowest resulting cost. Although any temperature within the band was allowed, the optimal temperature ending up matching the values used in occupancy-based setpoint control, signaling that for Case 1 setpoint control and MPC optimization yielded the same temperatures.

Table 12. Results for Case 1.

House	Energy Savings	Discomfort
2	2.1%	3.7 Kh
5	0.2%	0 Kh

Most hours of the simulation are within the comfort region, leading to high comfort for the occupants, but low energy savings. Hours in which the temperature was allowed to drift above comfortable temperatures were few due to the small number of hours when the prediction model accurately predicted the house to be vacant.

5.4.2. MPC Case 2A: Atlanta with Occupancy Prediction

In the second set of simulations, MPC was used with the individually-tuned hybrid models for Houses 1 and 2 in Atlanta in the summer. In this case, the maximum allowed temperatures during occupied hours were kept at the same values used during all unoccupied hours. By allowing a large temperature band at all times, the constraints within the cost function were reduced, and temperatures which produced the smallest cost function were used, rather than the being restricted by the temperature band. A summary of all the settings used for the simulation is shown in Table 13.

Results of these simulations are summarized in Table 14. For this simulation, the two houses saved an average of 9.0% in energy use. Exceedance is higher than Case 1 with an average of 35.3 Kh. Total occupied hours for both houses for the week was 185 h, with House 1 being occupied for 100 h and House 2 being occupied for 85 h. With the average exceedance of 35.3 Kh, an average distribution of thermal discomfort would yield 0.4 °C above the ideal temperatures.

Table 13. Settings used for MPC optimization in Case 2A.

Parameter	Value
City	Atlanta
Season	Summer
Houses	1 & 2
Prediction model	Individual Hybrid
Run period	1 week
Timestep	15 min
Planning horizon	24 h
Execution horizon	1 h
Occupied allowed temperatures	19 °C $\leq T_{optimal,k} \leq$ 27 °C
Unoccupied allowed temperatures	19 °C $\leq T_{optimal,k} \leq$ 27 °C
Temperature increments	0.5 °C
Comfort penalty slope (C)	1000
Optimization time per execution horizon	30 min

Table 14. Results for Case 2A.

House	Energy Savings	Discomfort
1	7.5%	30.8 Kh
2	10.4%	39.8 Kh

Figure 12 shows temperatures for House 1 for two days of the simulation. Chosen temperatures ranged from 19 °C to 27 °C, with the average setpoint temperature at 25.4 °C and 26.2 °C for House 1 and 2, respectively. With the expanded temperature range, temperature values selected did not always conform to setpoint temperatures as seen in Case 1. While setpoint values allowed an 8 °C range, ambient air temperatures occurring within the building had a 4.5 °C to 4.7 °C range. More extreme setpoint temperatures only lasted for an hour, preventing temperature within the building from reaching the setpoint and maintaining a comfortable space despite the setpoints used.

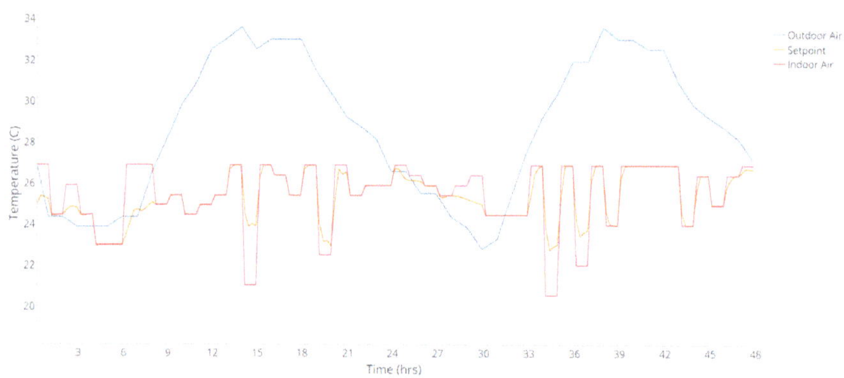

Figure 12. Case 2A temperatures for House 1 using individualized hybrid prediction model.

Figure 13 shows the electricity consumption resulting from the setpoint temperatures used. Due to changing setpoint temperatures, electricity consumption jumped in hours using low setpoint temperatures as more cooling occurred. In other hours, however, electricity was significantly less than the conventional constant temperature. Over the week simulated, electricity consumption was reduced to allow 7.5% and 10.4% energy savings for House 1 and 2, respectively.

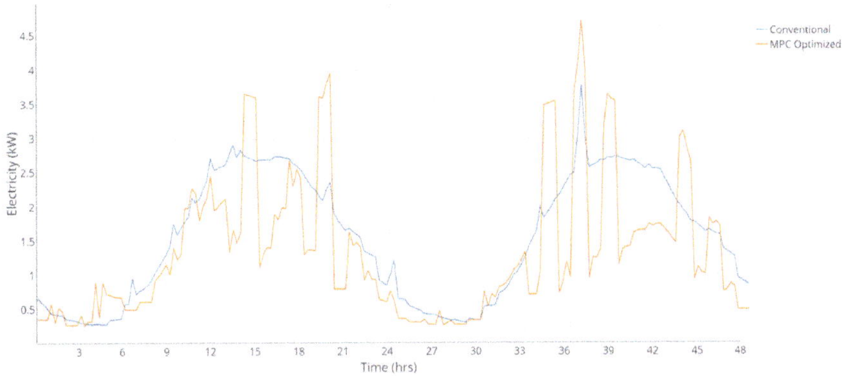

Figure 13. Case 2A electricity consumption for House 1 using individualized hybrid prediction model.

Figure 14 shows the duration curve of operative temperatures for House 1. The figure shows that allowed temperature deviation was higher for unoccupied hours. In hours that the house was occupied, operative temperature was kept closer to the center point temperature of 23.5 °C. With the chosen cost function, some temperature deviation was allowed to achieve higher energy savings. Unlike Case 1, which had tight occupied temperature constraints, temperature deviation in Case 2A is higher. Changes to the comfort penalty slope would change how much deviation is allowed and, in result, how much energy was saved.

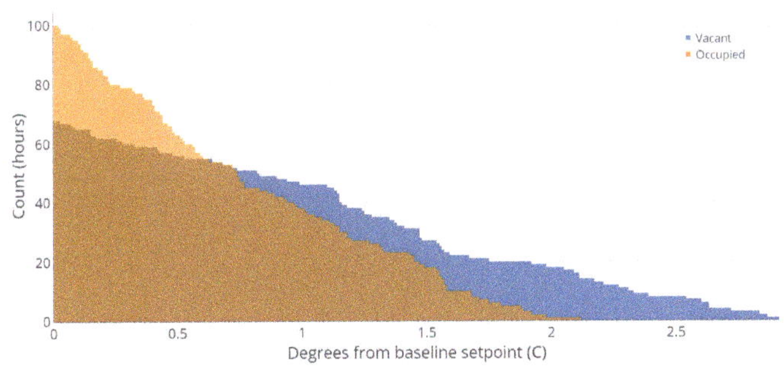

Figure 14. Case 2A duration curve of deviation from 24.5 °C operative temperatures.

5.4.3. MPC Case 2B: Atlanta with Perfect Occupancy Forecasting

In a third scenario, all settings used from Case 2A were kept the same except for the occupancy prediction model. In this scenario, actual occupancy data was used in Houses 1 and 2 to imitate perfect occupancy forecasting. This allows for an exploration of how imperfections in the occupancy prediction impact MPC results. A summary of all used settings are shown in Table 15.

Results from the MPC optimization with perfect occupancy predictions are shown in Table 16. Energy savings for Houses 1 and 2 increased by 5.4% and 2.9%, respectively, while comfort exceedance in both homes decreased (Figure 15). Like Case 2A, setpoint temperatures range from 19 °C to 27 °C. However, unlike Case 2A, low setpoint temperatures are used less often to achieve a quick temperature change. The resulting internal air temperature ranged from 22 °C to 27 °C.

Table 15. Settings used for MPC optimization in Case 2B.

Parameter	Value
City	Atlanta
Season	Summer
Houses	1 & 2
Prediction model	Perfect forecasting
Run period	1 week
Timestep	15 min
Planning horizon	24 h
Execution horizon	1 h
Occupied allowed temperatures	19 °C $\leq T_{optimal,k} \leq$ 27 °C
Unoccupied allowed temperatures	19 °C $\leq T_{optimal,k} \leq$ 27 °C
Temperature increments	0.5 °C
Comfort penalty slope (C)	1000
Optimization time per execution horizon	30 min

Table 16. Results for Case 2B.

House	Energy Savings	Discomfort
1	12.9%	21.0 Kh
2	13.3%	21.0 Kh

This experiment shows that with an accurate occupancy forecast, MPC optimization is able to allow less energy-intensive temperatures during vacant periods, without the penalty from discomfort when an occupant unexpectedly returns. This allows improvements in both energy savings and thermal comfort. Thus, with accurate occupancy forecasting, energy savings above 10% are possible with the occupancy patterns recorded. Accurate occupancy prediction, therefore, is essential in improving HVAC control.

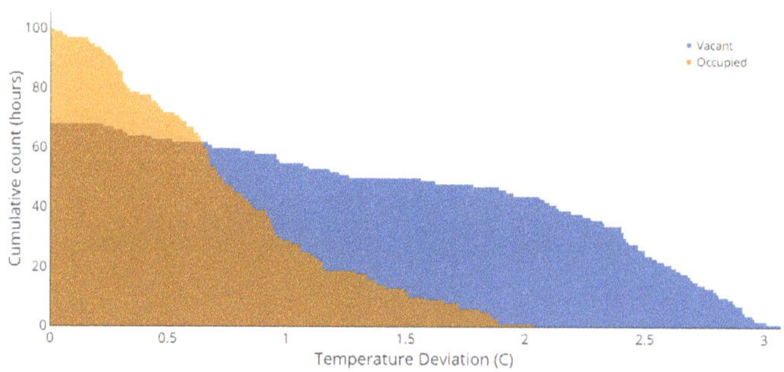

Figure 15. Case 2B duration curve of deviation from 24.5 °C operative temperatures.

6. Summary and Conclusions

Residential heating and cooling accounts for a large portion of annual energy consumption in the United States. Reducing energy use can contribute to furthering the goals of the Paris Agreement by reducing the burning of fossil fuels and thus reducing CO_2 emissions. Within the residential energy sector, energy savings can be realized by accounting for occupancy in HVAC control. This allows energy reduction without negatively impacting thermal comfort of occupants, which is essential to the widespread adoption of new HVAC control technologies.

A literature review revealed that non-probabilistic models historically have performed best for short-term occupancy prediction. By collecting real occupancy data from six different homes, individual non-probabilistic models were created and evaluated. Prediction inaccuracy in the models, termed state-matching error in the study, ranged from 8.0% to 48.7%. Model training data that used a moving, multi-week training set worked the best for all homes, with differences in occupancy patterns being the highest contributor to prediction inaccuracy. An examination of increased training time indicates that models can improve over time as more data is collected and included into the prediction model. Once the occupancy prediction models were generated, they were then incorporated in occupancy-based setpoint control and occupancy-based model predictive control in a building performance simulation. Five EnergyPlus home models were used to simulate the energy use and indoor temperatures for two-week periods in summer and in winter. Occupancy-based setpoint control showed possible energy savings from 0% to 50.0% over control methods that used a constant setpoint temperature depending on climate, occupancy pattern, and control strategy. Non-probabilistic prediction models achieved the highest energy savings, with an average of 10.0%, but with the disadvantage of high thermal discomfort for the occupants. By including an override, in which the occupancy prediction model can sense the actual occupant presence and react to it, thermal discomfort was reduced. In these hybrid occupancy models, the energy savings averaged 5.0%, while the number of hours that the space was deemed uncomfortable were low.

Model predictive control showed that energy savings is highly dependent on how the cost function and constraints are parameterized. In Case 1, where the temperature constraints were much stricter during occupied hours, little to no energy savings was achieved. However, in Case 2A, where temperature constraints were relaxed during occupied hours, energy savings increased with only a slight impact on discomfort. In Case 2B, where occupancy was perfectly predicted, both energy savings and thermal comfort improved, leading to two conclusions: First, a cost function that combines both energy consumption and thermal discomfort allows for flexibility for the user to determine what trade-off between energy savings and discomfort is appropriate for the them. Second, accurate occupancy prediction improves both performance aspects in the cost function. This shows that, as occupancy prediction improves, the ability for occupancy-aware HVAC control to maintain comfort and increase energy savings improves.

Author Contributions: Conceptualization, G.H.; methodology, G.H. and C.T.; software, C.T. and G.P.; validation, C.T., M.J., G.P. and G.H.; formal analysis, C.T.; investigation, C.T.; resources, M.J.; data curation, C.T. and M.J.; writing–original draft preparation, C.T.; writing–review and editing, C.T., M.J., G.P. and G.H.; visualization, C.T.; supervision, G.H.; project administration, G.H. All authors have read and agreed to the published version of the manuscript.

Funding: This research and the APC was funded by DOE ARPA-E grant number DE-AR0000938.

Conflicts of Interest: The authors declare no conflict of interest.

References

1. Tracking Buildings 2020—Analysis. 2020. Available online: https://www.iea.org/reports/tracking-buildings-2020 (accessed on 12 October 2020).
2. Pe, L. A review on buildings energy consumption information. *Energy Build.* **2008**, *40*, 394–398. [CrossRef]
3. Residential Energy Consumption Survey (RECS)-Data—U.S. Energy Information Administration (EIA). 2019. Available online: https://www.eia.gov/consumption/residential/data/2015/ (accessed on 19 September 2019).

4. Sun, K.; Hong, T. A simulation approach to estimate energy savings potential of occupant behavior measures. *Energy Build.* **2017**, *136*, 43–62. [CrossRef]
5. U.S. Energy Information Administration. *Drivers of U. S. Household Energy Consumption, 1980–2009*; U.S Department of Energy: Washington, DC, USA, 2015.
6. Fanger, P.O. *Thermal Confort Analysis and Applications in Environmental Engineering*; Mcgraw-Hill: New York, NY, USA, 1972.
7. ASHRAE. *ASHRAE Standard 55-2010 Thermal Environmental Conditions for Human Occupancy*; ASHRAE Inc.: Atlanta, GA, USA, 2010; Volume 2010, p. 42; ISSN 1041-2336.
8. Malinick, T.; Wilairat, N.; Holmes, J.; Perry, L. Destined to Disappoint: Programmable Thermostat Savings are Only as Good as the Assumptions about Their Operating Characteristics. In *ACEEE Summer Study on Energy Efficiency in Buildings*; American Council for an Energy-Efficient Economy: Washington, DC, USA, 2012; pp. 162–173.
9. Nägele, F.; Kasper, T.; Girod, B. Turning up the heat on obsolete thermostats: A simulation-based comparison of intelligent control approaches for residential heating systems. *Renew. Sustain. Energy Rev.* **2017**, *75*, 1254–1268. [CrossRef]
10. U.S. Department of Energy. *Overview of Existing and Future Residential Use Cases for Connected Thermostats*; Technical Report; Energetics Incorporated, Washington, DC, USA, 2016.
11. Afram, A.; Janabi-Sharifi, F. Theory and applications of HVAC control systems—A review of model predictive control (MPC). *Build. Environ.* **2014**, *72*, 343–355. [CrossRef]
12. Jia, M.; Srinivasan, R.S.; Raheem, A.A. From occupancy to occupant behavior: An analytical survey of data acquisition technologies, modeling methodologies and simulation coupling mechanisms for building energy efficiency. *Renew. Sustain. Energy Rev.* **2017**, *68*, 525–540. [CrossRef]
13. Aftab, M.; Chen, C.; Chau, C.K.; Rahwan, T. Automatic HVAC control with real-time occupancy recognition and simulation-guided model predictive control in low-cost embedded system. *Energy Build.* **2017**, *154*, 141–156. [CrossRef]
14. Chen, Z.; Soh, Y.C. Comparing occupancy models and data mining approaches for regular occupancy prediction in commercial buildings. *J. Build. Perform. Simul.* **2017**, *10*, 545–553. [CrossRef]
15. Lindelöf, D.; Afshari, H.; Alisafaee, M.; Biswas, J.; Caban, M.; Mocellin, X.; Viaene, J. Field tests of an adaptive, model-predictive heating controller for residential buildings. *Energy Build.* **2015**, *99*, 292–302. [CrossRef]
16. Afram, A.; Janabi-Sharifi, F. Supervisory model predictive controller (MPC) for residential HVAC systems: Implementation and experimentation on archetype sustainable house in Toronto. *Energy Build.* **2017**, *154*, 268–282. [CrossRef]
17. Lu, J.; Sookoor, T.; Srinivasan, V.; Gao, G.; Holben, B.; Stankovic, J.; Field, E.; Whitehouse, K. The smart thermostat: using occupancy sensors to save energy in homes. In Proceedings of the 8th ACM Conference on Embedded Networked Sensor Systems—SenSys '10, Zurich Switzerland, 2–5 November 2010; p. 211. [CrossRef]
18. Kleiminger, W.; Mattern, F.; Santini, S. Predicting household occupancy for smart heating control: A comparative performance analysis of state-of-the-art approaches. *Energy Build.* **2014**, *85*, 493–505. [CrossRef]
19. Southern California Edison. *Residential Smart Communicating Thermostat*; Work Paper SCE17HC054 Revision 0; California Public Utilies Commission: San Francisco, CA, USA, 2017.
20. O'Brien, W.; Gaetani, I.; Gilani, S.; Carlucci, S.; Hoes, P.J.; Hensen, J. International survey on current occupant modelling approaches in building performance simulation. *J. Build. Perform. Simul.* **2017**, *10*, 653–671. [CrossRef]
21. Socolow, R.H. The twin rivers program on energy conservation in housing: Highlights and conclusions. *Energy Build.* **1978**, *1*, 207–242. [CrossRef]
22. Al-Mumin, A.; Khattab, O.; Sridhar, G. Occupants' behavior and activity patterns influencing the energy consumption in the Kuwaiti residences. *Energy Build.* **2003**, *35*, 549–559. [CrossRef]
23. Andersen, R. The influence of occupants' behaviour on energy consumption investigated in 290 identical dwellings and in 35 apartments. In Proceedings of the Abstract from 10th International Conference on Healthy Buildings, Brisbane, Australia, 8–12 July 2012; pp. 1–3.

24. Gaetani, I.; Hoes, P.J.; Hensen, J.L. Occupant behavior in building energy simulation: Towards a fit-for-purpose modeling strategy. *Energy Build.* **2016**, *121*, 188–204. [CrossRef]
25. Best Directory|Building Energy Software Tools. 2019. Available online: https://www.buildingenergysoftwaretools.com/ (accessed on 19 September 2019).
26. Duarte, C.; Van Den Wymelenberg, K.; Rieger, C. Revealing occupancy patterns in an office building through the use of occupancy sensor data. *Energy Build.* **2013**, *67*, 587–595. [CrossRef]
27. Page, J.; Robinson, D.; Morel, N.; Scartezzini, J.L. A generalised stochastic model for the simulation of occupant presence. *Energy Build.* **2008**, *40*, 83–98. [CrossRef]
28. Tahmasebi, F.; Mahdavi, A. Exploring the Predictive Potential of Probabilistic Occupancy Models. In Proceedings of the BauSIM, Aachen, Germany, 22–24 September 2014. [CrossRef]
29. Richardson, I.; Thomson, M.; Infield, D. A high-resolution domestic building occupancy model for energy demand simulations. *Energy Build.* **2008**, *40*, 1560–1566. [CrossRef]
30. D'Oca, S.; Fabi, V.; Corgnati, S.P.; Andersen, R.K. Effect of thermostat and window opening occupant behavior models on energy use in homes. *Build. Simul.* **2014**, *7*, 683–694. [CrossRef]
31. Feng, X.; Yan, D.; Hong, T. Simulation of occupancy in buildings. *Energy Build.* **2015**, *87*, 348–359. [CrossRef]
32. Hong, T.; Taylor-Lange, S.C.; D'Oca, S.; Yan, D.; Corgnati, S.P. Advances in research and applications of energy-related occupant behavior in buildings. *Energy Build.* **2016**, *116*, 694–702. [CrossRef]
33. Li, X.; Yao, R. A machine-learning-based approach to predict residential annual space heating and cooling loads considering occupant behaviour *Energy* **2020**, *212*. [CrossRef]
34. Lee, Y.S.; Malkawi, A.M. Simulating multiple occupant behaviors in buildings: An agent-based modeling approach. *Energy Build.* **2014**, *69*, 407–416. [CrossRef]
35. Mahdavi, A.; Tahmasebi, F. Predicting people's presence in buildings: An empirically based model performance analysis. *Energy Build.* **2015**, *86*, 349–355. [CrossRef]
36. Tahmasebi, F.; Mahdavi, A. The sensitivity of building performance simulation results to the choice of occupants' presence models: A case study. *J. Build. Perform. Simul.* **2017**, *10*, 625–635. [CrossRef]
37. Yan, D.; Hong, T.; Dong, B.; Mahdavi, A.; D'Oca, S.; Gaetani, I.; Feng, X. IEA EBC Annex 66: Definition and simulation of occupant behavior in buildings. *Energy Build.* **2017**, *156*, 258–270. [CrossRef]
38. Mahdavi, A.; Taheri, M. An ontology for building monitoring. *J. Build. Perform. Simul.* **2017**, *10*, 499–508. [CrossRef]
39. Speake, A.; Wilson, E.; Salzman, M.; Hasz, A. *Cost Benchmarks for Residential 50% Energy Savings Packages*; National Renewable Energy Lab: Golden, CO, USA, 2012.
40. Oldewurtel, F.; Parisio, A.; Jones, C.N.; Gyalistras, D.; Gwerder, M.; Stauch, V.; Lehmann, B.; Morari, M. Use of model predictive control and weather forecasts for energy efficient building climate control. *Energy Build.* **2012**, *45*, 15–27. [CrossRef]
41. Roy, A.; Bai, L. Control of HVAC System in Residential Buildings using Model Predictive Control: Plant Model Selection. In Proceedings of the IISE Annual Conference, Pittsburgh, PA, USA, 20–23 May 2017; pp. 1538–1544.
42. Široký, J.; Oldewurtel, F.; Cigler, J.; Prívara, S. Experimental analysis of model predictive control for an energy efficient building heating system. *Appl. Energy* **2011**, *88*, 3079–3087. [CrossRef]
43. Bălan, R.; Cooper, J.; Chao, K.M.; Stan, S.; Donca, R. Parameter identification and model based predictive control of temperature inside a house. *Energy Build.* **2011**, *43*, 748–758. [CrossRef]
44. Corbin, C.D.; Henze, G.P.; May-Ostendorp, P. A model predictive control optimization environment for real-time commercial building application. *J. Build. Perform. Simul.* **2013**, *6*, 159–174. [CrossRef]
45. Pavlak, G.S.; Henze, G.P.; Cushing, V.J. Evaluating synergistic effect of optimally controlling commercial building thermal mass portfolios. *Energy* **2015**, *84*, 161–176. [CrossRef]
46. Henze, G.P.; Yuill, D.P.; Coward, A.H. Development of a Model Predictive Controller for Tankless Water Heaters. *HVAC R Res.* **2009**, *15*, 3–23. [CrossRef]
47. Tang, R.; Wang, S.; Xu, L. An MPC-Based Optimal Control Strategy of Active Thermal Storage in Commercial Buildings during Fast Demand Response Events in Smart Grids. *Energy Procedia* **2019**, *158*, 2506–2511. [CrossRef]

48. Woo, D.; Junghans, L. Framework for model predictive control (MPC)-based surface condensation prevention for thermo-active building systems (TABS). *Energy Build* **2020**, *215*. [CrossRef]
49. Coffey, B.; Haghighat, F.; Morofsky, E.; Kutrowski, E. A software framework for model predictive control with GenOpt. *Energy Build.* **2010**, *42*, 1084–1092. [CrossRef]
50. May-Ostendorp, P.; Henze, G.P.; Corbin, C.D.; Rajagopalan, B.; Felsmann, C. Model-predictive control of mixed-mode buildings with rule extraction. *Build. Environ.* **2011**, *46*, 428–437. [CrossRef]
51. Lawrence Berkeley National Laboratory; United States Department of Energy. *EnergyPlus Run Time Analysis*; Lawrence Berkeley National Laboratory: Oak Ridge, TN, USA; Environmental Energy Technologies Division: Berkeley, CA, USA, 2008.
52. Olsson, A.E. *Particle Swarm Optimization: Theory, Techniques and Applications*; Nova Science Publishers: New York, NY, USA, 2011.

energies

MDPI

Article

Efficiency of Energy Consumption between Reinforced Concrete Structure and Cross-Laminated Timber Based Hybrid Structure in East Asian Cities

Meng-Ting Tsai * and Wei-Ting Lin

Department of Architecture, National Taiwan University of Science and Technology, Taipei 10608, Taiwan; timtsai1005@hotmail.com
* Correspondence: Tsai@mail.ntust.edu.tw; Tel.: +886-2-27303212

Abstract: From the environmental perspective, wooden structures are favorable insulators that are suitable for carbon fixation and wooden-related products are considered the most sustainable material. Research has indicated that wooden structures have superior energy-saving performance compared to reinforced concrete (RC) structures. In this study, a CLT-based hybrid structure system that potentially improves the efficiency of energy consumption is proposed. The proposed hybrid structure system, which preserved original RC beams, columns and replaced CLT floors and walls, has less building weight compared to the original RC building. Additionally, less energy required for the manufacturing of building materials in the renovation of the aged building is achieved, compared to building a new CLT building. The energy consumptions for buildings with heights of 10 stories were compared. CLT and RC were selected as benchmark building materials to compare the energy-saving efficiencies with the proposed hybrid structure system. In addition, to examine the energy consumption differences at different latitudes, the energy consumptions in Taipei, Tokyo, Harbin, and Singapore were compared as well. The simulation results indicate the proposed hybrid structure system, which comprises RC beams and columns and CLT floors and walls, and has an energy-saving efficiency close to that of a CLT structure, by approximately 3–5% higher, however, had a superior energy consumption performance to the RC structure. In general, the proposed hybrid structure system can be effectively used for old building renewal in the selected Asian cities.

Keywords: energy consumption; carbon emission; Green Building-Rating System (GBS); cross laminated timber (CLT); hybrid structure system

Citation: Tsai, M.-T.; Lin, W.-T. Efficiency of Energy Consumption between Reinforced Concrete Structure and Cross-Laminated Timber Based Hybrid Structure in East Asian Cities. *Energies* **2022**, *15*, 165. https://doi.org/10.3390/en15010165

Academic Editor: Audrius Banaitis

Received: 28 October 2021
Accepted: 24 December 2021
Published: 27 December 2021

1. Introduction

Due to the environmental problems caused by global warming, energy consumption and carbon emissions must be reduced, and it is the only solution to stop the climate crisis [1]. The topic of environmental protection has drawn the attention of numerous countries. The building density in an area rapidly increases with increasing population. A report of the United Nations Environment Programme [2] indicates that the energy consumption of the construction sector accounts for 40% of the global energy consumption. In addition, the greenhouse gas emissions of the construction sector account for over 33% of the total global emissions, and the construction sector is considered the largest emission source. Consequently, to reduce their environmental impact, construction-related industries should aim to reduce their energy consumption and carbon dioxide emissions in the usage stage of the building life cycle.

The design and construction of wooden structures completely conform to the global trend of green building [3–7]. The material characteristics of cross-laminated timber (CLT), which has been rapidly developed recently, are different from those of traditional glue-laminated timber. CLT can provide high strength when used in vertical and horizontal combinations [8–11]. Theoretically, CLT can be used as the structural material for wooden structures over 10 stories. The fast construction of CLT structures has resulted in the rapid development of CLT structures in recent years [4–6]. From the environmental perspective, wooden structures are favorable insulators that are suitable for carbon fixation and wooden-related products are considered the most sustainable material [12]. From the Building Envelop point of view, the thermal conductivity of concrete is 1.4 w/m°k, which is 10 times the conductivity of wood. In addition, concrete has high thermal storage. In countries in cold zones, concrete walls are often placed in the direction of sunlight for thermal storage due to their high thermal storage characteristics. At night, the heat stored in concrete structures is released indoors; thus, the heating load in winter can be reduced [13]. However, in subtropical and tropical regions, the high thermal storage of concrete is one of the major reasons for increased air-conditioning consumption in the summer. Wood is a favorable insulator, and for the same thickness, the insulation value of wood is approximately 10 times that of concrete. Thus, wood can effectively block thermal conduction. Research has indicated that wooden structures have superior energy-saving performance to reinforced concrete (RC) structures [14–18]. Based on the analyses performed, it was indicated that wood-based buildings have a favorable energy balance in comparison to traditional buildings based on traditional materials [19]. Recent research concluded that CLT generally provides a significant improvement on energy efficiency, and studies on occupants' comfort in buildings have been examined [20,21]. However, its energy performance efficiency and occupants' comfort can be affected by weather, building size, internal loading, and HVAC control. Generally, CLT buildings provide sound insulation against cold weather as when the temperature outside is cold, the occupants can still find comfort in the house. The daily energy consumption of structures, such as their electricity generation and fuel consumption, influences the overall development direction of energy consumption for a country. Consequently, reducing energy consumption or adopting recyclable energy and thus reducing the overall energy consumption cost is a goal of many countries [22–25]. However, previous research has mostly focused on comparing the energy-saving efficiency between wooden and RC structures, as well as in high-latitude regions, energy-saving efficiency between hybrid structure systems and CLT or RC structures [14–18], as well as the performance of energy consumption in low-latitude regions, has not been focused.

For most East Asian cities, taking Taipei City as an example, RC is the primary building material for most structures. According to statistics, close to 400,000 residences were constructed between 1971 and 1990 in Taipei City [26]. These structures have a life expectancy of 30–50 years and currently require reconstruction or renovation. Thus, for urban renewal, a crucial aspect of this study is to understand how to use the advantages of wooden structures to develop efficient renewal models of energy consumption for the daily usage stage of the building life cycle. To propose a solution for the renovation of aged RC buildings, a new hybrid structure system that potentially improves the efficiency of energy consumption is proposed in this study. Hybrid structure systems by adopting CLT panels outside of original wall systems as a building's additional envelope has been studied [27–30]. However, adding the additional CLT panel increased the total weight of the building, potentially increasing the seismic load of the aged buildings, which had reduced seismic resistance ability, due to the decay of the materials. In this study, a renovation method that preserves the RC beams and columns in old buildings and uses CLT floors and walls for renewal was proposed. Replacing the original RC wall and floor with a CLT panel reduces the total weight of the building, hence helping reduce the potential seismic load [31,32]. Furthermore, the energy required for the manufacturing of building materials could represent in the near future, almost 400% of operational energy [33], it could cause less energy consumption if only part of the elements, such as wall and floor were replaced in the aged building instead of rebuilding a new building. The other objective of this study was to compare the energy-saving trends of these structures of different heights from high-latitude regions to low-latitude regions. The energy consumptions for buildings with heights of 4 and 10 stories were compared. CLT and RC were selected as benchmark building materials to compare the energy-saving efficiencies with hybrid structure systems.

Consequently, the Revit program was used with Green Building Studio (GBS) for simulating and analyzing the energy consumption of structures. The common residence structure type in Taipei City was selected, and its floor plan was used as the standard floor. The energy consumption efficiency of the hybrid structure system was compared with those of the other structure systems. Thus, a system with superior energy consumption efficiency could be identified for future urban renewal. In addition, to examine the energy consumption differences at different latitudes, the energy consumptions in Taipei City, Taiwan (taken as the standard); Tokyo, Japan (to the north of Taipei); Harbin city, China (to the north of Taipei), and Singapore (to the south of Taipei), which is close to the equator, were compared. Thus, the energy-saving efficiencies of wooden and RC structures in regions with different environmental conditions were compared.

2. Methodology

2.1. Assessment of Energy Consumption

Autodesk Green Building-rating System (GBS) was used for simulating the structure energy consumption in this study. GBS can be used as an independent cloud-service-based program or a plug-in component of the Revit program for energy analysis. GBS comprises the DOE-2.2 analysis core and can provide extremely detailed analysis. As a cloud-based tool, GBS can facilitate rapid computation on the Autodesk server. In general, the DOE-2.2 analysis core requires extremely detailed information on the building envelope and electromechanical system for computation. However, GBS presets numerous building envelope and electromechanical system parameters according to the ASHRAE standard. Thus, architects can focus more on the design factors that have decisive influences on the overall energy consumption of buildings and can ignore technical details. In addition to the building energy consumption, electricity consumption, and annual carbon emissions, GBS can calculate the Energy Star score of buildings. It can also evaluate the glass property and water usage efficiency scores according to the LEED evaluation system published by the U.S. Green Building Council. GBS can even determine the solar energy usage potential. In this study, a plug-in component of the Revit program 2019 version for energy analysis was used for simulation.

2.2. Assessment of Carbon Dioxide Emission

The Guidelines for National Greenhouse Gas Inventories published by the Intergovernmental Panel on Climate Change were used for the assessment of carbon dioxide emissions. In this study, the calculation method for the emissions of carbon dioxide, methane, and nitrous oxide was adopted to calculate the carbon emissions of different countries. GBS calculates the energy consumption of buildings according to two major parameters: electricity consumption and fuel consumption. The calculated carbon emission of electricity consumption differs according to the electricity carbon emission coefficients of different countries. The carbon emissions can be calculated using Equation (1). The emission coefficients of different countries are listed in Table 1. For calculating the carbon emission of fuel consumption, the fuel volume (m^3) is first converted into energy units. An energy of 38 MJ or 10.6 kWh can be generated by burning 1 m^3 of natural gas. The carbon emission of fuel consumption can then be calculated using Equation (2). The carbon emissions coefficient of fuel is listed in Table 2, and all the compared countries have the same carbon emissions coefficient of fuel. The carbon emissions of the total energy consumption of a building can be obtained by summing the carbon emissions of electricity and fuel.

Table 1. Electricity carbon emission coefficients of different countries.

Country	Electricity Carbon Emission Coefficient ($kg\ CO_2e/kWh$)
China (Harbin)	1.13 [34]
Japan (Tokyo)	0.47 [35]
Taiwan (Taipei and Kaohsiung)	0.53 [36]
Singapore	0.41 [37]

Table 2. Fuel carbon emission coefficient of all the compared countries.

Item	Carbon Emission Coefficient (CO_2e/m^3)
Natural gas	1.88 [38]

(A) Electricity consumption is converted into carbon emissions by using the following formula:

$$\begin{aligned} &\text{Electricity usage (kWh)} \times \text{electricity emissions coefficient (kg } CO_2e/\text{kWh)} \\ &= \text{electricity carbon dioxide emissions (kg)} \end{aligned} \quad (1)$$

(B) Fuel consumption is converted into carbon emissions by using the following equation:

$$\begin{aligned} &\text{Fuel usage } (m^3) \times \text{natural gas emissions coefficient (kg } CO_2e/\ m^3) \\ &= \text{fuel carbon dioxide emissions (kg),} \end{aligned} \quad (2)$$

(C) The carbon emission of total energy consumption is calculated as follows:

$$\begin{aligned} &\text{Carbon emission of total energy consumption (kg)} \\ &= \text{electricity carbon dioxide emissions (kg)+ fuel carbon dioxide emissions (kg)} \end{aligned} \quad (3)$$

3. Simulation Modeling

To determine the energy usage efficiencies under different conditions, the energy consumption and carbon emission were compared for different numbers of stories (4 and 10), different construction materials (RC and CLT), and cities with different latitudes (from north to south, Harbin, Tokyo, Taipei, and Singapore). The related structure usage situations and air admission timing of the air-conditioning system were set. In addition, the energy simulation was only conducted for the daily usage stage in the structure life cycle. When GBS was used for structure energy simulation, the basic settings, simulation parameters, weather data, electromechanical system, indoor load, and operation schedules had to be input in the simulation process. The basic parameters and settings of this study are as follows:

1. According to the descriptions on the official website of Autodesk, the data on weather stations were obtained from the World Meteorological Organization.
2. For concrete materials, the pre-existing data in the program were used. The parameters of CLT walls were obtained from relevant research [39,40].
3. For the electromechanical system and indoor load, detailed data were required from the DOE-2.2 analysis core to the electromechanical system for the operation. However, the building envelope and electromechanical system parameters were preset in GBS according to the ASHRAE standards. Different air-conditioning systems are used in different countries. The preset parameters were used for the mechanical system and indoor load in this study. The preset parameters of heating, ventilation, and air-conditioning (HVAC) systems were a central variable air volume system, hot-water heating, a performance coefficient of 5.96 for the freezer, and a boiler efficiency of 84.5.

Because this study was a preliminary study, a 24/7 operation schedule was set. Thus, simulations were conducted 7 days a week and 24 h a day annually. Then, GBS was used to analyze the simulated building energy consumption. In the comparison of the energy consumptions of different building materials (RC and CLT) in cities at different latitudes, only the energy consumption in the operation stage was considered. In addition, the following assumptions were made in the simulation process for the energy consumption:

1. The window positions remained the same when switching from an RC structure to a CLT structure; thus, the illumination demands remained the same.
2. Except for the balcony, the indoor temperatures of all the rooms were controlled between 18 and 26 °C.
3. No heating or cooling was conducted in the stair areas.
4. Electricity was used for the air-conditioning system and illumination, and fuel was used for heating.

3.1. Target Building

To determine the energy usage efficiencies for different numbers of stories and different weather conditions, the standard floor of social housing in Taipei City was used as the standard floor in this study. The layouts and basic information of a four-story building and a 10-story building are listed in Table 3 and displayed in Figure 1.

Table 3. Basic information on the target buildings.

Building Type		**4F**	**10F**
Single-story floor area		192.8 m^2	702.6 m^2
Total floor area		771.2 m^2	7026.0 m^2
floor-to-ceiling height	others	3.0 m	3.2 m
	Ground level	3.2 m	3.6 m
Total surface area		1161.2 m^2	3906.6 m^2
Exterior window ratio		16.4%	19.9%
User number per unit area		3 people/100 m^2	3 people/100 m^2
Total user number		23 people	211 people
Average illumination power		6.6 W/ m^2	6.6 W/ m^2

(**a**) four-story building (**b**) 10-story building

Figure 1. Standard floor plans of the target buildings.

3.2. Building Materials

The targeted building that was selected was a typical social housing building surrounded by buildings in central Taipei city, and the orientation of the building is with the main entrance facing southward. This building type is adopted and simulated in other selected cities, and the surrounding condition of the building is assumed the same. The simulated building was an RC structure with fixed floor plans on different floors. In this study, the building height, building direction, total area, and opening, such as windows and doors were fixed for the building simulation. The material of major structures, such as beams and walls, was replaced with CLT. Thus, the influence of the floor height on energy-saving efficiency was examined. In addition, the energy consumption efficiencies of the RC and CLT structures were compared. Moreover, only the energy consumption in the daily life stage of the building was examined; the structural properties of the structure were not determined.

The physical properties of RC and CLT, including their heat transfer coefficient, specific heat, and density, were also determined for complete analysis. As presented in Table 4, RC has a higher heat transfer coefficient, specific heat, and density than CLT does. Thermal resistance and heat loss are inversely correlated. Thus, an increase in the thermal resistance of the wall material between the interior and exterior of a building can reduce the heat loss of the building (Equation (4)). As presented in Table 4, CLT walls have a higher thermal resistance than RC walls do. The thermal resistance of a 300-mm-thick external CLT wall is up to 3.3 $m^2 \cdot K/W$ (U-Value = 0.3 $W/m^2 \cdot K$), while with a variable thickness of insulation applied, the U-Value ranges from 0.13 to 0.45 $W/m^2 \cdot K$ [41]. The thermal resistance of a 300-mm-thick RC external wall is only 0.25 $m^2 \cdot K/W$, which is approximately 1/13th of the thermal resistance of a 300-mm-thick CLT external wall. Table 5 lists the physical properties of RC and CLT walls of different thicknesses. Figure 2 displays the models of numerical analysis. Four- and Ten-story CLT and RC buildings were constructed for comparing the energy consumptions in their daily life stages. In this paper, the buildings in different cities with different latitude were studied, to emphasize the influence of the building envelope on total energy consumption, the reflected U-Values of different cities for their roof and floor are assumed as follow, based on the different climate zone located. U-value of the roof was 0.28, 0.45, 0.50, 0.50 for Harbin, Tokyo, Taipei, and Singapore respectively. Moreover, the U-value of the floor was 0.38, 0.50, 0.80, 1.50 for Harbin, Tokyo, Taipei, and Singapore respectively. In order to simplify the simulation, the U-value of the window and door were set to be the same in order to clarify the major influence of the exterior wall.

$$\text{Heat loss} = (A/R) \times (\text{Tindoor} - \text{Toutdoor}) \tag{4}$$

A: external surface area of the building.
R: thermal resistance (R value).
Tindoor: indoor air temperature.
Toutdoor: outdoor air temperature.

Table 4. Physical properties of different walls.

Wall Material	Wall Purpose	Total Thickness (mm)	Heat Transfer Coefficient $W/(m^2 \cdot K)$	Thermal Resistance $(m^2 \cdot K)/W$	Thermal Mass kJ/K
RC	Exterior wall	150	3.87	0.25	60.98
	Interior wall	120	3.46	0.29	22.59
CLT	Exterior wall (10F)	300	0.30	3.3	4.66
	Exterior wall (4F)	215	0.38	2.59	3.91
	Interior wall	150	0.59	1.67	2.48

Table 5. Materials of different wall types.

Wall Material	Wall Purpose	Total Thickness (mm)	Wall Schematic	Interior Material	Thickness (mm)
RC	Exterior wall	150	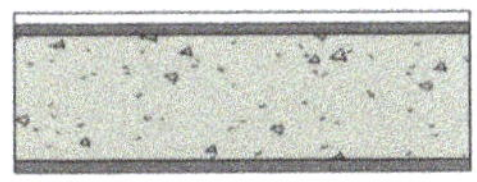	Ceramic tile Cement mortar Concrete Cement mortar	10 10 120 10
	Interior wall	120	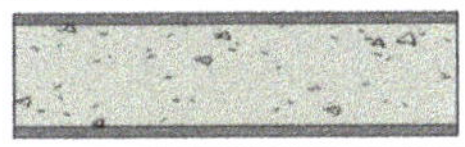	Cement mortar Concrete Cement mortar	10 100 10
CLT	Exterior wall (10F)	300	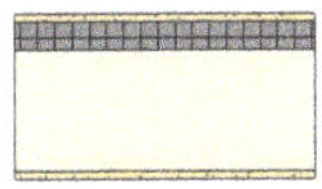	Plasterboard Rigid insulation wall CLT Plasterboard	15 50 220 15
	Exterior wall (4F)	215	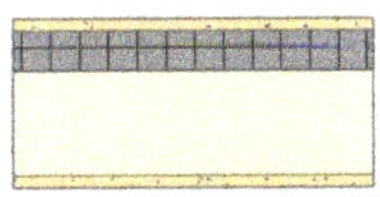	Plasterboard Rigid insulation wall CLT Plasterboard	15 50 135 15
	Interior wall	150	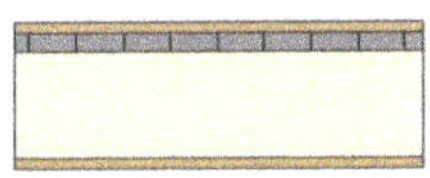	Plywood Rigid insulation wall CLT Plywood	10 20 110 10

3.3. Cities in Different Latitude

Cities at different latitudes have different climates, which influences the energy consumption of buildings. The climate includes the highest and lowest outdoor temperatures and humidity. In this study, the energy consumption of buildings in cities at different latitudes was also simulated to determine how the overall energy usage efficiency varied with the latitude. As depicted in Figure 3, the latitude of Taipei was selected as the standard. The other cities selected for the comparison were Tokyo and Harbin, which are located to the north of Taipei, as well as Singapore, which is located close to the equator and to the south of Taipei. These cities were selected for comparing the energy-saving efficiencies of RC and CLT buildings of different heights in different environmental conditions.

The monthly average temperatures of the aforementioned cities in 2019 are presented in Table 6 and Figure 4. The lowest monthly average temperature in Harbin, which is located at a high latitude, was −17.2 °C in January, and the highest monthly average temperature in Harbin was 23.6 °C in July. Thus, the largest monthly average temperature difference was 40.8 °C in Harbin. The lowest monthly average temperatures in Singapore, which is located at a low latitude, were 26.8 °C and 27 °C in December, and the highest monthly average temperatures in Singapore were 30.6 °C and 28.9 °C in April and March, respectively. The highest monthly average temperature difference in Singapore was only 3.8 °C. The highest monthly average temperatures of Tokyo and Taipei, which are located in the temperate zone, were 28.6 °C in August and 30.6 °C in July, respectively. These temperatures were comparable to the high temperatures in Singapore, which is located in the tropics.

Table 6. Weather information in the considered cities.

City	Temperature (°C)		Relative Humidity (%)		Wind Speed (m/s)		Radiation (Wh/m^2)	
	Hottest	Coldest	Highest	Lowest	Highest	Lowest	Highest	Lowest
Harbin	33	−30	81	45	4.2	1.9	209	106
Tokyo	34	−1	77	52	3.9	2.6	198	132
Taipei	35	7	84	76	4.6	2.2	166	44
Singapore	34	24	88	82	3.9	1.2	96	49

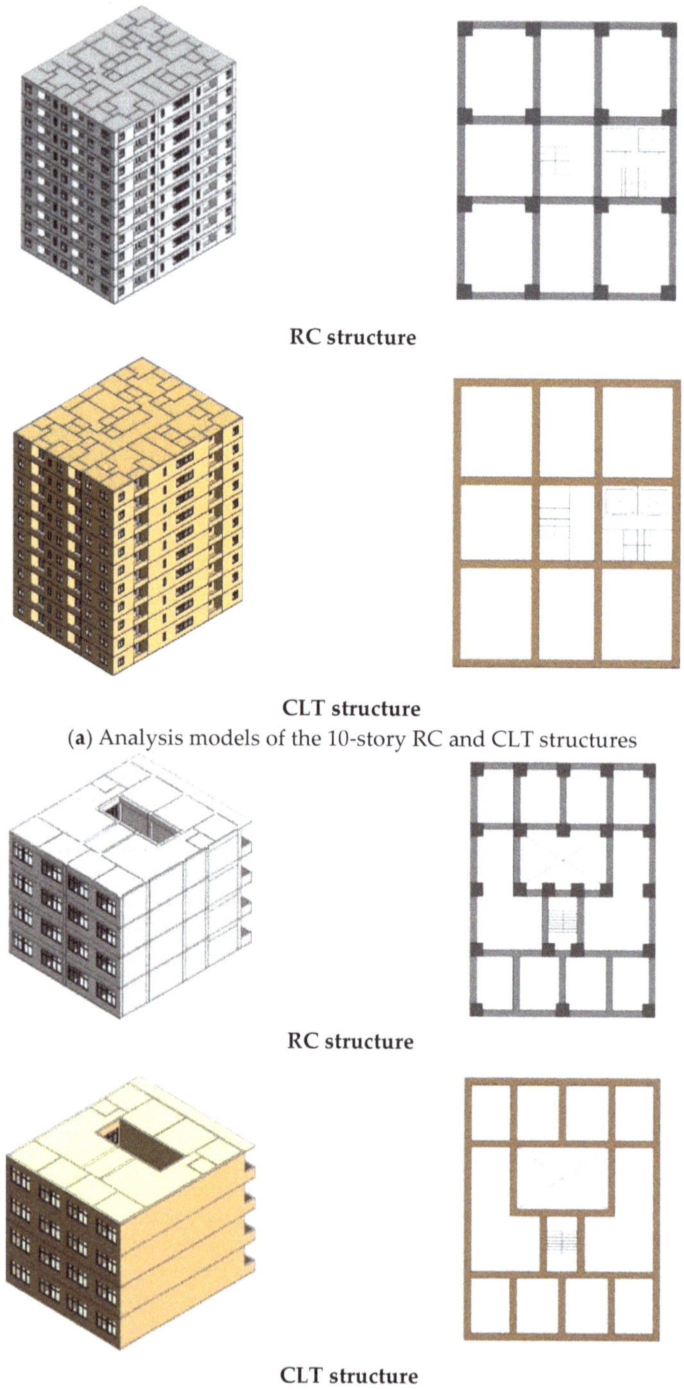

Figure 2. Analysis models of numerical simulation.

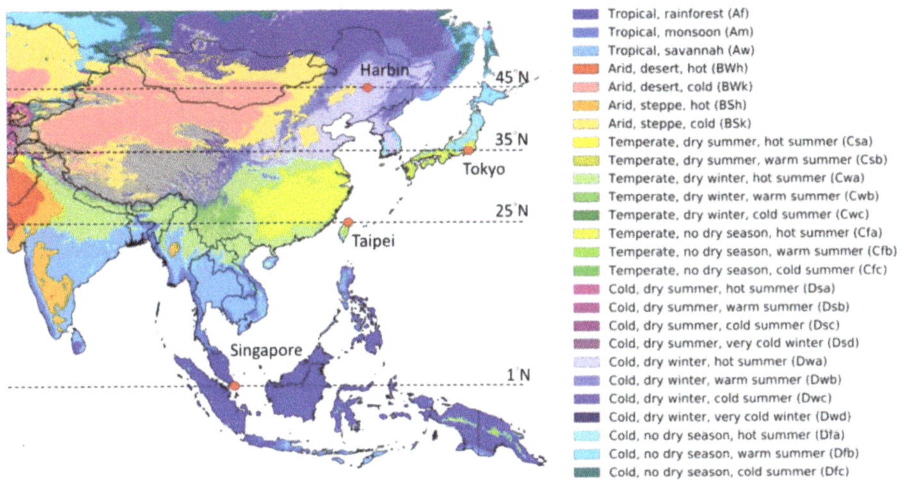

Figure 3. Latitudes of the considered cities.

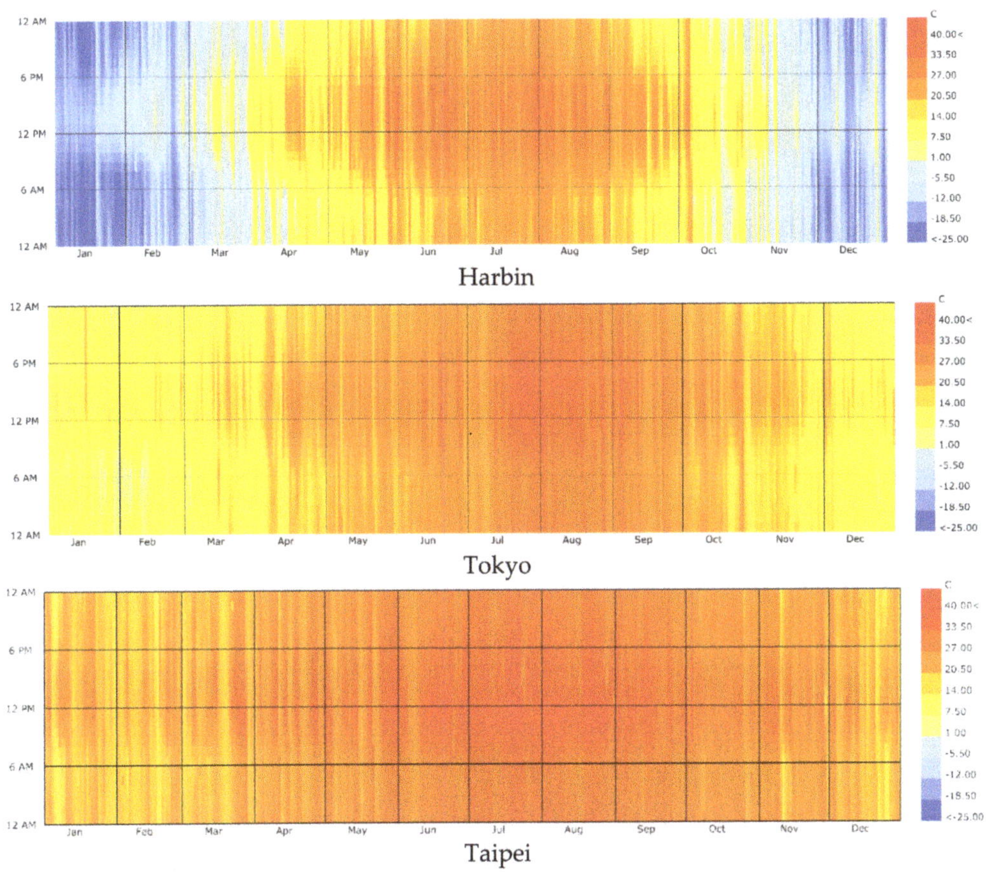

Figure 4. *Cont.*

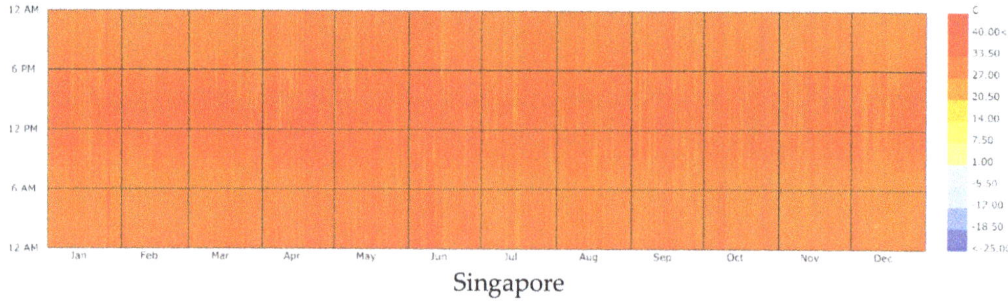

Figure 4. Annual temperatures of the considered cities.

4. Results and Discussion

4.1. Energy Consumption

In the simulation, electricity was used by the HVAC system, illumination equipment, and other equipment. Fuel was used by the HVAC and domestic water heating systems. Table 7 presents the simulation results for the 4- and 10-story structures made of different building materials and located in cities at different latitudes.

Table 7. Energy consumption comparison of structures with different heights in different cities.

		Harbin		Tokyo		Taipei		Singapore	
		RC	**CLT**	**RC**	**CLT**	**RC**	**CLT**	**RC**	**CLT**
(A) Electricity (kWh/m^2/yr)	4 storys	244	148	165	148	175	179	232	227
	10 storys	195	105	143	119	157	151	202	190
(B) Electricity (MJ/m^2/yr)	4 storys	877	534	594	533	644	631	833	818
	10 storys	702	377	517	426	566	543	726	682
(C) Fuel (MJ/m^2/yr)	4 storys	4021	2251	1597	1020	761	603	390	389
	10 storys	3075	1917	1304	899	701	583	382	382
Sum (B) + (C) (MJ/m^2/yr)	4 storys	4898	2785	2191	1553	1392	1247	1223	1207
	10 storys	3777	2294	1821	1325	1267	1126	1108	1064

For Harbin, which is located at the highest latitude among the considered cities, the annual electricity consumptions of the RC and CLT structures were 244 and 148 kWh/ m^2, respectively. For Singapore, which is at the lowest latitude, the annual electricity consumptions of the RC and CLT structures were 232 and 227 kWh/ m^2, respectively. For both 4- and 10-story structures, the simulation results indicated that the electricity consumption of RC structures was considerably higher than that of CLT structures at high and low latitudes. However, at low latitudes, the difference in the electricity consumption of RC and CLT structures was relatively small (Figure 5). The same trend was observed for fuel energy usage. Thus, the energy-saving efficiencies of CLT structures were higher than those of RC structures. The differences in the energy-saving efficiencies of CLT and RC structures were higher at higher latitudes (Figure 6).

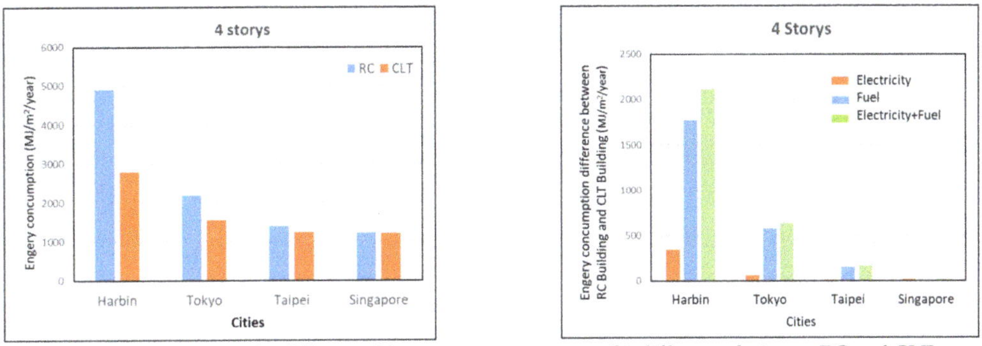

(**a**) Energy consumption in different cities (**b**) difference between RC and CLT

Figure 5. Energy consumption of 4 story building in different cities.

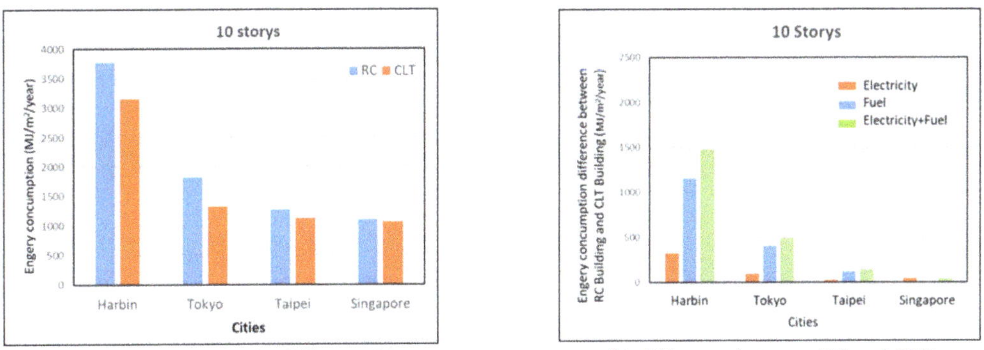

(**a**) Energy consumption in different cities (**b**) difference between RC and CLT

Figure 6. Energy consumption of 10 story buildings in different cities.

A comparison of the total energy consumptions of RC and CLT structures is presented in Table 8 and Figure 7. For four-story structures, the total energy consumptions of the CLT structures were approximately 57%, 71%, and 88% of those of the RC structures in Harbin, Tokyo, and Taipei, respectively. No significant difference was observed in the total energy consumptions of the four-story CLT and RC structures in Singapore. For 10-story structures, the total energy consumptions of the CLT structures were approximately 61%, 73%, and 89% of those of the RC structures in Harbin, Tokyo, and Taipei, respectively. For Singapore, the difference between the total energy consumptions of the 10-story CLT and RC structures was marginally higher than that of the four-story CLT and RC structures.

Table 8. Energy consumption Ratio between RC Building and CLT Building.

		Harbin	Tokyo	Taipei	Singapore
Electricity (Ratio of CLT/RC)	4 storys	0.61	0.90	0.98	0.98
	10 storys	0.54	0.82	0.96	0.94
Fuel (Ratio of CLT/RC)	4 storys	0.56	0.64	0.79	1.00
	10 storys	0.62	0.69	0.83	1.00
Total Consumption (Ratio of CLT/RC)	4 storys	0.57	0.71	0.88	0.99
	10 storys	0.61	0.73	0.89	0.96

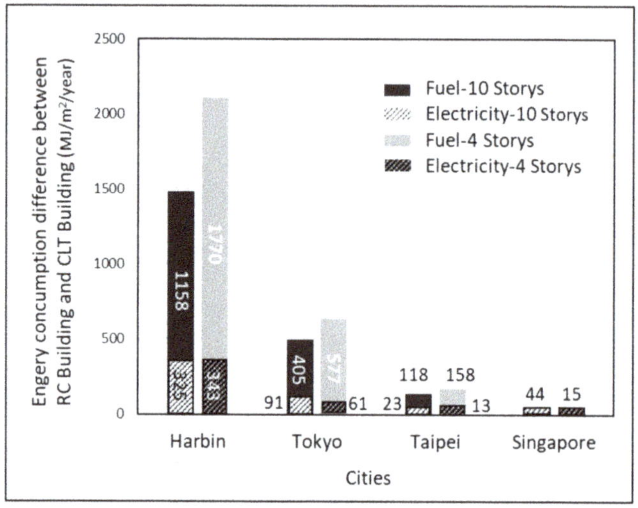

Figure 7. Comparison of different types of energy consumption in different cities.

The electricity and fuel energy consumption ratios of the CLT and RC structures were compared. Electricity consumption mainly originates from rooms with air-conditioning. Thus, a higher number of stories caused a lower ratio of the floor area being directly heated by sunlight. Consequently, the electricity-energy-saving efficiency of the 10-story structures was higher than that of the four-story structures (Figure 7). However, this energy-saving efficiency decreased with decreasing latitude. The consumption of fuel energy mainly originates from rooms with heating. Thus, the larger the surface area of a structure, the higher is the heat exchange. Table 8 and Figure 7 indicate that all the fuel energy consumption ratios of 10-story structures were higher than those of four-story structures in the different cities. In addition, CLT structures at high altitudes (Harbin) had a high fuel energy consumption efficiency. The energy consumption ratios of 4- and 10-story CLT structures were approximately 56% and 62% of those of 4- and 10-story RC structures, respectively. For Singapore, which is at a low latitude, indoor heating demands were low. Thus, the fuel energy consumption ratios did not exhibit a significant difference.

4.2. CO_2 Emission

Table 9, Figures 8 and 9 present the carbon dioxide emissions per unit area for 4- and 10-story RC and CLT structures in cities at different latitudes. The carbon emissions of electricity energy consumption mainly originate from air-conditioning systems, illumination systems, and basic facilities of the structure. The carbon emissions of fuel energy consumption mainly originate from the use of heating systems. Electricity carbon emissions were calculated according to the electricity emission coefficients of the countries in which the considered cities are located (Tables 1 and 2). Thus, the carbon emissions from the same electricity energy consumption were different in different countries.

The electricity carbon emissions of four-story RC and CLT structures were 276 and 167 kg/ m^2·yr, respectively, in Harbin; thus, the difference in the electricity carbon emissions of the two types of four-story structures was 109 kg/ m^2·yr in Harbin. In Singapore, the electricity carbon emissions of four-story RC and CLT structures were 96 and 93 kg/ m^2·yr, respectively, which represents an electricity carbon emission difference of 3 kg/ m^2·yr. The difference in carbon emissions was small because the monthly average temperature in every month in Singapore was higher than 26 °C, which was the temperature set in this study for the air-conditioning system to be turned on. The air-conditioning system demands were high. Thus, the demands for carbon emissions from electricity energy consumption were high for both RC and CLT structures. For Harbin, in addition to the air-conditioning system demands, the heating system demands were high. Thus, significant differences were observed in the carbon emissions of electric energy consumption for RC and CLT structures in Harbin. In addition, the electricity emissions of Harbin were higher than those in other cities due to the higher electricity carbon emission coefficient in Harbin (1.13 kg CO_2e/kWh). The carbon emissions of fuel energy consumption for four-story RC and CLT structures in Harbin were 199 and 111 kg/ m^2·yr, respectively, which represents a fuel carbon emission difference of 88 kg/ m^2·yr. In Singapore, the fuel carbon emissions of four-story RC and CLT structures were both 12 kg/ m^2·yr. In cities at higher latitudes, the heating system needs were higher. Thus, the fuel carbon emissions and fuel carbon emission differences increased considerably with the latitude.

Table 9. Carbon emissions comparison of structures with different heights in different cities.

		Harbin		Tokyo		Taipei		Singapore	
		RC	CLT	RC	CLT	RC	CLT	RC	CLT
(A)Electricity (kg/m^2/yr)	4 storys	276	167	79	71	95	93	96	93
	10 storys	220	119	69	57	84	81	83	78
(C)Fuel (Kg/m^2/yr)	4 storys	199	111	79	50	38	30	12	12
	10 storys	152	95	69	65	35	29	12	12
Sum (A)+(B) (Kg/m^2/yr)	4 storys	474	278	158	121	133	123	108	105
	10 storys	372	213	133	102	118	109	95	90

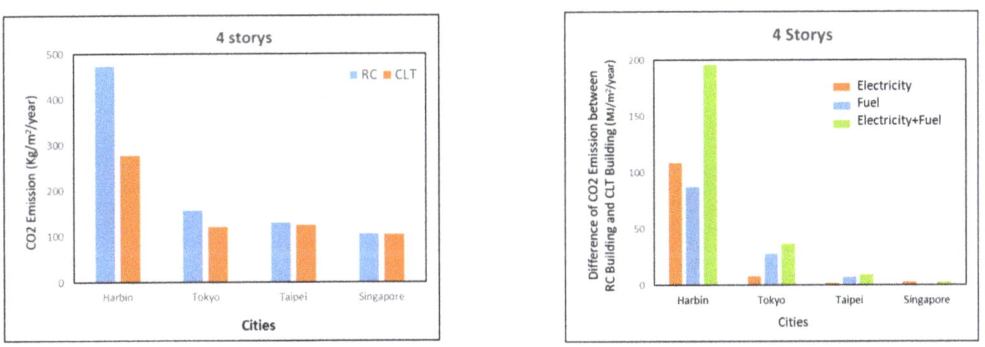

(**a**) Carbon emissions in different cities (**b**) Difference between RC and CLT

Figure 8. Carbon emissions of 4 story building in different cities.

Figure 10 indicates that the 10-story structures had superior carbon emission reduction effects to those of the four-story structures for electricity and fuel energy consumption. For regions at high latitudes (Harbin), the differences in electricity carbon emissions of 10- and 4-story structures were not significant. However, the carbon emissions reduction efficiencies of structures with more stories could be shown in the differences in fuel carbon emissions.

4.3. Potential Hybrid Structure System for Renovation in Asian Cities

Previous results of this study indicate that CLT structures have higher energy-saving and carbon reduction efficiencies than RC structures do at different latitudes. These efficiencies increase with the number of floors. Moreover, the differences in the energy consumption efficiencies of CLT and RC structures increase with latitude. In this section, CLT and RC structures that were studied as benchmark building materials were furthermore compared with the energy-saving efficiencies of a hybrid structure system.

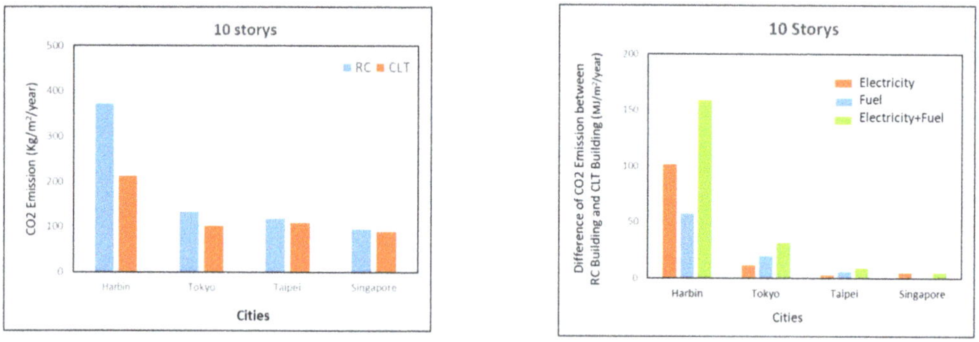

(**a**) Carbon emissions in different cities

(**b**) Difference between RC and CLT

Figure 9. Carbon emissions of 10 story building in different cities.

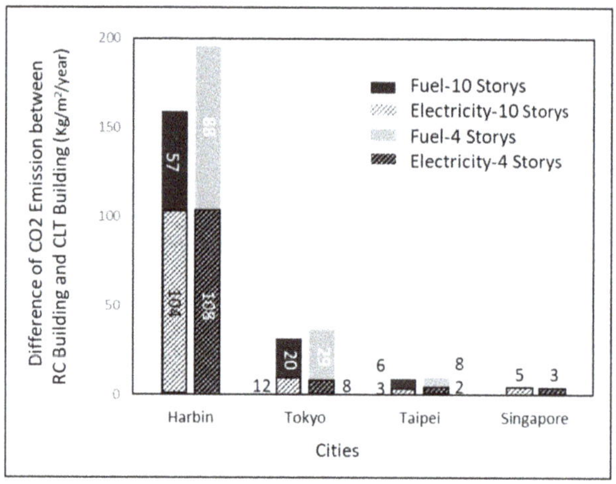

Figure 10. Comparison of different type of Carbon emissions in different cities.

In Taipei, most buildings are RC structures that are approximately 30–50 years old. These structures are still usable; however, their overall energy consumption is high because of the RC building material. Consequently, a strategy is proposed for the renewal of these buildings in the future. The proposed strategy mainly involves preserving existing RC beams and columns and renewing the floors and walls with CLT. On the basis of this concept, a hybrid structure analysis model was established in this study. Figure 11 illustrates the model of preserving the RC beams and columns and replacing the floors and walls with CLT. The energy consumptions of CLT and RC structures of the same size were also compared. The analysis and simulation conditions were the same as stated previously as shown in Section 3.

The analysis results presented in Table 10 indicate that the 10-story hybrid structure system had a superior electricity energy consumption performance to the RC structure. The electricity energy consumption performance of the hybrid structure was only marginally worse (3–5% lower) than that of the CLT structure. Thus, the proposed hybrid structure system is close to a CLT structure in terms of electrical energy consumption. For fuel energy consumption, no significant difference was observed between the different structures in Singapore, which is located at a low latitude. In Taipei, Tokyo, and Harbin, which are located at relatively high latitudes, the fuel energy consumption of the hybrid structures was higher than that of the RC structures and not significantly different from that of the CLT structures. The CLT and hybrid structures exhibited no significant difference in their total energy consumption. The aforementioned structures exhibited lower total energy consumptions than the RC structures did. The energy-saving efficiency of the proposed hybrid structure system, which comprises RC beam structures and CLT floors and walls, is close to that of CLT structures. Taking energy consumption in RC structure as 100%, the relative consumption ratio is illustrated and as shown in Table 11, indicating that using CLT as building skin, such as hybrid structure proposes in this study, performs as well as the CLT structure. For all the East Asian cities selected in this study, the proposed hybrid structure system can be used effectively for old building renewal.

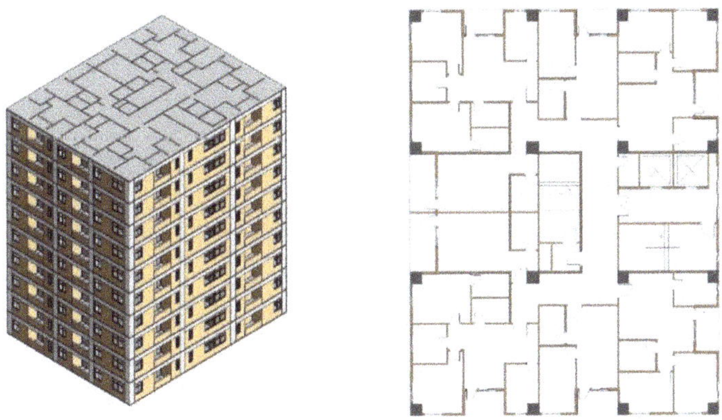

Figure 11. Analysis models of hybrid structure system.

Table 10. Energy consumption comparison of hybrid structures with RC and CLT structures in different cities.

	Harbin			Tokyo			Taipei			Singapore		
	RC	**CLT**	**Hybrid**	**RC**	**CLT**	**Hybrid**	**RC**	**CLT**	**Hybrid**	**RC**	**CLT**	**Hybrid**
(A) Electricity ($MJ/m^2/yr$)	702	377	389	517	426	436	566	543	549	726	682	692
(B) Fuel ($MJ/m^2/yr$)	3075	1917	1874	1304	899	882	701	583	570	382	382	382
Sum (A) + (B) ($MJ/m^2/yr$)	3777	2294	2263	1821	1325	1318	1267	1126	1119	1108	1064	1074

Table 11. Comparison of energy consumption ratio of hybrid structures with RC and CLT structures in different cities.

City	RC Structure Ratio (RC/RC)	Hybrid Structure Ratio (Hybrid/RC)	CLT Structure Ratio (CLT/RC)
Harbin			
Tokyo			

Table 11. *Cont.*

City	RC Structure Ratio (RC/RC)	Hybrid Structure Ratio (Hybrid/RC)	CLT Structure Ratio (CLT/RC)
Taipei	100	88	88
Singapore	100	97	96

5. Conclusions

In this study, the energy consumption efficiency of the proposed hybrid system was compared with those of the RC and CLT structure systems. Detailed energy consumption and carbon emissions were compared for different numbers of floors (4 and 10), different building materials (RC and CLT), and cities at different latitudes (from north to south, Harbin, Tokyo, Taipei, and Singapore) to understand the energy usage efficiencies under different conditions. In order to clarify the major influence of exterior walls on energy consumption, part of the parameters for the simulation model are simplified, such as doors and windows, further detailed study needs to be conducted in order to obtain a deep understanding. For the preliminary study, the following conclusion is drawn, and worthy to provide to local authorities for the policymaking regarding urban renewal issues. For the proposed hybrid structure system, the electricity consumption performance of 10-story hybrid structures was superior to that of RC structures but marginally inferior to that of CLT structures (by approximately 3–5%). Thus, the electrical energy consumption of the hybrid structures was close to that of the CLT structures. No significant differences were observed in the fuel energy consumptions of the different structures in Singapore. In Taipei, Tokyo, and Harbin, which are located at higher latitudes than Singapore is, the fuel energy consumption of the hybrid structures was higher than that of the RC structures. No significant difference was observed in the fuel energy and total energy consumptions of the hybrid structure system and CLT structures. The fuel energy and total energy consumptions of the aforementioned structures were lower than those of the RC structures, indicating the advantage of the hybrid structure system used for old building renovation. In conclusion, the proposed hybrid structure system, which comprises RC beams, columns and CLT floors and walls, has less building weight compared to the original RC building, and less energy required for the manufacturing of building materials [31] in the renovation of the aged building, has an energy-saving efficiency close to that of a CLT structure. For all the East Asian cities selected in this study, the proposed hybrid structure system can be effectively used for old building renewal.

Author Contributions: Conceptualization, M.-T.T.; methodology, M.-T.T.; software, W.-T.L.; validation, W.-T.L. and M.-T.T.; formal analysis, W.-T.L. and M.-T.T.; investigation, W.-T.L. and M.-T.T.; resources, W.-T.L. and M.-T.T.; data curation, W.-T.L. and M.-T.T.; writing—original draft preparation, W.-T.L.; writing—review and editing, M.-T.T. All authors have read and agreed to the published version of the manuscript.

Funding: This research was financially supported by the Ministry of Science and Technology of Taiwan, R.O.C. under Grant No. MOST 109-2221-E-011-032—and the National Taiwan University of Science and Technology.

Institutional Review Board Statement: Not applicable.

Informed Consent Statement: Not applicable.

Data Availability Statement: Not applicable.

Conflicts of Interest: The authors declare no conflict of interest.

References

1. Pierrehumbert, R. There Is No Plan B for Dealing with the Climate Crisis. *Bull. At. Sci.* **2019**, *75*, 1–7. [CrossRef]
2. Guidelines on Green Houses of Worship, UN Environment Programme, July. 2020. Available online: https://www.unenvironment.org/resources/publication/guidelines-green-houses-worship (accessed on 5 October 2020).
3. Green, M.; Taggart, J. *Tall Wood Buildings: Design, Construction and Performance*, 2nd ed.; Birkhäuser: Basel, Switzerland, 2020.
4. Popovski, M.; Mohammad, M.; Ni, C.; Rezai, M.; Below, K.; Malczyk, R.; Sherstobitoff, J. Design and Construction of Tall Wood Buildings: Input Data, Testing and Advanced Analysis. In Proceedings of the 13th World Conference on Timber Engineering (13WCTE), Quebec City, QC, Canada, 10–14 August 2014.
5. Connolly, T.; Loss, C.; Iqbal, A.; Tannert, T. Feasibility Study of Mass-Timber Cores for the UBC Tall Wood Building. *Buildings* **2018**, *8*, 98. [CrossRef]
6. Connolly, T.; Moudgil, M.; Loss, C.; Iqbal, A. Feasibility of Cross-Laminated Timber Cores for the UBC Tall Wood Building. In Proceedings of the 15th World Conference on Timber Engineering (15WCTE), Seoul, Korea, 20–23 August 2018.
7. Švajlenka, J.; Kozlovská, M. Evaluation of the Efficiency and Sustainability of Timber-Based Construction. *J. Clean. Prod.* **2020**, *259*, 120835. [CrossRef]
8. C. P. *CLT Handbook: Cross-Laminated Timber*; Gagnon, S., Ed.; FPInnovations: Vancouver, BC, Canada, 2011.
9. Alsayegh, G.; Mukhopadhyaya, P.; Wang, J.; Zalok, E.; van Reenen, D. Preliminary Characterization of Physical Properties of Cross-Laminated-Timber (CLT) Panels for Hygrothermal Modelling. *Adv. Civ. Eng. Mater.* **2013**, *2*, 472. [CrossRef]
10. Laguarda Mallo, M.F.; Espinoza, O. Outlook for Cross-Laminated Timber in the United States. *BioResources* **2014**, *9*, 7427–7443. [CrossRef]
11. Weissensteiner, J.; Barbu, M.C. Cross Laminated Timber—European Experiences. *Pro Ligno* **2013**, *9*, 69–72.
12. Balasbaneh, A.T.; Sher, W. Comparative Sustainability Evaluation of Two Engineered Wood-Based Construction Materials: Life Cycle Analysis of CLT versus GLT. *Build. Environ.* **2021**, *204*, 108112. [CrossRef]
13. Nässén, J.; Hedenus, F.; Karlsson, S.; Holmberg, J. Concrete vs. Wood in Buildings—An Energy System Approach. *Build. Environ.* **2012**, *51*, 361–369. [CrossRef]
14. Haibo, G.; Ying, L.; Yiping, M.; Haoyu, H.; Cheng, S.; Yu, S. A Comparison of the Energy Saving and Carbon Reduction Performance between Reinforced Concrete and Cross-Laminated Timber Structures in Residential Buildings in the Severe Cold Region of China. *Sustainability* **2017**, *9*, 1.
15. Haibo, G.; Ying, L.; Wen-Shao, C.; Yu, S.; Cheng, S. Energy Saving and Carbon Reduction in the Operation Stage of Cross Laminated Timber Residential Buildings in China. *Sustainability* **2017**, *9*, 292. [CrossRef]
16. Khavari, A.M.; Pei, S.; Tabares-Velasco, P.C. Energy Consumption Analysis of Multistory Cross-Laminated Timber Residential Buildings: A comparative study. *J. Archit. Eng.* **2016**, *22*, 04016002. [CrossRef]
17. Ying, L.; Haibo, G.; Cheng, S.; Wen-Shao, C. Assessing Cross Laminated Timber (CLT) as an Alternative Material for Mid-Rise Residential Buildings in Cold Regions in China—A Life-Cycle Assessment Approach. *Sustainability* **2016**, *8*, 1047. [CrossRef]
18. Dong, Y.; Cui, X.; Yin, X.; Chen, Y.; Guo, H. Assessment of Energy Saving Potential by Replacing Conventional Materials by cross Laminated Timber (CLT)—A Case Study of Office Buildings in China. *Appl. Sci.* **2019**, *9*, 858. [CrossRef]
19. Švajlenka, J.; Kozlovská, M.; Badida, M.; Moravec, M.; Dzuro, T.; Vranay, F. Analysis of the Characteristics of External Walls of Wooden Prefab Cross Laminated Timber. *Energies* **2020**, *13*, 5974. [CrossRef]
20. Adekunle, T.O. Occupants' Comfort and Stress Indices in a Structural Timber School Building in the Northeast US in Different Seasons. *Build. Res. Inform.* **2020**, *48*, 331–348. [CrossRef]
21. Adekunle, T.O. A Field Survey on Thermal Comfort of Occupants and Cold Stress in CLT School Buildings. In *Smart and Sustainable Cities and Buildings*; Roggema, R., Roggema, A., Eds.; Springer: Cham, Switzerland, 2020.
22. Amitha, J.; Satheeskumar, N.; Tuan, D.; Priyan, M.; Nick, H.; Lu, A. Life Cycle Performance of Cross Laminated Timber Mid-rise Residential Buildings in Australia. *Energy Build.* **2020**, *223*, 110091. [CrossRef]
23. Laleman, R.; Albrecht, J. Estimations of the Reserve Margin during the Nuclear Phase-out. *Int. J. Electr. Power Energy Syst.* **2016**, *81*, 416–426. [CrossRef]
24. Gülen, G.; Bellman, D.K. Scenarios of Resource Adequacy in ERCOT: Mandated Reserve Margin, Impact of Environmental Regulations and Integration of Renewables. *Electr. J.* **2015**, *28*, 89–100. [CrossRef]
25. Lindner, B.; Brits, R.; van Vuuren, J.; Bekker, J. Tradeoffs between Levelling the Reserve Margin and Minimising Production Cost in Generator Maintenance Scheduling for Regulated Power Systems. *Int. J. Electr. Power Energy Syst.* **2018**, *101*, 458–471. [CrossRef]
26. National Statistics (Taiwan). Housing Conditions of 2010 Statistical Tables. Population and Housing Census. 2010. Available online: https://census.dgbas.gov.tw/PHC2010/english/rehome.htm (accessed on 25 September 2020).
27. Margani, G.; Evola, G.; Tardo, C.; Marino, E.M. Energy, Seismic, and Architectural Renovation of RC Framed Buildings with Prefabricated Timber Panels. *Sustainability* **2020**, *12*, 4845. [CrossRef]

28. Barbagallo, F.; Margani, G.; Marino, E.M.; Moretti, A.; Tardo, C. Impact of Retrofit of RC Frames by CLT Panels and Friction Dampers. In Proceedings of the 8th International Conference on Computational Methods in Structural Dynamics and Earthquake Engineering, CompDyn 2021, Streamed from Athens, Athens, Greece, 28–30 June 2021.
29. Pelaz, B.; Blanco, J.M.; Cuadrado, J.; Egiluz, Z.; Buruaga, A. Analysis of the Influence of Wood Cladding on the Thermal Behavior of Building Façades; Characterization through Simulation by Using Different Tools and Comparative Testing Validation. *Energy Build.* **2017**, *141*, 349–360. [CrossRef]
30. Cambiaso, F.; Pietrasanta, M.V. Innovative Timber Construction: Sustainability and High Performance Building Skin. *Int. J. Eng. Technol.* **2014**, *6*, 47–54. [CrossRef]
31. Haba, R.; Kitamori, A.; Mori, T.; Fukuhara, T.; Fukuhara, T.; Isoda, H. Development of CLT Panels Bond-in Method for Seismic Retrofitting of RC Frame Structure. *J. Struct. Constr. Eng. (Trans. AIJ)* **2016**, *81*, 1299–1308. [CrossRef]
32. Khajehpour, M.; Pan, Y.; Tannert, T. Seismic Analysis of Hybrid Steel Moment Frame CLT Shear Walls Structures. *J. Perform. Constr. Facil.* **2021**, *35*, 04021059. [CrossRef]
33. Pacheco-Torgal, F.; Faria, J.; Jalali, S. Embodied Energy versus Operational Energy. Showing the Shortcomings of the Energy Performance Building Directive (EPBD). *Mater. Sci. Forum* **2013**, *730*, 587–591. [CrossRef]
34. Taipower Company Sustainabilty Report. 2008. Available online: https://www.taipower.com.tw/upload/85/2018010916401984784.pdf (accessed on 16 January 2021).
35. CO_2 Emissions, CO_2 Emissions Intensity and Electricity Sales. 2020. Available online: https://www.tepco.co.jp/en/corpinfo/illustrated/environment/emissions-co2-e.html (accessed on 22 January 2021).
36. Bureau of Energy, Ministry of Economic Affairs, Taiwan. 2019. Available online: https://www.moeaboe.gov.tw/ECW/populace/news/Board.aspx?kind=3&menu_id=57&news_id=17638 (accessed on 16 January 2021).
37. EMA: Singapore Energy Statistics | Energy Transformation. 2020. Available online: https://www.ema.gov.sg/Singapore-Energy-Statistics-2019/Ch02/index2 (accessed on 18 January 2021).
38. Ministry of Economic Affairs, Taiwan. 2017. Available online: https://reurl.cc/pd3oj8 (accessed on 16 January 2021).
39. Peng, C. Calculation of a Building's Life Cycle Carbon Emissions Based on Ecotect and Building Information Modeling. *J. Clean. Prod.* **2016**, *112*, 453–465. [CrossRef]
40. The Technical Characteristics of KLH Cross-Laminated Timber Panels. Available online: http://www.klhuk.com/media/29233/technical%20characteristics.pdf (accessed on 5 May 2020).
41. Stora Enso. CLT. Technical Brochure. Available online: https://www.storaenso.com/-/media/documents/download-center/documents/product-brochures/wood-products/clt-by-stora-enso-technical-brochure-en.pdf (accessed on 11 May 2020).

energies

MDPI

Article

Impact of Enclosure Boundary Patterns and Lift-Up Design on Optimization of Summer Pedestrian Wind Environment in High-Density Residential Districts

Zhengrong Jiang [1,*] and Weijun Gao [2]

[1] College of Urban Construction, Zhejiang Shuren University, Hangzhou 310015, China
[2] Faculty of Environmental Engineering, The University of Kitakyushu, Fukuoka 8080135, Japan; gaoweijun@me.com
* Correspondence: jiangzr@zjsru.edu.cn; Tel.: +86-0571-88285810

Abstract: A comfortable wind environment favors the sustainable development of urban residential districts and public health. However, the rapid growth of high-rise urban residential districts leads to low wind velocity environments in summer. This study examines the influence of enclosure boundary patterns and lift-up design on the wind environment and proposes an optimization strategy to improve the low wind velocity environment in residential districts in summer. A typical residential district in Hangzhou was selected; the average wind velocity, calm wind zone ratio and comfortable wind zone ratio were selected as the evaluation indexes. The wind environment for different enclosure boundary patterns and lift-up designs were obtained via computational fluid dynamics (CFD) simulations. The results indicate that the pedestrian wind environment is greatly improved in residential districts by reducing the height/width of the enclosure boundary, increasing the permeability rate and adopting a lift-up design in all buildings within residential districts. A combination of permeable railings and lift-up design is recommended; this can increase the average wind velocity and the ratio of comfortable wind zones by 70% and 200%, respectively. This study provides practical guidelines for the optimization of a low wind velocity environment in Chinese high-density residential districts in summer.

Keywords: high-density residential district; pedestrian wind environment; computational fluid dynamics; enclosure boundary; lift-up design

Citation: Jiang, Z.; Gao, W. Impact of Enclosure Boundary Patterns and Lift-Up Design on Optimization of Summer Pedestrian Wind Environment in High-Density Residential Districts. *Energies* **2021**, *14*, 3199. https://doi.org/10.3390/en14113199

Academic Editors: Giorgio Ficco, Mark Luther, Igor Martek and Mehdi Amirkhani

Received: 8 April 2021
Accepted: 26 May 2021
Published: 30 May 2021

1. Introduction

With the rapid development of urbanization in China, land resources are becoming increasingly scarce, and building density and height in urban residential districts are rapidly increasing [1,2]. High-density residential districts exhibit reduced ventilation and heat dissipation conditions, thus leading to low wind velocity environments and poor thermal comfort in residential districts [3]. This problem is particularly serious in summer, especially in the densely populated eastern coastal areas of China [4]. These related phenomena may negatively affect the sustainable development of urban residential districts. For instance, air stagnation due to a low wind velocity environment may prevent the evacuation of airborne pathogens [5,6]. During epidemic periods of infectious diseases, this issue is particularly prominent, and the concentration of pollutants poses a clear threat to public health safety [7]. Especially during the development of the COVID-19 epidemic, it is essential to improve the pedestrian-height wind environment in urban high-density residential districts for public health safety purposes [8].

A series of studies has been conducted to investigate effective methods for improving the pedestrian wind environment. Field measurements, wind tunnel experiments, and computational fluid dynamics (CFD) simulations are often employed to analyze outdoor wind environments [9,10]. Du et al. adopted the overall mean velocity rate (OMVR) as an evaluation criterion, and through a wind tunnel experiment on the campus of the Hong Kong Polytechnic University, they demonstrated that the area of acceptable wind comfort increases from 20% to 50% via the adoption of building opening and lift-up design [4]. Tsanga et al. performed wind tunnel experiments to study the influence of building separation on pedestrian wind environments under low and high wind conditions, and found that when the building separation is smaller than half of the building width, it imposes adverse effects on the natural ventilation conditions for pedestrians [11].

Although wind tunnel experiments are reliable, they are relatively expensive and insufficiently flexible to be applied to architecture design on a large scale. Instead, the CFD simulation method has the advantages of high speed, easy implementation, high accuracy and effectiveness, and low cost [12,13]. Therefore, with the development of computer technology and grid generation techniques, increasing numbers of researchers have performed CFD simulations to study urban wind environments [14–16]. For instance, Hang et al. used CFD simulations to numerically investigate the air age and air exchange efficiency of an ideal high-density building complex under the effect of parallel winds, and proposed that wide streets and small building array lengths could help improve ventilation [17]. Based on a highly simplified model, Li and Chen adopted the CFD technique to simulate the wind environment around buildings and found that when the area of the opening remains the same, the larger the number of openings, the better the distribution of the vertical wind velocity behind the building [18].

In recent years, an increasing number of studies have focused on the pedestrian wind environment in residential districts. For example, Wang et al. introduced dimensionless layout parameters of rank-and-file building groups through wind tunnel experiments, and they proposed that changing the building layout and orientation can help produce a good wind environment [19]. Hadavi and Pasdarshahri quantified the effects of the wind velocity, urban planning district density, and urban form on the infiltration rate via CFD simulations, and concluded that building tightness enhancement could reduce the infiltration rate and augment the exfiltration rate [20]. Hu et al. analyzed the relationship between the residential building density and wind environment and reported that when the residential building density increases from 0.18 to 0.32, the outdoor average wind velocity ratio decreases by 0.18 and the average air age increases by 58.63 s [21]. Hong and Lin compared the wind environments of several residential areas with different layout patterns and tree arrangements and concluded that orienting the long facades of buildings parallel to the prevailing wind direction can yield satisfactory thermal comfort [22]. Ghadikolaei et al. analyzed the effects of wing wall at the balcony on the natural ventilation performance in medium-rise residential buildings [23]. Besides, Tes et al. explored the impact of lift-up building models with different core heights and widths, and found that the lift-up core height is the most influential parameter and governs the area and magnitude of high and low wind velocity zones around such buildings [24]. Guo et al. compared the natural ventilation performance of different building morphologies via CFD simulation, and indicated that enclosed city blocks are unfavorable to natural ventilation [25]. Kubota et al. carried out wind tunnel tests on 22 residential neighborhood districts selected from actual Japanese cities, and concluded that an increase in the gross building coverage ratio decreases the mean wind velocity ratio [26].

However, most of these studies were concerned with the effects of different building layout patterns, densities and heights on the wind environment in residential districts. The influence of different enclosure boundary patterns on residential wind environments has rarely been considered in studies to date. Several studies have analyzed the influence of the lift-up building model on the wind environment, but few have analyzed the influence of the combination of the enclosure boundary and lift-up design on the wind comfort of a residential district. The research objects in most of the literature are highly simplified, ideal models such that planning and architectural practice constraints are neglected. The results of related studies do not reflect the complexity and diversity of urban construction projects.

Chinese residential districts exhibit their own characteristics due to local urban guidelines, customs and culture [22]. They are distinguished by enclosed boundaries, lift-up designs and regular layout patterns. Closed boundaries are mainly divided into four patterns: low-rise podiums, main buildings, solid walls and permeable railings. Appropriate types of enclosure boundaries and the combined design of boundaries and lift-up options have been noted as being highly important for achieving a comfortable outdoor environment in residential districts. However, existing studies generally neglect these aspects. Moreover, there remains a lack of quantitative analysis and research on the effect of different combined designs on the wind environment in residential areas. It is essential for architects to develop a design method to determine the optimum arrangement for creating a comfortable wind environment.

In this paper, the characteristics of 16 high-density residential districts in the Yangtze River Delta region of China were analyzed. Then, a typical residential district in Hangzhou was selected to investigate the low wind velocity environment problem in summer. With regard to the CFD simulation method, the average wind velocity, calm wind zone ratio and comfortable wind zone ratio were selected as the evaluation indexes to quantify the results of the wind environment under different enclosure boundaries and lift-up forms, and the impacts of different improvement measures on the low wind velocity environment were determined. The research framework is shown in Figure 1. The results of this study provide practical guidelines for the optimization of low wind velocity environments in high-density residential districts in summer.

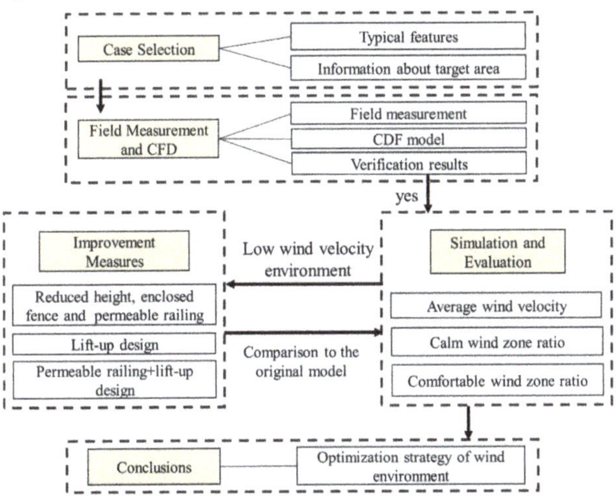

Figure 1. Framework and workflow of this study.

2. Methodology

2.1. Case Selection

2.1.1. Survey of Residential Districts in the Yangtze River Delta

The Yangtze River Delta urban agglomeration is one of the most economically developed and urbanized areas in China. According to Chinese specifications [27], in this study, a high-density residential district is defined as a site area ranging from 2–10 ha containing 11 layers or more with a floor area ratio higher than 2. Sixteen high-density residential districts are randomly selected in typical cities. Figure 2 shows the locations of these residential districts, and Figure 3 illustrates the details of these residential districts with the depicted information sourced from Google Earth images. Due to the geographic environmental constraints, sunlight norms, and urban planning requirements in the Yangtze River Delta region, certain common characteristics are observed in these residential districts.

Figure 2. Distribution map of the case locations.

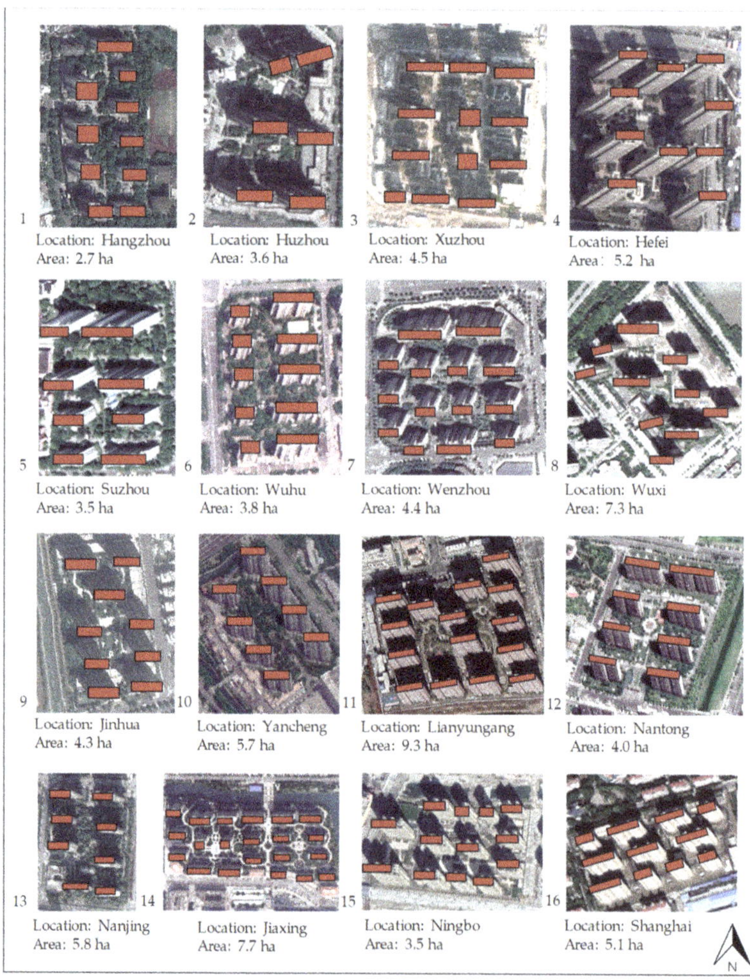

Figure 3. Site layouts of 16 high-density districts in the Yangtze River Delta region of China.

2.1.2. Common Characteristics of High-Density Districts

- The Yangtze River Delta is situated in the hot summer and cold winter region. The best orientation is toward the south or slightly toward the southeast or southwest to attain sufficient sunlight and ventilation. The layout of buildings largely indicates a combination layout mode of the row type or the row type + the point type (see Figure 4).
- The residential districts are gated communities [28]. The boundaries of the residential districts are physically enclosed, leaving only a few pedestrian and vehicular access entrances. In addition to natural boundaries (e.g., lakes), the enclosed boundaries of these residential districts usually include four types of enclosures: low-rise podiums, main buildings, solid walls, and permeable railings (as shown in Figure 5).
- In certain newly constructed high-density residential districts, some or all buildings adopt a lift-up design to provide public spaces for people to rest or parking for non-motorized vehicles.

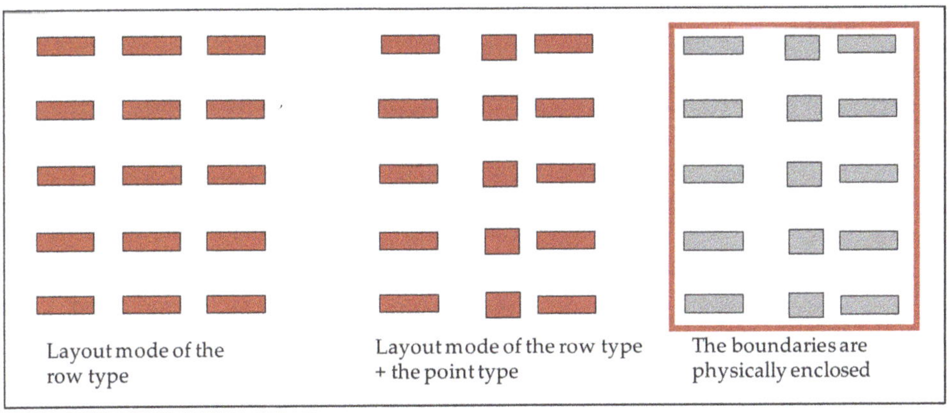

Figure 4. Characteristics of high-density residential districts in the Yangtze River Delta region of China.

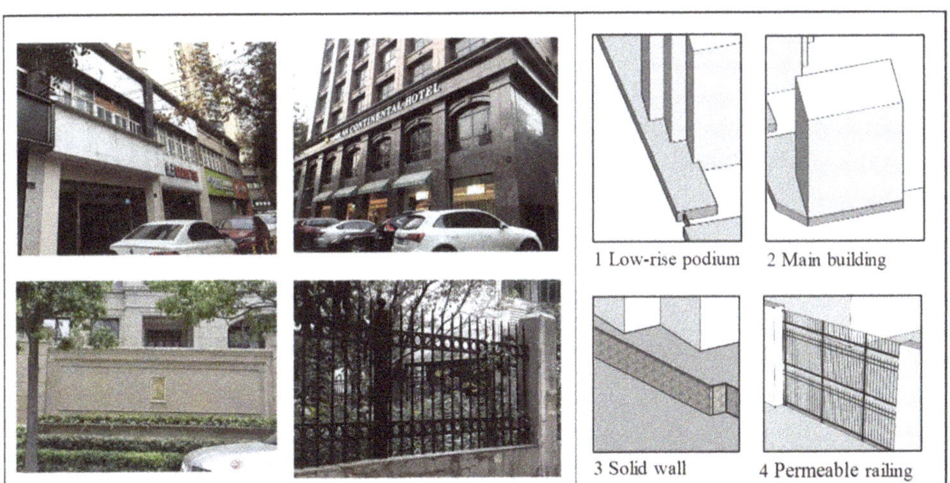

Figure 5. Four typical patterns of enclosed boundaries in residential districts.

2.1.3. Selection of a Typical Case

Based on the common characteristics of residential districts, Case 1 (i.e., a typical high-density residential district in the central urban area of Hangzhou city) was chosen as the research object. The residential district mainly consists of ten high-rise buildings A–J (refer to Figure 6a), among which buildings A–E contain lift-up areas with a height of 3 m. The main structure of the lift-up buildings is elevated from the ground by a combination of columns, shear walls and a central core. Figure 6a shows the locations of the lift up areas and the vertical structural elements. Figure 6b shows the schematic diagrams of the lift-up building E. The core area percentage is expressed as the percentage of area covered by the core of the plan area of the building at the ground level, which is termed the "area percentage" [24]. The area percentage value in this study was 18.1–27.4%.

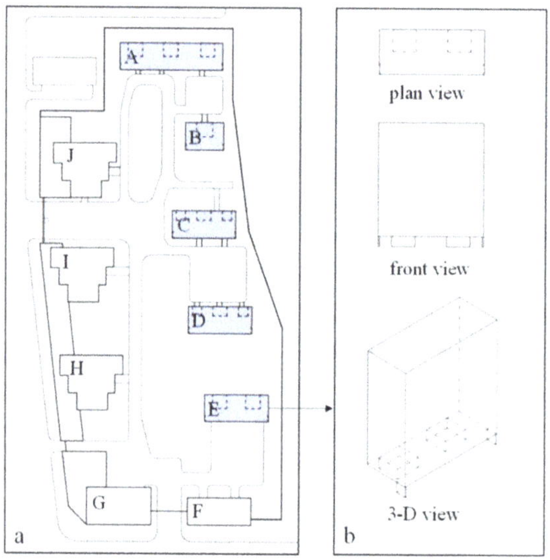

Figure 6. Schematic diagram of lift-up buildings: (**a**) The locations of the lift-up areas and the vertical structural elements, (**b**) schematic diagrams of the lift-up building E.

City roads border the western and southern sides of the residential district. The western side is enclosed by a low-rise podium that is 8 m in height and 11 m in depth, while the northern and eastern sides are separated from the neighboring areas via permeable railings. The dimensions of the area are 240 m at its longest point and 120 m at its widest point, with a total area of 2.7 ha and a maximum building height of 54 m.

2.2. CFD Simulation Setup

2.2.1. Geometric Model and Domain Size

According to existing research [29,30], to accurately reproduce the flow field in this residential district, the geometric model considers the building information in the target area (i.e., the blue area shown in Figure 7a), and the surrounding area is three times larger than the target area (see Figure 7a). Detailed information was retrieved from field measurements and Google Earth images.

The CFD calculation area in the present study satisfies a distance of 4D from the entrance to the nearest building boundary, a distance of 6D to the outflow boundary, a distance of 4W to the side boundary and a distance of 5H to the top boundary, where D is the total length of the building complex (D = 720 m), H is the maximum building height of the target complex (H = 72 m), W is the total width of the building complex (W = 360 m) and the obstruction rate is 2.4%, which is lower than the recommended maximum obstruction ratio of 3% [31].

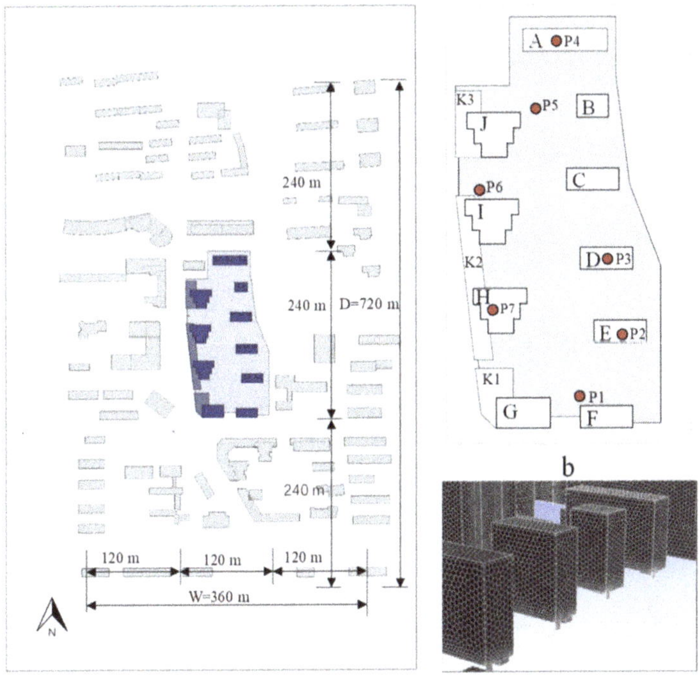

a c

Figure 7. CFD model: (**a**) typical residential district and surrounding buildings, (**b**) measurement point locations, (**c**) grid arrangement.

2.2.2. CFD Model and Grid Arrangement

A commercial CFD program, ANSYS Fluent 19.0, was adopted in the present study. The unsteady Reynolds-averaged Navier–Stokes (RANS) equations with the RNG k-e turbulence model were used to solve the turbulent wind flow in the computational domain [32]. This model was verified by simulating wind flows at the urban scale, with an average relative error of 12% with respect to the measured data [33].

A polygonal grid scheme was adopted in this study, which is adaptable to different geometric shapes. The grid spacing was set to 20 m in the surrounding open area, which was gradually reduced to 3 m in the object model. To accurately represent key simulation areas (e.g., lift-up floors and pedestrian heights), the grid size was thus refined to 1 m. This resulted in a total grid number of 3 million, and the corresponding local grid division is shown in Figure 7c.

2.2.3. Boundary Conditions

As the research target is located in the city, it was difficult to obtain the boundary data in the city by field measurement. Moreover, it was hard to gain enough meteorological data in the large area, and difficult to conduct field measurement to verify the large-scale simulation results of the wind environment. Hence, the wind data from the Hangzhou Meteorological Bureau were used as the inflow boundary conditions. The Hangzhou Meteorological Bureau used the interpolation method to get the wind field based on the real-time data of each meteorological monitoring point. The average wind velocity near the inflow boundary from 10:40 a.m. to 11:40 a.m. on 22 July 2020 was 1.7 m/s and the wind direction was south-southwest. The upstream vertical boundary in the calculation domain was set as the inflow boundary, and the outflow boundary was set as the pressure outlet. The vertical velocity profile along the inflow boundary was modeled as a power law based on the structural building code [34].

$$V_h = V_0 \left(\frac{h}{h_0} \right)^{\alpha} \tag{1}$$

where V_h is the wind velocity at a height of h, m/s; V_0 is the wind velocity at height h_0 m/s; h_0 = 10 m; and α is the roughness index. The target area is located in a dense urban area of a large city with a ground topography of category C, and the exponent of the power law α = 0.22 [34]. The computations were performed on an 8-core workstation (Intel Core i7-9700, 3.0 GHz) with 32 GB DDR of system memory. When the scaled residuals reached 10^{-4} for mass conservation, U, V, W, k, ε, the solution was considered to be converged.

2.3. CFD Validation

To verify the reliability of the CFD simulation results, this study carried out field measurements in the target residential district. The field measurements were conducted from 10:40 a.m. to 11:40 a.m. on 22 July 2020, with seven anemometers (405i, Testo, Testo AG). The field measurement net covered seven locations for pedestrian activities in the district, as shown in Figure 7b. Velocity measurement points 1–6 were located 1.5 m above the ground, measurement points 2–4 were situated in the lift-up area, and measurement point 7 was located 1.5 m above the roof.

The wind velocity results obtained with the CFD simulation method at a height of 1.5 m were compared to the measured results, as shown in Figure 8. The average error between the CFD model simulation data and the measured data is 20.4%. The trends at the different positions are the same, which verifies the reliability of the CFD simulation results [29]. Therefore, the CFD model established in this study and the obtained simulation results are credible.

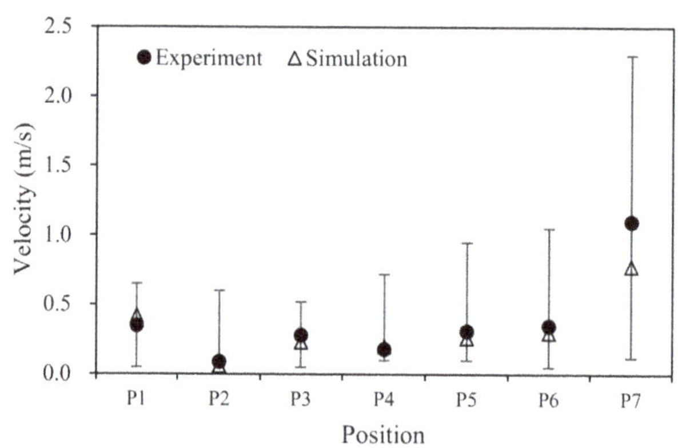

Figure 8. Comparison of the average measured and simulated wind velocity values at points 1–7.

2.4. Case Design

Considering engineering practices, on the basis of not altering the layout of residential districts, this study adopted the enclosure boundaries and the lift-up design as the parameter design, and 6 cases were obtained (refer to Figure 9).

Case O represents the original state of the research target. The northern and eastern enclosure boundaries are permeable iron railings, and the southern enclosure boundary is the main building. The dominant wind direction in Hangzhou in summer is south-

southwest (SSW), and the changes in enclosure boundary patterns largely involve the optimization of the western boundary.

Case A1, as shown in Figure 9, indicates that the existing lift-up areas of buildings A–E are blocked. Cases B1, B2, and B3, as shown in Figure 9, reveal the changes in the western enclosure boundary pattern. Case B1 reduces the enclosure boundary height and changes the podium height from 8 to 4 m. In Case B2, the enclosure height and depth are reduced. The podium is changed to a solid wall with a height of 3 m and a thickness of 0.3 m. In Case B3, the degree of the enclosure boundary is changed, and the podium is replaced by a permeable railing with a height of 3 m. Case C1 (see Figure 9) considers the combined design of a permeable railing and lift-up, i.e., the western podium is changed to a 3-m-high permeable railing. All high-rise buildings A–J contain a lift-up area that is 3 m high.

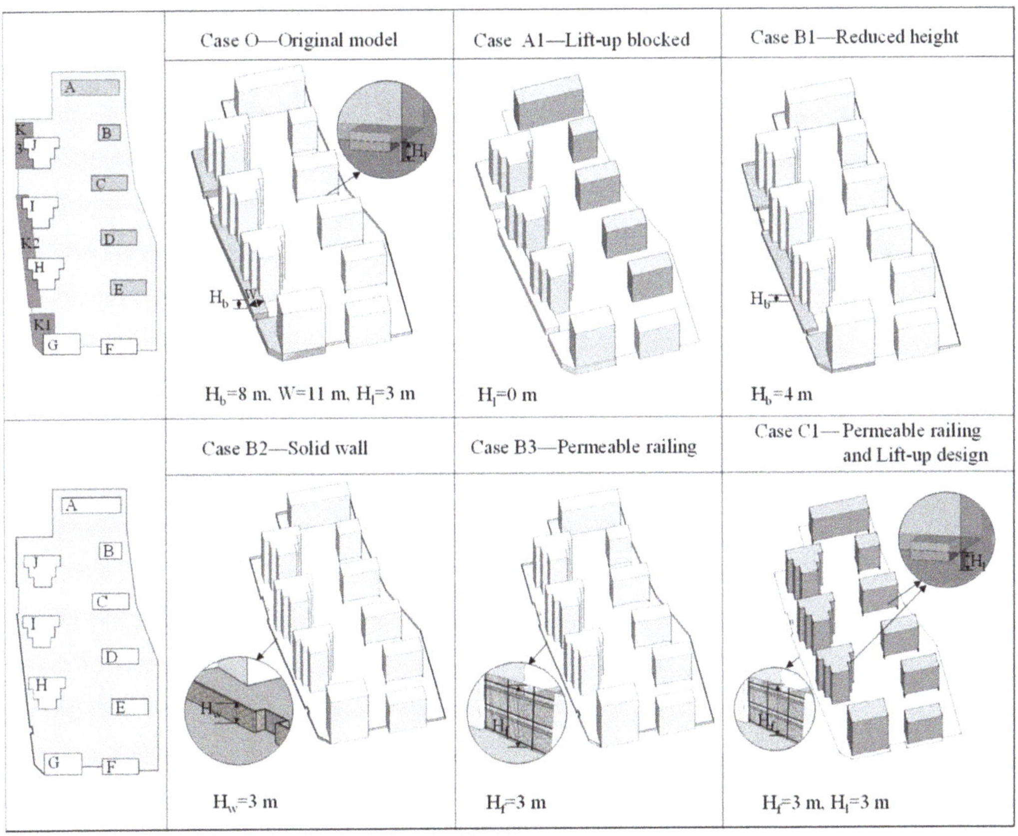

Figure 9. Design of the six cases.

2.5. Wind Environment Evaluation Index

According to the study of Du et al., summer wind comfort criteria are typically classified into five groups [35]: (i) an unfavorable wind environment for pedestrian activities due to low wind conditions; (ii) a favorable wind environment for pedestrian activities, including light, gentle and moderate breezes; (iii) a tolerable wind environment for any activity; (iv) an intolerable wind environment for any activity; and (v) a dangerous outdoor environment.

According to the research of Lawson [36], the NEN 8100 standard [37], and the air ventilation assessment (AVA) scheme [38] regarding wind comfort indicators, the summer low wind comfort indicators are summarized in Table 1. Based on previous studies [39,40], the following indicators were selected as evaluation indexes for the summer low wind velocity environment:

- Average wind velocity: The average wind velocity considers all velocity measurement points at a height of 1.5 m above the ground surface in the residential district.
- Calm wind zone ratio: The area with a wind velocity lower than 1 m/s is defined as a calm wind zone. Considering comfort and pollutant diffusion, the lower the calm wind zone ratio at the measurement points, the better the wind environment.
- Comfortable wind zone ratio: The area with a wind velocity between 1 and 5 m/s is a comfortable wind zone. The higher the comfortable wind zone ratio, the better the wind environment.

Table 1. Summer low wind comfort criteria in terms of the average wind velocity.

Type	Description	Average Wind Velocity (m/s)
Unfavorable	Calm	0–1 [36,38]
	Light breeze	1–1.8 [36]
Favorable	Gentle breeze	1.8–3.6 [36]
	Moderate breeze	3.6–5.0 [37]
Tolerable	Fresh breeze	5.0–7.6 [37]
Intolerable	Strong breeze	7.6–15 [37]
Dangerous	Gale	>15 [37]

3. Results and Discussion

Wind data for the target residential district were retrieved from the design standards for wind and thermal environmental design of residential buildings [41]. In July, the dominant wind direction in Hangzhou is SSW at a wind velocity of 2.5 m/s. The summer wind rose map is shown in Figure 10. Within the outdoor pedestrian activity range of the residential district, 70 measurement points were selected at a uniform height of 1.5 m (refer to Figure 11). The wind environment results in the different cases were obtained through CFD simulations, and the wind environment evaluation indexes in the different cases were summarized and compared. Moreover, axes X1, X2, and X3, corresponding to the three typical boundaries, were selected to evaluate the change trends in the wind velocity at typical axial velocity measurement points.

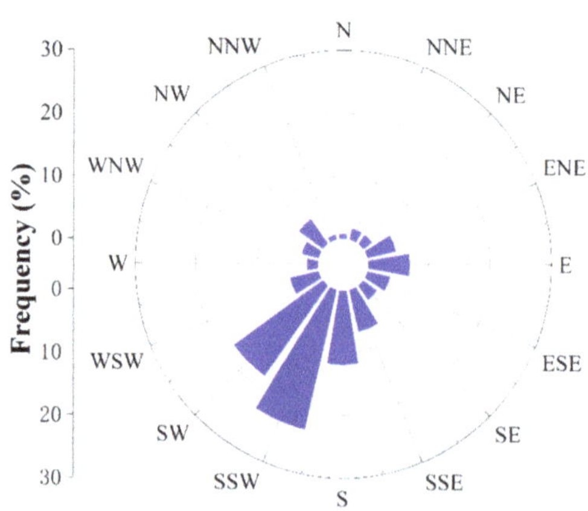

Figure 10. Wind rose map for Hangzhou in July, information used from reference [41].

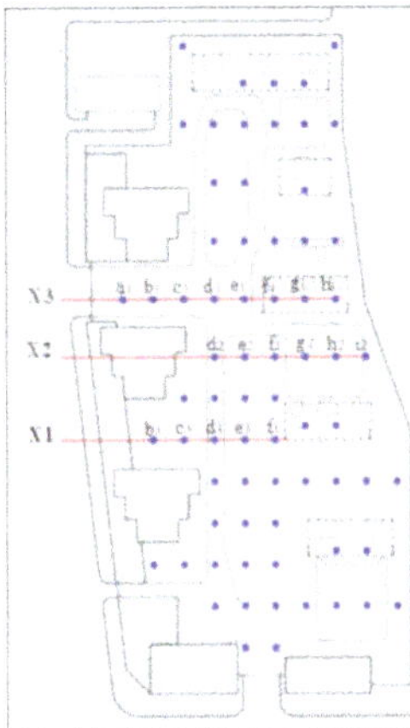

Figure 11. Location of the 70 velocity measurement points.

3.1. Original Model

Figures 12–14 depict the wind environment simulation results for the six cases in summer. Case O, as shown in Figure 12, indicates that the wind velocities in the western and southern entrance plaza areas are higher than those in the other areas and that the wind velocities in most of the residential district are low. Table 2 shows that the average wind velocity in the whole residential district is 0.63 m/s. Eighty-three percent of the outdoor spaces throughout the entire residential district are found in the calm wind zone, which is not conducive to pedestrian activities, and only 17% of the outdoor spaces are located in the comfort zone. In general, the overall wind comfort in the residential district is unsuitable for pedestrian activities during the hot and humid summer months.

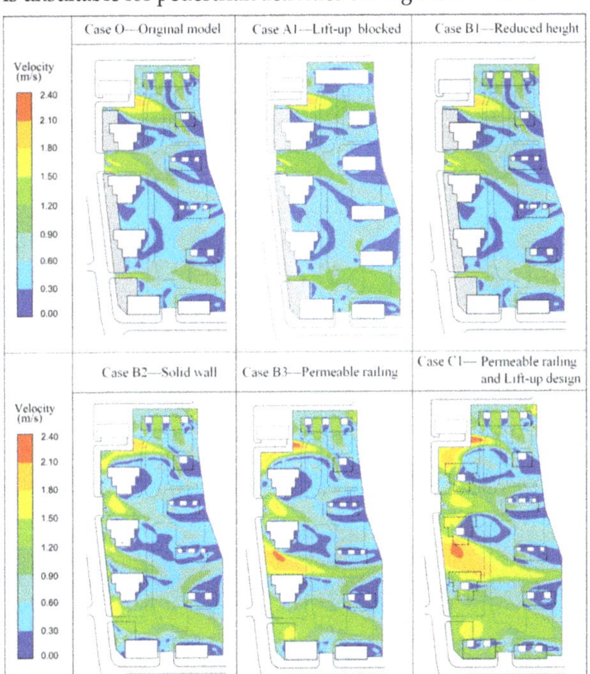

Figure 12. Wind velocity simulation results corresponding to the six cases.

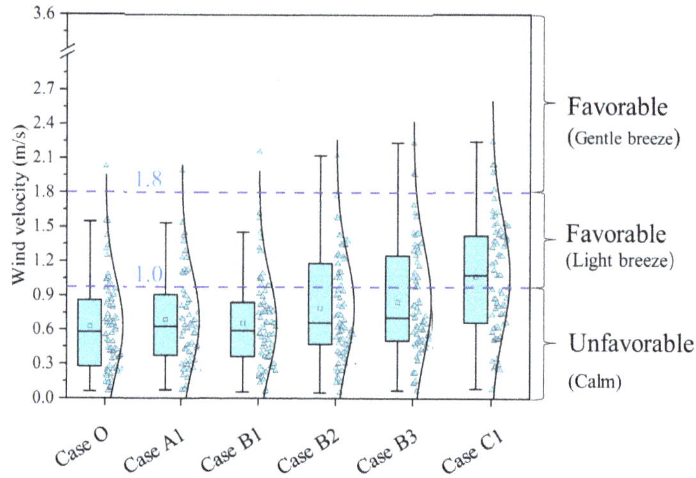

Figure 13. Wind velocity box plots for the six cases.

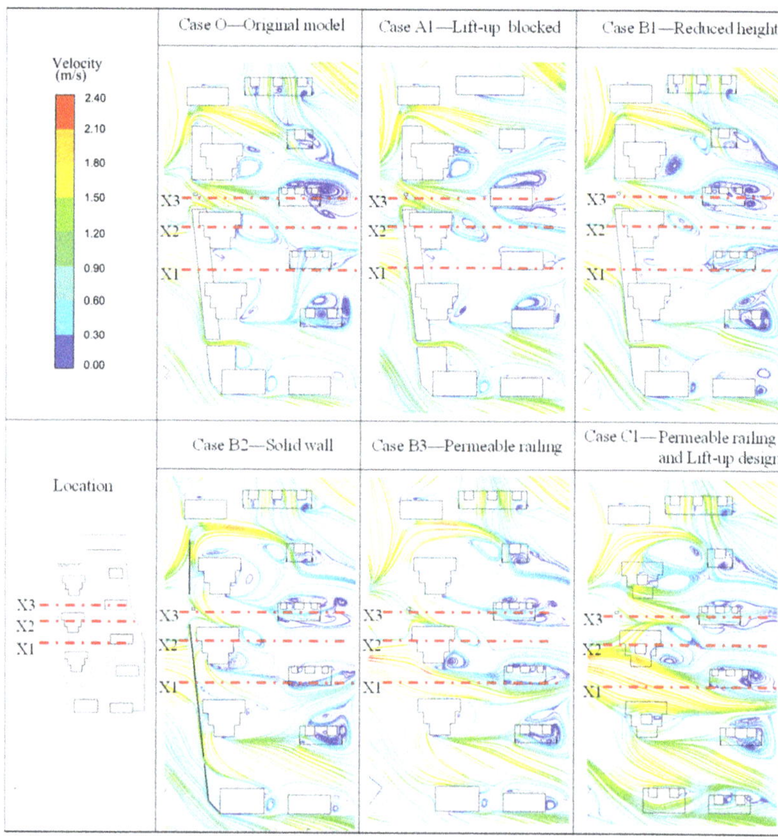

Figure 14. Streamline patterns of X1, X2, and X3 at pedestrian height for the six cases.

Table 2. Results of the wind environment simulation for the six cases.

Type	Case O	Case A1	Case B1	Case B2	Case B3	Case C1
Average wind velocity (m/s)	0.63	0.69	0.66	0.79	0.85	1.07
Number of measurement points	70	58	70	70	70	70
Calm wind zone ratio	83%	78%	83%	69%	66%	49%
Comfortable wind zone ratio	17%	22%	17%	31%	34%	51%
Ratio of wind velocity increase	/	*	5%	25%	35%	70%
Ratio of calm wind zone reduction	/	*	0	17%	20%	41%
Ratio of comfortable wind zone increase	/	*	0	82%	100%	200%

* There are no wind velocity measurement points in the lift-up area in Case A1.

3.2. Lift-Up Design

In Figures 12–14, the results of wind environment simulation show that winds beyond the lift-up area are blocked for Case A1. Comparing Case A1 with Case O reveals that the lift-up buildings within the residential district increase the wind velocity in nearby areas. Axis X2 is located 15 m downwind to the north of lift-up building D (refer to Figure 11). Figure 15 shows the wind velocity at the monitoring points along axis X2, i.e., at g2, h2, and i2, and the wind velocities are 0.12, 0.15 and 0.09 m/s higher in Case O than those in Case A1, respectively.

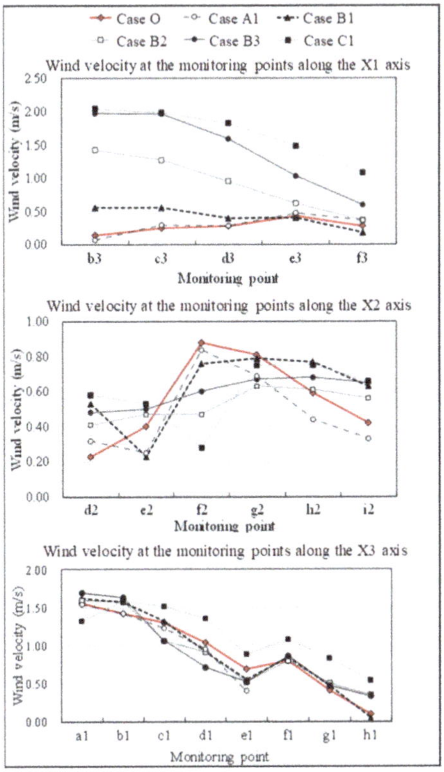

Figure 15. X1, X2, and X3 velocity measurement point wind velocities for the six cases.

However, the lift-up design in Case O does not improve the low wind velocity environment throughout the entire residential area. Table 2 indicates that the overall average wind velocity in Case A1 is 0.69 m/s, and the comfortable wind zone ratio is 22%. The overall wind comfort zone in the residential area remains limited. There are no velocity measurement points in the lift-up area in Case A1, so the average wind velocities cannot be directly compared. When the lift-up area velocity measurement points are removed, the average wind velocity in Case A1 is slightly higher (3%) than that in Case O.

3.3. Patterns of Enclosure Boundaries

Case B1 in Figures 12–14 shows the wind environment simulation results after podium height reduction. The wind velocity behind the podium greatly increases, but the wind velocity change in the rest of the residential district is small. The average wind velocity reaches 0.66 m/s, which is 5% higher than that in the original case. Furthermore, the comfortable wind zone ratio is 17%, and 83% of the outdoor spaces occurring in the calm wind zone are suitable for pedestrian activities. In general, the improvement in residential wind comfort is not obvious.

Case B2 in Figures 12–14 depicts the wind environment simulation results after adoption of a solid wall. Figure 13 and Table 2 demonstrate that the average wind velocity reaches 0.79 m/s, which is 25% higher than that in the original case. The comfortable wind zone ratio is 31%, which is 82% higher than that in the original case. Overall, the wind environment comfort is notably improved.

Case B3 in Figures 12–14 shows the wind environment simulation results after the adoption of a permeable railing. Table 2 indicates that the average wind velocity in the residential district reaches 0.85 m/s, i.e., 35% higher than that in the original case. The comfortable wind zone ratio reaches 34%, and the increase in the ratio of the wind comfort zone is 100%.

According to Cases B1 to B3, the higher the permeability of the enclosure boundary is, the smaller the wind barrier, and the higher the proportion of the wind comfort zone. The wind comfort increases from Cases B1 to B3 and is optimal in Case B3.

3.4. Combination of the Permeable Railing and Lift-Up Design

Figure 12, specifically Case C1, and Figure 14 show the wind environment simulation results based on the permeable railing and lift-up design. All buildings in the residential area adopt the lift-up design. As indicated in Table 2, the average wind velocity in the residential area reaches 1.07 m/s, and the wind comfort zone ratio is 51%. Compared to Case B3, Case C1 also lifts the boundary buildings, and the average wind velocity increases by 26%. The ratio of the calm wind zone decreases by 26%, and the ratio of the comfortable wind zone increases by 50%. The lift-up design is applied to all buildings in the residential district, which notably improves the wind environment.

Compared with Case O, Case C1 has an average wind velocity that increases by 70%; more than half of the outdoor areas are located in the comfort zone, the ratio of the calm wind zone decreases by 41%, and the ratio of the wind comfort zone increases by 200%. The comfortable wind environment in the residential area is greatly improved. Among these six cases, the combined case exhibits the most apparent improvement effects on the wind environment.

Through a wind tunnel experiment on the campus of the Hong Kong Polytechnic University, Du and Mak demonstrated that pedestrian-level wind comfort is greatly improved with the adoption of measures such as opening and lift-up [3]. Tes et al. analyzed the impact of lift-up building models with different core heights and widths on surrounding wind environments [24].

In this study, we confirmed that the lift-up design affects wind comfort on hot summer days and further examined the difference between the partial lift-up design and the total lift-up design. In addition, the effect of the enclosure boundary patterns on wind comfort in residential districts was examined.

3.5. Effects of Different Enclosure Boundaries on the Wind Environment along the Horizontal Direction

To further study the influence of the different enclosure boundary patterns on the surrounding wind velocity environment, three enclosure boundary patterns were selected based on Case O: the low-rise podium, main building, and permeable railing. Along the leeward direction of the boundary, horizontal lines at a height of 1.5 m were selected as axes X1, X2, and X3. Velocity measurement points were selected at 11-m intervals along the horizontal lines. Figure 10 shows the locations of the axes.

As shown in Figures 14 and 15, the wind velocity along the X1 axis is most notably affected by different boundary conditions. In case O, the 8-m podium provides the most obvious barrier to wind flow. In case B1, the 4-m-high podium provides a weaker barrier to wind. In case B2, the solid wall provides a smaller barrier to wind. The wind velocity at the various measurement points exhibits relatively large variations. Compared to that for the original case, the wind velocity at point b3 nearest to the boundary (11 m) indicates the most notable increase, with the maximum increase occurring in Case C1 (i.e., an increase of 1.9 m/s). The wind velocity at point f3 (55 m) also increases, with a maximum increase of 0.8 m/s in Case C1. The wind velocity is directly affected by the conditions in these cases. The main reason for this phenomenon is that the wind velocity of the horizontal boundary corresponding to the X1 axis depends on the enclosure boundary patterns.

The changes in the X2 axis wind velocity vary in different cases. Compared with those in Case O, the maximum increase in the wind velocity occurs at d2 in Case C1, with an increase of 0.4 m/s, while the maximum decrease occurs at f2 in Case C1, decreasing by 0.6 m/s. The horizontal boundary corresponding to the X2 axis is the main building. As shown in Figure 15, the wind velocity is not greatly affected by the change in cases but is affected mainly by the corresponding wind shadow area of the main building.

As shown in Figures 14 and 15, the variation in the range of the wind velocity along the X3 axis is very limited in the different cases. The maximum wind velocity change occurs at h1 in Case C1, and the wind velocity increases by 0.4 m/s. The horizontal boundary corresponding to axis X3 is the permeable fence, which is the residential area entrance. The wind velocity is slightly affected by the change in cases.

In general, the different enclosure boundary patterns directly influence the horizontal wind field along the downwind direction, and the wind gradually decreases with increasing distance from the boundary. This exerts a slight effect on the wind velocity along the nonadjacent axes.

4. Conclusions

The purpose of this study was to analyze the impacts of various enclosure boundary patterns and the lift-up design on the pedestrian wind environment in a high-density residential district and to improve the low wind velocity environment in summer. A typical residential district in Hangzhou was selected, and the wind environment results for different enclosure boundary patterns and the lift-up design were obtained via CFD simulations. The average wind velocity, calm wind zone ratio and comfortable wind zone ratio in different cases were quantitatively analyzed, and optimization strategies for the pedestrian low wind velocity environment in residential districts were determined. The main conclusions are as follows.

1. Reducing the height and width of the enclosure boundary and increasing the permeability rate of the enclosure boundary increase the overall average wind velocity at the pedestrian level in the residential district. Compared with that in Case O, the average wind velocities in Cases B1, B2, and B3 increase by 5%, 25% and 35%, respectively.
2. The lift-up design in all buildings of residential districts can notably improve the wind environment. Compared to that in Case B3, the average wind velocity in Case C1 increases by 26%, and the ratio of the comfortable wind zone increases by 50%. When the buildings inside the residential district adopt the lift-up design, the wind velocity is only improved in nearby areas. However, the low wind velocity environment throughout the whole residential district is not enhanced.
3. The combination of permeable railings and the lift-up design should be recommended because this combination has the most notable improvement effect. Compared with that in Case O, the average wind velocity in Case C1 increases by 70%, and the ratio of comfortable wind zone increases by 200%.

4. The enclosure boundary pattern has an obvious influence on the wind field along the horizontal downwind direction but only a slight influence on the wind speed along the nonadjacent axes. Moreover, the influence gradually decreases with increasing distance from the boundary.

In this study, we focus on the influence of the enclosure boundary patterns and the lift-up design in the hot summer and cold winter region of China. However, the characteristics of residential districts vary greatly in different areas of China. We will investigate the effects of different enclosures for various climates in future studies. Simulations for larger areas will be performed to gain more accurate CFD inflow boundary conditions. Besides, we will also try to use the Weather Research and Forecasting (WRF) and CFD methods to obtain more complete data for the inflow boundary.

Author Contributions: Z.J. developed the original idea, designed the methodology, and wrote the original draft, which was revised by W.G. Both authors have read and agreed to the published version of the manuscript.

Funding: This research received no external funding.

Institutional Review Board Statement: Not applicable.

Informed Consent Statement: Not applicable.

Data Availability Statement: Not applicable.

Acknowledgments: The authors are thankful to anonymous reviewers for their helpful comments.

Conflicts of Interest: The authors declare no conflict of interest.

References

1. Xiao, M.; Simon, S.; Pregger, T. Energy System Transitions in the Eastern Coastal Metropolitan Regions of China—The Role of Regional Policy Plans. *Energies* **2019**, *12*, 389. [CrossRef]
2. Zhong, S.; Qian, Y.; Zhao, C.; Leung, R.; Wang, H.; Yang, B.; Fan, J.; Yan, H.; Yang, X.; Liu, D. Urbanization-induced urban heat island and aerosol effects on climate extremes in the Yangtze River Delta region of China. *Atmos. Chem. Phys.* **2017**, *17*, 5439–5457. [CrossRef]
3. Du, Y.; Mak, C.M. Improving pedestrian level low wind environment in high-density cities: A general framework and case study. *Sustain. Cities Soc.* **2018**, *42*, 314–324. [CrossRef]
4. Peng, L.; Liu, J.P.; Wang, Y.; Chan, P.W.; Lee, T.C.; Peng, F.; Wong, M.S.; Li, Y. Wind weakening in a dense high-rise city due to over nearly five decades of urbanization. *Build. Environ.* **2018**, *138*, 207–220. [CrossRef]
5. Vervoort, R.; Blocken, B.; Van Hooff, T. Science of the Total Environment Reduction of particulate matter concentrations by local removal in a building courtyard: Case study for the Delhi American Embassy School. *Sci. Total Environ.* **2019**, *686*, 657–680. [CrossRef] [PubMed]
6. Qin, H.; Lin, P.; Siu, S.; Lau, Y.; Song, D.X. Influence of site and tower types on urban natural ventilation performance in high-rise high-density urban environment. *Build. Environ.* **2020**, *179*, 106960. [CrossRef]
7. Lin, G.; Zhang, S.; Zhong, Y.; Zhang, L.; Ai, S.; Li, K.; Su, W.; Cao, L.; Zhao, Y.; Tian, F.; et al. Community evidence of severe acute respiratory syndrome coronavirus 2 (SARS-CoV-2) transmission through air. *Atmos. Environ.* **2021**, *246*, 118083. [CrossRef]
8. Coccia, M. How (Un)sustainable Environments Are Related to the Diffusion of COVID-19: The Relation between Coronavirus Disease 2019, Air Pollution, Wind Resource and Energy. *Sustainability* **2020**, *12*, 9709. [CrossRef]
9. Huang, T.-L.; Kuo, C.-Y.; Tzeng, C.-T.; Lai, C.-M. The Influence of High-Rise Buildings on Pedestrian-Level Wind in Surrounding Street Canyons in an Urban Renewal Project. *Energies* **2020**, *13*, 2745. [CrossRef]
10. Blocken, B.; Stathopoulos, T.; Van Beeck, J.P.A.J. Pedestrian-level wind conditions around buildings: Review of wind-tunnel and CFD techniques and their accuracy for wind comfort assessment. *Build. Environ.* **2016**, *100*, 50–81. [CrossRef]
11. Tsang, C.W.; Kwok, K.C.S.; Hitchcock, P.A. Wind tunnel study of pedestrian level wind environment around tall buildings: Effects of building dimensions, separation and podium. *Build. Environ.* **2012**, *49*, 167–181. [CrossRef]
12. Zhen, M.; Zhou, D.; Bian, G.M.; Yang, Y.; Liu, Y. Wind environment of urban residential blocks: A research review. *Archit. Sci. Rev.* **2019**, *62*, 66–73. [CrossRef]
13. Blocken, B. 50 years of Computational Wind Engineering: Past, present and future. *J. Wind Eng. Ind. Aerodyn.* **2014**, *129*, 69–102. [CrossRef]
14. Liu, S.; Pan, W.; Zhang, H.; Cheng, X.; Long, Z.; Chen, Q. CFD simulations of wind distribution in an urban community with a full-scale geometrical model. *Build. Environ.* **2017**, *117*, 11–23. [CrossRef]

15. Park, B.; Lee, S. Investigation of the Energy Saving Efficiency of a Natural Ventilation Strategy in a Multistory School Building. *Energies* **2020**, *13*, 1746. [CrossRef]
16. Ying, X.; Wang, Y.; Li, W.; Liu, Z.; Ding, G. Group Layout Pattern and Outdoor Wind Environment of Enclosed Office Buildings in Hangzhou. *Energies* **2020**, *13*, 406. [CrossRef]
17. Hang, J.; Li, Y. Age of air and air exchange efficiency in high-rise urban areas and its link to pollutant dilution. *Atmos. Environ.* **2011**, *45*, 5572–5585. [CrossRef]
18. Li, Y.; Chen, L. Study on the influence of voids on high-rise building on the wind environment. *Build Simul.* **2020**, *13*, 419–438. [CrossRef] [PubMed]
19. Wang, X.; Chen, Y.; Sun, N.; Chen, S.; Li, Q. Wind tunnel experiment and optimization design of wind environment around buildings with parallel layout. *J. Zhejiang Univ. Sci. A* **2012**, *46*, 454–462. (In Chinese) [CrossRef]
20. Hadavi, M.; Pasdarshahri, H. Quantifying impacts of wind speed and urban neighborhood layout on the infiltration rate of residential buildings. *Sustain. Cities Soc.* **2020**, *53*, 101887. [CrossRef]
21. Hu, K.; Cheng, S.; Qian, Y. CFD Simulation Analysis of Building Density on Residential Wind Environment. *J. Eng. Sci. Technol. Rev.* **2018**, *11*, 35–43. [CrossRef]
22. Hong, B.; Lin, B. Numerical studies of the outdoor wind environment and thermal comfort at pedestrian level in housing blocks with different building layout patterns and trees arrangement. *Renew. Energy* **2015**, *73*, 18–27. [CrossRef]
23. Ghadikolaei, F.M.; Ossen, D.R.; Mohamed, M.F. Effects of wing wall at the balcony on the natural ventilation performance in medium-rise residential buildings. *J. Build. Eng.* **2020**, *31*, 101316. [CrossRef]
24. Tse, K.T.; Zhang, X.; Weerasuriya, A.U.; Li, S.W.; Kwok, K.C.S.; Mak, C.M. Adopting 'lift-up' building design to improve the surrounding pedestrian-level wind environment. *Build. Environ.* **2017**, *117*, 154–165. [CrossRef]
25. Guo, F.; Zhu, P.; Wang, S.; Duan, D.; Jin, Y. Improving Natural Ventilation Performance in a High-Density Urban District: A Building Morphology Method. *Proc. Eng.* **2017**, *205*, 952–958. [CrossRef]
26. Kubota, T.; Miura, M.; Tominaga, Y.; Mochida, A. Wind tunnel tests on the relationship between building density and pedestrian-level wind velocity: Development of guidelines for realizing acceptable wind environment in residential neighborhoods. *Build. Environ.* **2008**, *43*, 1699–1708. [CrossRef]
27. GB 50180-2018. In *Standard for Urban Residential Area Planning and Design*; Ministry of Housing and Urban-Rural Development of the People's Republic of China: Beijing, China, 2018.
28. Deng, F. Comparative study on gated communities of China and the western private community. *Urban Probl.* **2011**, *11*, 2–8. (In Chinese) [CrossRef]
29. Liu, S.; Pan, W.; Zhao, X.; Zhang, H.; Cheng, X.L.; Long, Z.; Chen, Q. Influence of surrounding buildings on wind flow around a building predicted by CFD simulations. *Build. Environ.* **2018**, *140*, 1–10. [CrossRef]
30. Zhuang, Z.; Yu, Y.; Ye, H.; Tan, H.; Xie, J. Review on CFD Simulation Technology of Wind Environment around Buildings. *Build. Sci.* **2014**, *30*, 108–114. (In Chinese) [CrossRef]
31. Tominaga, Y.; Mochida, A.; Yoshie, R.; Kataoka, H.; Nozu, T.; Yoshikawa, M.; Shirasawa, T. AIJ guidelines for practical applications of CFD to pedestrian wind environment around buildings. *J. Wind Eng. Ind. Aerodyn.* **2008**, *96*, 1749–1761. [CrossRef]
32. Shih, T.H.; Liou, W.W.; Shabbir, A.; Yang, Z.; Zhu, J. A new k-ε eddy-viscosity model for high Reynolds number turbulent flows–model development and validation. *Comput. Fluids* **1995**, *24*, 227–238. [CrossRef]
33. Blocken, B.; Persoon, J. Pedestrian wind comfort around a large football stadium in an urban environment: CFD simulation, validation and application of the new Dutch wind nuisance standard. *J. Wind Eng. Ind. Aerodyn.* **2009**, *97*, 255–270. [CrossRef]
34. GB50009-2012. In *Code for Loads of Building Structures*; Ministry of Housing and Urban-Rural Development of the People's Republic of China: Beijing, China, 2012.
35. Du, Y.; Mak, C.M.; Kwok, K.; Tse, K.T.; Lee, T.; Ai, Z.; Liu, J.; Niu, J. New criteria for assessing low wind environment at pedestrian level in Hong Kong. *Build. Environ.* **2017**, *123*, 23–36. [CrossRef]
36. Lawson, T.V. The Wind Content of the Built Environment. *J. Wind Eng. Ind. Aerodyn.* **1978**, *3*, 93–105. [CrossRef]
37. Willemsen, E.; Wisse, J.A. Design for wind comfort in The Netherlands: Procedures, criteria and open research issues. *J. Wind Eng. Ind. Aerodyn.* **2007**, *95*, 1541–1550. [CrossRef]
38. Ng, E. Policies and technical guidelines for urban planning of high-density cities–Air ventilation assessment (AVA) of Hong Kong. *Build. Environ.* **2009**, *44*, 1478–1488. [CrossRef]
39. Ma, T.; Chen, T. Classification and pedestrian-level wind environment assessment among Tianjin's residential area based on numerical simulation. *Urban Clim.* **2020**, *34*, 100702. [CrossRef]
40. Janssen, W.D.; Blocken, B.; van Hooff, T. Pedestrian wind comfort around buildings: Comparison of wind comfort criteria based on whole-flow field data for a complex case study. *Build. Environ.* **2013**, *59*, 547–562. [CrossRef]
41. DB33/1111-2015. In *Design Standards for Wind and Thermal Environment of Residential Buildings*; Zhejiang Housing and Urban-Rural Development Department: Beijing, China, 2015.

energies

MDPI

Article

Bushfire: Retrofitting Rural and Urban Fringe Structures—Implications of Current Engineering Data

Glenn P. Costin

School of Architecture & Built Environment, Faculty of Science, Engineering and Built Environment, Deakin University, Geelong, VIC 3220, Australia; g.costin@deakin.edu.au

Abstract: Since the 2009 Black Saturday bushfires in which 173 lives were lost, two-thirds of whom died in their homes, the question of what a home prepared for bushfire looks like has been repeatedly raised. The 2019/2020 fires saw us not much further advanced. This paper seeks to consolidate what is known about bushfire behavior, its influence upon structures, and, through this data, infer improved standards of practice for retrofitting rural and urban fringe homes. In particular, the prevention of ember and smoke incursion: the data suggesting the prior as the main mechanism of home destruction; the latter as high risk to sheltering occupant health. The article is framed around a comprehensive literature review, and the author's own experiences and observations from fire impacted structures in Victoria's northeast. The article's import lies in demonstrating how embers and smoke may enter homes otherwise seen to be appropriately sealed prior to the fire's approach. Included in the findings are developed hypotheses based on thermal expansion, pressure differentials and backdraft; offering defined paths towards future research. In addition, the work provides practical advice towards mitigating the identified issues using retrofit practices based upon the author's practical experience as a tradesperson and building designer.

Keywords: bushfire; retrofit; ember attack; pressure differential; urban fringe; rural housing; pyro-tornadogenesis; backdraft

Citation: Costin, G.P. Bushfire: Retrofitting Rural and Urban Fringe Structures—Implications of Current Engineering Data. *Energies* **2021**, *14*, 3526. https://doi.org/10.3390/en14123526

Academic Editors: Mark Luther, Igor Martek and Mehdi Amirkhani

Received: 19 April 2021
Accepted: 24 May 2021
Published: 14 June 2021

1. Introduction

Prior to the Black Saturday fires of 2009 Australian bushfire policy could be summed up by the phrase 'stay or go'. The premise underlying this approach suggests most homes lost to bush fire succumb to ember attack, not the fire front [1–8]. Received wisdom held, holds today, that ember attack can be defended against, and thus many homes saved. The policy, however, was based upon another premise: that homes were prepared, and residents mentally and physically capable of such defense. The year 2009, and the loss of 173 lives—two-thirds of whom died in their homes—changed that perspective radically [2,3].

Whilst conceptually 'stay or go' still exists, active defense risks to homeowners, discussed by many including state and territory fire authorities [2,3,9–12] have altered the underpinning message. Today, Australians are advised to prepare property before the bushfire season, then leave early should a fire start. In conditions categorized as Catastrophic or Code Red (state dependent categories) the advice is to prepare the home and leave *before* a fire event begins [13,14]. In emergencies, with or without a 'state of emergency' declaration, mandatory evacuations may be ordered, though the legalities of forced removal from a home property are debatable, state specific and unclear [15,16].

Occasionally evacuation is not possible; on others, the indicators of potential fire are low, and communities are taken by surprise. In December of 2015, over 100 homes were lost in the Victorian coastal community of Wye River [6]. On that occasion, the McArthur Forest Fire Danger Index (FFDI)—the measure by which Australia's fire danger levels of High, Very High, Severe, etc., are identified—was only 49 or 'Very High' [6]. Code Red or Catastrophic is 100+. Fortunately, due to the fire's approach direction, no lives were lost.

Both before [17] and after [18] that event, the FFDI as a sole indicator of extreme bushfire risk has been considered questionable. This aside, what remains clear is that homes were not prepared adequately. The 2019/2020 fires again exposed this truth, with fires raging, and homes lost, in all states and territories of the country except the capital. NSW and Victoria took the brunt of losses, with over 2800 houses destroyed in those two states alone. Sadly, on this occasion, 34 human lives were also directly lost [19]. Indirectly, a further 417 lives were lost and 4456 hospitalized due to smoke inhalation [20].

Such deaths give rise to another purpose behind retrofitting: air quality. Occasionally the house is the only shelter available; though Dengate [9] notes that some people decide to stay, others have no such option [2,3,10,21]. Retrofitting will not turn an older home into a bushfire bunker, but it will improve its potential for occupant survival. Much retrofitting discussion is about sealing the home against embers, however as identified above, smoke is just as important, studies also identifying in utero growth restriction of babies through the mother's bushfire smoke inhalation [22,23]. Examples of new homes purposefully built to withstand bushfire attacks without smoke incursion are promising [24]: improving indoor air quality should likewise be a consideration in retrofitting [25].

So what does a home prepared for bushfire actually look like? Specifically, how might we retrofit a home to improve it, and its resident's, preparedness? When constructing a new home in Australia, performance requirements are found in the National Construction Code (NCC) [26] and the standard *AS3959 Construction of buildings in bushfire-prone areas* [27]. However, that standard's veracity is questioned both in this article and by others [28], due to its currently limited interpretation of ember attack mechanisms. Further advice for new and existing homes is given by state and territory fire agency guides on land and property preparedness. Yet, whilst commentary on defensible Australian homes began as early 1945 [1], only recently has discussion focused on retrofitting.

This paper explores the efficacy of these approaches, and AS3959, by drawing upon extant literature, current engineering modeling, reported experiences of homeowners and the author's own experiences. Improved retrofitting actions are proposed based upon these findings; particularly issues arising from potential pressure differentials in extreme bushfire events.

2. Researching Bushfire and Its Influence Upon Structure

What causes a structure to burn is best studied through careful analysis of burnt buildings [1,4–6,29], practical experimentation [30–32], or, when such is problematic (missing evidence, cost, risk), through computer modeling [33,34]. Much of this work has been, or is being, undertaken in facilities around the globe. This article consolidates that material, combining a comprehensive literature review and anecdotal evidence. In reviewing the literature a broad range of terms and phrases were chased through online search engines, academic libraries, Springer, Scopus, Google Scholar and the like. To this was added news reports, blogs, drilled down for quality evidence such as photographs or events witnessed by multiple parties. The main terms used derived from previously understood retrofitting actions and basic bushfire defense systems such as: bushfire sprinklers; intumescent paints; water tanks and bushfire; glazing; ember attack and the like. These terms were followed by those surrounding the hypothesis of pressure differentials, cyclonic winds, pyrogenic winds and such, prior to turning to the specifics of structural components and materials under temperature extremes.

Within this core of articles, the hypothesis of pressure differential was explored; seeking more information as the data exposed questions, potential answers, new terminology: testing the hypothesis in light of known data and reported experiences. Future, targeted, research was thus developed in pursuit of insights to aid retrofitting of older homes, guide the construction of new homes, improve legislated standards and, perhaps, solve mysteries surrounding why seemingly secure homes are destroyed by bushfire, whilst others, much older and more decrepit, survive. To interpret the insights gained from this exploration, an understanding must first be had of the structure, design and context of the typical Australian home; be it rural, or within the rural-urban fringe interface.

The Australian Home: A Structural Description

The rationale for retrofitting existing dwellings against bushfire attack is clear. Also clear, within the literature, is that ember ingress is the main concern. To understand how ingress occurs requires appreciation of the typical Australian home's structure. Generally, this is a one or two story timber or steel frame building clad with brick veneer, timber weatherboard, cement sheet or rendered panels of autoclaved aerated concrete or polystyrene. Roof cladding is either corrugated metal sheeting or tiles of concrete or glazed terracotta. Windows are usually timber or light aluminum framing, frequently with single panes of glass as little as 3 mm in thickness. More recent homes will have thicker glass (5 mm) and or be double glazed. The whole structure, with rare architectural exceptions, will be constructed upon either timber, concrete or steel stumps or posts, brick piers or a concrete slab. Raised floor structures may have this underfloor space enclosed by a vented brick dwarf wall, be partially enclosed with timber battens, or be fully open.

Common to homes is some form of verandah. In addition, there may be a deck, generally of timber, though on occasions steel-framed, with timber or polymer strip flooring. Eaves of 450–600 mm are typical, though recent trends towards 'eave-less' homes have reduced this to gutters only in many cases. Fascias are frequently timber, or more recently, light steel. Eave soffits are usually cement sheet lined, though older homes may have timber strips venting the roof space. Further roof venting may be available through grills or slats in gables or via static or spinning metal vents. Guttering is usually of galvanized or pre-painted steel, though aluminum and PVC systems exist. Generally, gutters feed to above or below ground water tanks made of steel, fiberglass, plastic or concrete. On occasions, water may be stored under raised floor houses.

In older homes doors and windows seldom seal well, particularly door sills. Weatherboard homes commonly have gaps where external architraves overlap boards. Others may not be flashed correctly at the heads, leaving gaps for embers to settle or be driven into the framework. Some windows will simply not close properly, or have openings in bathrooms or toilets to prevent condensation or odors.

The above demonstrates that most homes under consideration provide ample opportunity for ember ingress. At first glance, retrofitting these homes against bushfire is not a complex task. However, even at a basic level, it requires an understanding of their weaknesses and an appreciation of how bushfires compound these weaknesses through dynamic, high velocity, winds, extreme temperatures and pressure differentials.

3. What Is Known of How Houses Burn in Bushfires

Recent research [4–7] strongly supports Barrow's 1945 [1] suggestion that embers were the most significant cause of house destruction. The fires at Tathra, NSW, in March 2018 furthered this understanding. In that instance, only 32 of the 69 homes destroyed were in a decreed bushfire prone area [35]. The rest were within the seaside township where neither flames nor fire front heat flux had influence. Research has suggested that ember penetration into urban or suburban areas is typically less than 700 m [36], however the Canberra fires of 2003 destroyed homes in the suburb of Lyons over 2 km distant from the fire front [29]. Notably, the Australian Institute for Disaster Resilience (AIDR) Canberra Fires Field Report [29] documents bushfire generated winds damaging homes well before flame arrival; leading to an extremely high percentage (91%) of totally destroyed homes: i.e., once fire took hold, invariably the house was lost.

However, stating ember attack alone is insufficient as it fails to account for the randomization of its effect. In research [1,37,38], news reports [39] and blogs [40], there are multiple examples of homes that 'should' have been destroyed, yet survived intact despite abutting seemingly secure but destroyed homes. Whilst chance must be conceded, there are potentially other factors in play.

During a bushfire, residents are advised to seal homes as much as possible: citing Barrow's proposition ' ... a house should be as air-tight as practicable ... ' [1] (p. 1). Windows, doors, vents, everything should be blocked to prevent ember incursion. With modern homes, this is reasonably achievable: potentially to the point of fault. An untested hypothesis held by this author, to some degree suggested by Ghaderi et al.'s modeling [33], is that of pressure differentials in bushfire contexts: Differentials caused by pyrogenic winds—fire generated winds high in both temperature and velocity—impacting homes incapable of rapid pressure equalization. In such instances at least two things may challenge the integrity of a building: extreme uplift forces; and rapid, potentially destructive, pressure equalization.

The first, uplift, is well understood and engineered for, particularly in cyclonic zones, through AS 1684.3 [41] and AS 4055 [42]. Unfortunately, bushfires also occur in non-cyclonic zones where homes are designed for significantly lower wind pressures. That extreme bushfires can create their own weather systems is also well understood. Pyrocumulonimbus clouds and their associated fire thunderstorms bring about fierce winds, downbursts, and in extreme cases, pyro-tornadogenesis—fire tornadoes generated by the rotating convective winds of the cloud [43]. Video evidence of bushfires assaulting Canberra suburbs Kambah and Chapman in 2003 [43,44] show roof sheeting falling kilometers from the fire front, and the fire tornado's enormity—approximated at 500 m base diameter [45].

Pyro-tornadoes aside, Ghaderi et al.'s investigation [33] into wind-driven surface fires demonstrates that the presence of a structure alters the intensity and dynamics of that fire, supporting earlier work by Honey and Rollo [34]. A range of factors deriving from this study have great import into how homes are challenged by fire, and hence, may be defended. One notable outcome [33] (p. 12) shifts our interpretation of fire behavior from steady heat action (the premise behind AS3959) to dynamic pulsation, uplifting vortices, extremely low pressures and high velocity reverse airflows (reflective of Sharples et al. 2012 wind-terrain modeling [46]).

Figures 1–3 below, evidence these key influences upon a standard flat plane structure (effectively a 6 m × 6 m × 6 m box replicating a house) downstream from an oncoming fire. The main points of interest are as follows:

- The fire produces a low pressure zone immediately downstream from its source which draws the flames forward, (colored red) at velocities significantly greater than the wind (inlet velocity) driving it.
- That ground hugging behavior is significantly foreshortened when confronted by a building, at which point the winds and flames flow upwards.
- Immediately behind the structure, there are fast moving reverse air flows (blue zones) that also drive the plume upwards.
- The fire's momentum immediately in front of the building slows, becomes intermittent; pulsating more rapidly (twice the frequency, P1 circled in Figure 3b 0.93 Hz) than when moving over open ground (P2 & P3 circled at 0.42 Hz and 0.46 Hz respectively).
- This higher frequency raises the convective heat load through increased periods of surface contact, whilst limiting any cooling potential.

Additionally, radiative heat flux was determined to focus upon upper wall portions facing the fire, whilst the sides, rear and roof are significantly less affected.

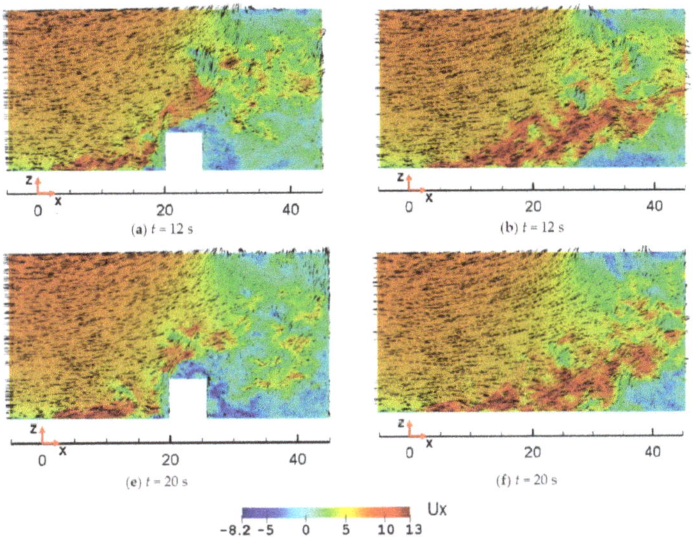

Figure 1. In subfigures (**a**) and (**e**) the white box instances the presence of a 6 m cubic building. Subfigures (**b**) and (**f**) show the fire passing by the building 9 m distance from its center. I.e., 6 m from the side [33] (p. 9). Colorbar represents velocity in m/s along x axis.

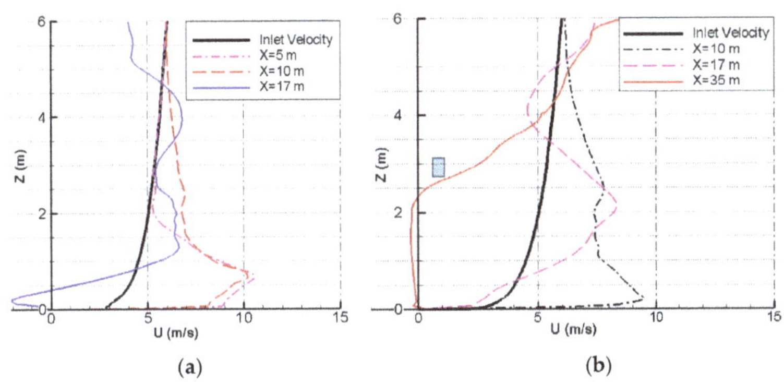

Figure 2. Subfigures (**a**,**b**) show the vertical profile of the airflow's changing longitudinal velocity at different longitudinal positions relative to the inlet location. Subfigure (**a**) shows airflow aligned with the building's center line. Subfigure (**b**) shows airflow passing 9 m on either side of center line [33] (p. 10).

Of particular interest is the identified wind velocities. The fire front wind velocity is seen to be significantly greater than the driving input. I.e., the fire is drawn forward by self-generated low pressures, not simply driven forward by high winds. The input wind velocity in this model is only 6 km/h. Thunderstorms commonly generate wind gusts of 90 km/h, more damaging storms, 160 km/h; cyclones may exceed 360 km/h [47]. The velocity the fire itself travels at in such conditions is discussed in Sharples et al.'s study of the Canberra 2003 fires [46]. Such dynamic low-hi-low heat, wind velocity and pressure fluctuations place inordinate stress upon a structure.

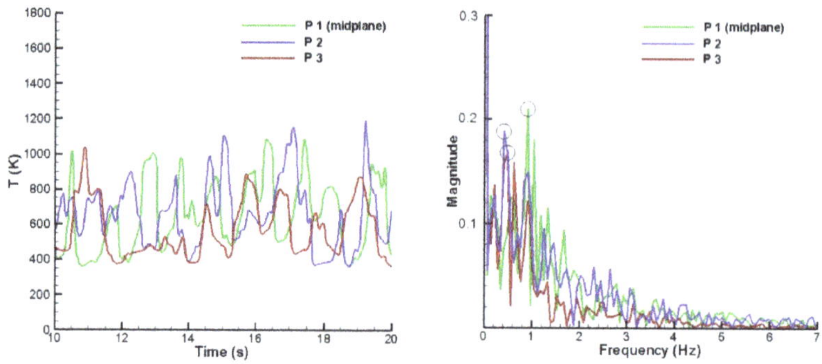

(**a**) (**b**)

Figure 3. Subfigure (**a**) shows the instantaneous surface temperature as measured in Degrees Kelvin at 3 points located on the surface 5 m from the building front. P1 is directly on the building's center line, whilst P2 and P3 are located 9 m on either side of the center. Subfigure (**b**) shows the frequency of peak temperature signals at these same sensors [33] (p. 12).

The significance of the above regarding ember attack is multifaceted: particularly ember disposition being the rear or downstream side of the structure, not the surface facing the fire front. This is concerning for homes with verandahs or other semi enclosed spaces on this side. Also, embers attack both the front and rear of the structure at an upwards trajectory, exposing any weakness under eaves and verandahs.

Figure 4 describes high velocity winds over and around a structure with (b) and without fire (a). Figure 4b demonstrates that in the presence of fire, myriads of small vortices initially hug the ground, then buoyancy inducing forces dominate, creating larger vortices that randomly cross hatch the structural zone generating significant uplift: The structure is attacked simultaneously by embers, heat flux, and wind loads from multiple directions, introducing dynamic loading at numerous points and angles.

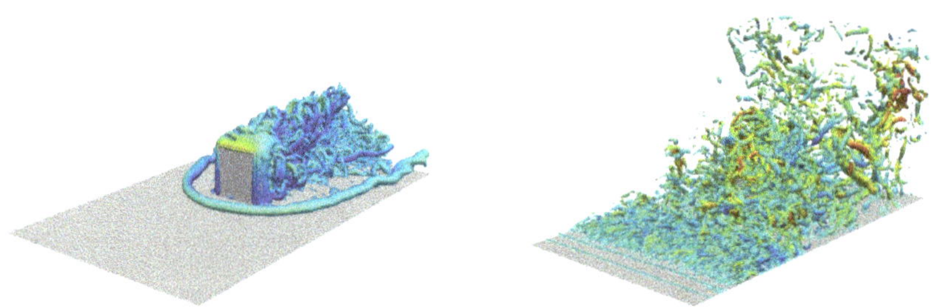

(a) (b)

Figure 4. The vertical airflow structures in the presence of fire (**b**) and without (**a**) [33] (p. 11).

A defensively well sealed building now works against itself as rising temperatures and, to a lesser degree, pressures, occur within the home. What the internal temperature and internal-external pressure differential might be is not addressed in this modeling—nor in any modeling to date. However, it may be assumed that this differential at least adds behind surface loading to uplift forces already existent through high velocity winds.

Coupled with the dynamic fire front heat shock inflicted upon windows, this pressure may suffice to fracture or shatter glass. In either case, cladding and windows of the structure may be compromised: creating gaps and openings, allowing ember ingress and potentially an explosive backdraft action.

The temperatures modeled are also concerning for exposed structural steel elements. It is well argued [48] that at around 600 °C—steel's yield strength drops by approximately

60%, Young's modulus falling by 70%. Given high wind loads, and Figure 3a showing surface temperatures fluctuating between 400 °K and 1000 °K (125 °C to 725 °C), exposed structural steel failure is potentially high. In relation to ember incursion, failure need not involve collapse—columns, beams or lintels need only flex—creating gaps sufficient to allow ember ingress; ingress being considered possible through gaps as little as 2 mm [27].

Understanding the forces affecting structures as dynamic, not stable is therefore crucial to both new homes and retrofitting. Buildings must not be viewed as static entities: rather, as vibrating, pulsating structures attempting to 'breathe', whilst swaying, rocking, being twisted, even if not visibly so.

3.1. *Inference: Structural and Cladding Integrity; Ember Incursion*

Current modeling does not reflect an actual house with its complexities of form. However, the aforementioned studies provide sufficient data to highlight key areas of risk from ember attack; suggesting useful actions. Rising internal pressures would at first seem a positive defense outcome. I.e., positively pressurizing a room, corridor, or stairwell is a typical fire defense strategy in high-rise buildings [26]. The problem comes, however, with the principles of backdraft and flashover (the latter discussed by Caird Ramsay et al. [37,38]). Flashover occurs when energy trapped within a space cycles upwards such that all materials reach their ignition point—a 'fuel-dependent phenomenon' [49] (p. 55). Backdraft occurs when a closed compartment, bordering on combustion but low on oxygen, is suddenly provided oxygen through an opening—such as broken windows [49]. Add fumes inside a home from furnishings, finishes and fittings [50], frequently volatile organic compounds (VOCs)—Fleischmann et al.'s unburned hydrocarbons [51]—and a rapid combustion event may occur acting explosively outwards.

Not enough is known about VOC off-gassing in homes experiencing extreme bushfire temperatures to state authoritatively that these could amount to an explosive event. Despite multiple anecdotal and media accounts of buildings 'exploding' [52], evidence after the fact is insufficient to support, or refute, such accounts [4,6,37,38]. However, the data above suggests the potential should be conceded. Regardless, the potential for cladding breaches and backdraft remains. Finding means by which to reduce internal air pressures safely, therefore, forms part of the discussion on refitting strategies that follows. Before developing that discussion, an outline of applicable Australian Standards [27,41,42,53] is requisite to understanding contemporary approaches to construction and retrofitting of homes in bushfire zones.

3.2. *Australian Standards and the Bushfire Attack Level (BAL)*

Whilst not required for all home retrofits, AS3959 [27] remains the preeminent guide for Australian housing in bushfire zones. Integral to this standard is the Bushfire Attack Level (BAL) rating system. Briefly, the 'BAL' is a statement of the likely severity of a site or building's exposure to bushfire and ember attack. A radiant heat flux range statement in kW/m^2, it is an evaluation of a structure's context. The evaluation includes:

- Identified Fire Danger Index (FDI) for a given location in Australia
- Vegetation classification
- Slope of the land under this classified vegetation
- Distance the classified vegetation is from the building

Derived ratings are either BAL-LOW, 12.5, 19, 29, 40, or FZ (Flame Zone). The higher the BAL, the more threatened a structure. Once identified, guidance on acceptable construction is given to each BAL level through the major sections of the standard. A typical visual guide is given in Figure 5 below:

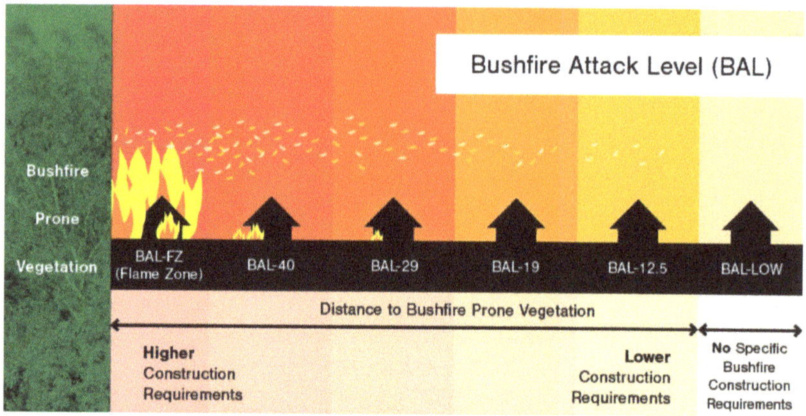

Figure 5. Pictorial description of Bushfire Attack Level (BAL) ratings [54].

AS3959 becomes challenged when dynamic, not static, heat flux is considered. As others have found [55], even these static values are questionable when detailed modeling is applied. Add the previously identified vortices, uplift and negative pressure loadings and the construction provisions lose relevance. This does not negate the standard's total relevance, however, BAL must now be understood as indicative rather than definitive—potentially a significant understatement. When retrofitting or designing and constructing new buildings far greater attention to detailing, hold down and material choices should apply.

The BAL system of identifying risk is further challenged when Figure 5 is reflected upon in light of documented home losses; many destroyed despite BAL-LOW ratings. It fails to account for ember attack many kilometers ahead of the fire front, frequently within town or urban fringe areas many streets back from classified vegetation. Yet the BAL system remains valuable in identifying radiant heat and flame proximity for given locations. It allows homeowners, when considering retrofitting, to better prepare land around buildings to reduce heat and flame levels. But with regards to embers, distance from the flame front is not a safeguard.

3.3. *Australian Standards and Wind Ratings*

Wind ratings for Australian houses are defined by AS 4055 [42], or AS1170.2 [56]. For the purposes of this paper, outlining AS4055's rating system will suffice for exampling wind speeds in a later discussion. Ratings within AS4055 are given as either N1 through to N6, or C1 through to C4. N or C ratings depend upon regional location; 'C' referencing areas likely to experience cyclonic conditions. The individual ratings suggest limit state (serviceability and ultimate) design wind gust speeds. Stated in m/s, classes N3 through to N6 have the same design gust speeds as C1 through to C4, being: 32 (50 ultimate), 39 (61), 47 (74) and 55 (86) respectively. The difference between the N and C becomes evident when calculating positive and negative pressures applicable to given surfaces or building elements (roof, walls or the like). For example, a C2 rated roof surface has a general ultimate limit state pressure of −3.21 kPa (a net uplift), whereas N4, with the same gust speed of 61 m/s, is deemed to have a net uplift pressure of only −2.21 kPa. These derive from a table of pressure coefficients recognizing the dynamic low-hi-low pressures sustained in cyclonic events. Ghaderi et al.'s [33] study suggests that such dynamic pressures evolve from bushfires as well.

4. Retrofitting: Applied Implications

Conceding that embers cause most bushfire home losses, the ambition of retrofitting is to prevent their ingress and eliminating or reducing ember deposition zones of high ignition risk. The previous section's guidance support's most government or fire authority recommendations. However, the material allows review of existing retrofitting publications more informatively, several of which are outlined briefly below. The following then discusses key areas of bushfire attack—pressure differentials, uplift, angles of attack, counter flowing winds and the like—relative to those structural elements most influenced, seeking to mitigate such attack and inhibit ember and smoke ingress. In making recommendations it is conceded that retrofit activity is framed by what is 'reasonably practicable'. Available time, financial resources, structural element access and emotive connection to home and contents, all influence perceptions of achievability.

A recently published bushfire resource for the construction and retrofitting of homes in Queensland [57], has relevance beyond that state. Particularly useful is the Level of Protection system that informs builders/owners of the protection achievable with a particular range of actions. An extensive document, it also directs builders to particular sections of the relevant standards [27,41,58].

Another developing Queensland publication [59] lays the ground work for retrofitting against extreme winds. Yet despite being a Geoscience Australia and James Cook University's Cyclone Testing Station (JCU CTS) collaboration, under the banner of the Bushfire and Natural Hazards CRC, there is no linkage between bushfires and high wind loads. Despite this lack, it promises to be a valuable resource in identifying levels of protection gained against 0.2 s peak wind gusts gained from different actions undertaken [60].

The Victorian Building Authority (VBA) and Country Fire Authority (CFA) have also produced valuable guides specific to retrofits [61,62]. One deals with domestic homes whilst the other focuses upon commercial buildings defined as Class 9 under the NCC—those that might be used as last resort shelters for vulnerable communities—it holds advice relevant to domestic structures, particularly subfloor venting, piping and roof penetrations. As with the recommendations being promoted here, these documents frequently promote methods exceeding AS3959.

Reflecting on these documents in light of research implications presented earlier, the following frequently references Australian Standards and building codes to higher than 'normal' standards.

4.1. *Pressure Differentials, Dynamic Loadings and Cladding*

These are issues least addressed by current retrofitting advice and AS 3959 generally. Of particular import is the dynamic loading bushfires impose: loads suggesting cyclonic standards (such as cladding fixing AS 1562.1 [63]) whereby materials must satisfy a Low-Hi-Low testing regime that 'rattles' elements over a seven stage sequence of varying pressures. The only guide to this procedure, and any claims of compliance, is a draft document published in April 2009 by JCU CTS [64].

The following 'grounds' this discussion in practice: i.e., how does the preceding data inform the typical Australian home retrofit? Though only common building elements are addressed, the principle of increased attention to tie down may be applied more generally. At the section's end, the discussion turns to idealized means by which internal-external pressure differentials may be alleviated.

4.1.1. Metal Sheet Roof Cladding

In older homes roof sheeting is commonly nailed, not screwed; whilst capping and flashings to many relatively new homes are pop-riveted rather than screwed—Australian standards [63] only requiring screws from 2018. Both are problematic regarding tie-down—the identified potential for high uplift pressures and dynamic wind loadings is likely to exceed their withdrawal capacity. Even if sheeting or capping is not torn away, sufficient gaps may be created at sheet ends and overlaps allowing ember ingress. Sheet roofs should thus be check and fixings upgraded to screw fastenings using cyclonic fixing patterns following AS 1684.3. Gaps between flashings and corrugated profiles should be filled with non-combustible material such as rock wool.

4.1.2. Tile Roofs (Terracotta or Cement)

In low wind areas (AS 4055 ratings N1, N2, N3), tiles tend to be tied down minimally or not at all. Given higher and dynamic wind loads associated with bushfires, tiles should be mechanically fastened to a minimum of N4 or C2 wherever there is the possibility to do so. Appropriate sarking should be provided under the tiles, though this may be considered impractical in many instances due to the costs associated. When retrofitting incorporates new cladding, appropriate sarking must be installed, and sheeting rather than tiles considered.

4.1.3. Further Roofing Considerations

In the above cases, the tie-down has only been achieved for the cladding. This remains inadequate if tie-down of battens to rafters, rafters to wall plates and studs, is inadequate. Hence the concept of 'reasonably practicable'; i.e., it depends upon owner willingness or financial capacity. Yet, where possible, tie-down fixings should be upgraded at these points. AS 1684.3 provides appropriate guidance in this area. It is important to work to these higher wind loads, not standard contextual loads which may be as low as N1—which the previous information has shown inadequate in extreme bushfire scenarios.

4.1.4. Wall Cladding

Fixings to sheet cladding (steel, ply, cement sheet), should be checked for security and upgraded anywhere movement is detected. Gaps and cracks should be filled. Likewise for weatherboards; replacing damaged boards. Older homes may have no sarking, with the upward trajectory of embers, gaps in board overlaps become susceptible, allowing embers to settle within the wall cavity. Eave/wall junctions, re-entrant and external wall corners, should be carefully inspected; loose or damaged stops or cover straps replaced, gaps filled. Intumescent paints may be considered to protect against fire front heat flux, or low areas were embers may settle, however, the service life of such coatings remains doubtful [65]. Closing gaps to prevent ember penetration is the main aim.

4.2. Subfloor Protection

As per existing retrofitting guides, these areas should be shielded by steel mesh and close fitting battens. In the case of brick dwarf walls, gaps or cracks should be filled, vents should be upgraded if vent openings are greater than 2 mm. The junction between dwarf walls and main wall cladding should be inspected and gaps filled.

4.3. Fascia, Eave, and Guttering

This area is particularly open to ember attack due to wind approach angles—upwards at 45–60 degrees from horizontal. Generally, soffits are non-combustible cement sheet, and gutters of galvanized or painted steel. However, fascias are frequently either light steel or timber. Under heat loads, steel may deflect, allowing soffit sheets to drop from groves; joiners between sheets are generally light PVC which may melt, gaps for embers now becoming present. The soffit/wall cladding junction should also be inspected. Fixing should be checked as per previous elements.

4.4. Gutter Guards

This is an area of contention amongst homeowners, builders and suppliers/manufacturers. This author's experiences and anecdotal reflections of homeowners and architects, suggests that no available system works effectively. Metal guards are considered the most effective in being less susceptible to damage from birds and small marsupials, however, the mesh tends to trap leaf ends and grass still grows under the guard. Many such guards cannot be easily removed for cleaning. When installed, systems for flushing gutters without removal should be integrated, and regularly used.

4.5. Sarking

Sarking has been listed in several of the above recommendations. Leonard et al. [57] recommend against conventional sarking, even flammability index 5 rated (NCC requires 5 or less). That document recommends the use of flame resistant sarking, but fails to identify what that actually means. Kempster [66], however, describes bushfire roof and wall systems incorporating fireproof blankets, sarking and insulation elements tested to AS 1530 [67] offering a flammability index of 1. The materials within these systems can be individually applied as part of a retrofit program but require cladding removal to do so.

4.6. Structural Steel

Structural steel is common to Australian homes as beams, columns and light gauge wall and roof framing. In normal conditions, these elements are stable, serviceable and non-combustible. However, suffering extreme bushfire temperatures components may buckle under load having lost potentially 60–70% structural capacity. Alternatively, tie-down capacity of joints may become compromised or gaps in cladding generated: the latter through steel's linear expansion coefficient of 1.2×10^{-6}, meaning with a temperature rise of as little as 60 °C (from 40 °C ambient to heat flux induced 100 °C material temperature), a 2.4 m or 2.7 m steel wall will gain height by approximately 2 mm.

Exposed posts and beams undergo greater elongation. An 8 m long beam, may expand linearly when raised by 100 °C, as much as 10 mm. Though non-failure temperatures, issues arise when components are restrained, either by other steel components, or elements such as brick or concrete columns. Baetu et al. [48] demonstrate that constrained structural steel is more likely to fail early—before yield temperatures and loads are reached—due to constrained liner expansion causing premature buckling. This was particularly notable amongst lightweight steel components. The import here is not that structural steel will inevitably fail in bushfire conditions, rather that it may, due to even minor deflection, be a factor in exposing the internal elements of walls and roofs to ember ingress through the generation of gaps as the fire front passes.

Retrofit actions to counter the issue of steel's expansion may be in the form of:

- Intumescent coatings to exposed steel elements (columns and beams).
- Fireproof mastic sealants with high (40% or greater) expansion characteristics to any areas where expanding steel elements may lead to gap generation.
- Fireproof mastic at junction of metal fascia and soffit lining to limit lining sag should the fascia buckle outwards.
- For steel frames, similar mastics around window and door frames where they meet claddings, particularly at window heads.

4.7. Of Pipes and Penetrations

Contemporary plumbing invariably involves PVC components, often with long exposed lengths under suspended subfloors. Rigid PVC's flash and sustained ignition temperatures of approximately 400 °C and 450 °C respectively [68] are achievable in a fire front. Though unlikely to carry fire directly through a wall, roof or floor element, PVC components will melt, droop or compact as the melting range is only 115–245 °C (manufacturing process dependent) [69]. It is by this mechanism that embers and flame may gain entry into or through structural and protective elements either during the initial assault or after the fire front has passed.

Fireproof collars are available for retrofitting that hinge around PVC piping. Likewise, there are metal fittings that couple around a pipe and backfilled with fireproof mastics. Where piping is more exposed, consideration should be given to applying intumescent coatings or flameproof insulating wrap, shielding the PVC from heat to prevent unnecessary replacement.

4.8. Windows and Doors

Glazing represents a weak point in the defense against ember attack and the bushfire front itself. Many early homes may have large panes of ordinary annealed gas only 3 mm thick: totally inadequate against a fire front. Likewise, they are easily shattered by flying debris from trees or neighboring properties. More modern homes may have thicker glass, though still generally ordinary annealed.

Upgrading of windows against bushfire needs careful consideration outside the scope of this article as solutions must be highly contextualized; reflecting requirements of AS 3595, the NCC, and through this code, multiple further standards. However, Bowditch et al.'s [32] research offers some guidance. Though reporting on a range of glass and framing types exposed to bushfire temperatures, the simulation was of steady heat flux as against the rapid fluctuations observed by Ghaderi et al. [33]. Likewise, no high velocity winds were brought to bear and the glazing units stood in an open frame. The latter factors meaning that internal-external pressure differentials were not considered. These issues aside, the findings suggest that improved glazing should include 5 mm toughened glass as a minimum (particularly the exterior pane is part of a double glazed unit).

A cheaper, or more easily applied, means of upgrading window security against heat, flame and impact, is shielding. Multiple systems are available, from proprietary roller shutters to cost effective hinged panels, shutters or screens [70]. In either case, 'non-corrosive steel' (generally taken to mean stainless steel) or bronze should be used and installed such that no gaps greater than 2 mm remain.

Doors and windows, designed to open, mean small gaps are common to ensure free movement. It is imperative that these gaps have high temperature resistant seals that will prevent smoke ingress, not just embers. Standard rubber seals are generally inadequate for this purpose, likewise brush seals. High quality neoprene seals should be installed wherever possible.

5. Pressure Equalization Strategies—Requisite or Not?

Notably lacking attention, this area needs applied research and theoretical modeling. The modeling previously discussed [33] is suggestive of pressure differentials affecting bushfire attacked homes but does not confirm it. After the 1983 Ash Wednesday fires this author noted examples of untouched weatherboard homes beside destroyed brick dwellings. Similarly, following the 2008 Black Saturday fires, a grass fire destroyed a new, seemingly secure but evacuated, brick veneer home and steel shedding. Nearby, stood an old weatherboard house unscathed. This difference between the two outcomes could be 'chance': Alternatively, old cottages leak air with little resistance; whilst the new home, appropriately closed and sealed against ember incursion, failed to quickly equalize the internal-external pressure. Barrow's [1] documenting of timber clad homes with mesh covered windows and vents surviving bushfires led to our understanding of ember attack. Potentially there is another message here too.

Whilst awaiting further research it is worth considering potential remedies, particularly from a retrofitting perspective. One solution is release valve venting. Associated with every room, they must allow rapid release of pressure but disallow incursion of embers or smoke. Notably, the hypothesis is less about high pressures forming suddenly within the house, and more about sudden low pressures—relative to the interior—becoming dominant outside the structure. It is about reducing that pressure differential sufficiently to limit structural flex and component creep, or the fracturing of already heat shocked glazing, either of which may allow ember incursion and destructive backdraft ignition.

6. Defense through Water

Dousing a structure with water before a bushfire arrives has long been considered a reliable defense strategy. Several studies have been undertaken [71–74] reporting on the effectiveness of bushfire defense sprinkler systems. To this may be added news reports and testimony of those who have saved or lost homes where sprinkler systems have been deployed [40]. However, discussion remains as to its potential for heightening risk through over expenditure of water—leaving inadequate supply for fire front defense.

Water storage and supply are crucial to active fire defense systems; particularly for those who remain and defend a property believing themselves equipped to do so. Wilkinson & Eriksen [75] outline water scarcity, pump failures, power outages and melting componentry—pipes lines, hoses, fittings and even water storage tanks—as the 'weak link' in seemingly well prepared systems. They also note that reticulated supply cannot be guaranteed due to power cuts to pumping stations, or through drought depleted reserves being drawn down prior to fire front emergence. Some cities or towns may impose restrictions on reticulated water supply to conserve storage, small town or farm homes use tank or bore water and are not required to comply, making conservation discretionary.

Current water storage policy, though varying state to state, requires maintaining dedicated firefighting reserves of between 10,000 L to 22,000 L to be maintained on site depending upon block size [76–78]. This raises the question of what a viable storage container looks like. Previous research [4,6,79] suggests that polyethylene (PE) tanks are inadequate in the face of significant heat sources, even when full. However, when appropriately sited, shielded or installed underground, they may offer low cost alternatives. No data is available regarding the effectiveness of intumescent coatings to such tanks. Concrete or metal tanks were shown to have no such limitations. Of importance in all cases, however, is that connections and above ground piping should match those of the local fire authority and be of metal.

Spray and Sprinkler Systems

When appropriately designed and deployed, spray systems have shown high value in protecting homes [40,80]. What is appropriate remains widely discussed in a range of reports in Australia and internationally [40,57,61,71–75,81]. A recent study into the efficacy of airborne droplets for attenuating the effects of bushfire offers useful insights, whilst raising further doubts [73]. The key findings from these studies show that:

- Water sprayed directly onto upper wall surfaces are more than twice as effective at lowering heat flux than spraying way from the building.
- Air flows significantly influence the sprinkler effectiveness (varying almost 40%).
- Sprinklers are effective when using very high flow rates, but these rates were beyond the storage capacity of most domestic settings.
- Further investigation required to determine the efficacy of small droplets (0.1 mm) at moderate flow rates.

Post bushfire blogs and news reports tell of sprinkler efficacy in some instances, and failure in others [40,80]. The main questions arise around spray head location, pressures, flow volumes and spray direction. In the 2020–2021 Australian fires, a rural property was actively defended by the occupants with the aid of preinstalled sprinkler systems [80]. The design of the home and immediate landscape aiding their success (a green roofed structure). However, the most important asset was the water supply: more than 200,000 L held in concrete tanks. Saving the property took over 150,000 L, most expended through the sprinklers. Such volumes of water are seldom available. Arguments against sprinklers have focused on this expenditure of water prior to fire front arrival. Yet another commentary [40] argues that sprinklers did little to prevent the destruction of the home despite plentiful water supply.

Most researchers and manufacturers concur that spray head type and location are the keys to a system's success [71,82,83]. Recommendations include static low volume, high pressure, spray heads that generate a 'screen' of fine water droplets. Others use rotating heads to similar effect, deploying heads around eaves and fascia. Systems located on the roof ridge line are considered ineffective due to the high uplift winds taking the water away from the property: a premise supported by the modeling discussed earlier [33]. The modeling of Green [71], suggests spray systems should be close to a structure's walls for efficacy, directed in such a manner that water is carried back to the upper walls, wall bases and under eaves by the encroaching winds and vortices.

Most importantly, spray systems and service lines must be fireproof. Whilst CPVC piping is fire resistant and has been successfully tested for unshielded interior fire suppression systems, questions remain about their resilience after extensive exposure to UV [84,85]. The recommendation for metal systems thus remains.

7. Retrofitting and the Context: Landscaping

In 2011 Honey & Rollo [34] analyzed the influence of structural form, proximity, and the arrangement and alignment of multiple structures upon the transport, passage and deposition of embers. They also investigated the significance of air pressure and velocity, with findings not dissimilar to Ghaderi et al.'s modeling. In dealing with the form they experimented with alignment and, through raising structures off ground, introduced interrupting airflows limiting ember build up on the lee side. Potentially the most important aspect of this study is their observationally derived hypothesis concerning land sculpting in front of buildings in order to confuse wind flows and dictate preferred ember deposition [34].

A useful CFA [86] document speaks to appropriate landscape design for bushfire, however, the focus is upon materials, layouts and plant/tree selection. Sculpting of the landmass itself is not raised. In practice, such sculpting may be problematic for many homes due to block size restrictions, suggesting collaborative sculpting across multiple blocks, or council-assisted sculpting of government or state forest lands. Further studies need to be undertaken, however, before such works could be assured of success; particularly in light of Sharples et al.'s modeling and experimentation demonstrating powerful reverse airflows driving fires laterally upon ridges and rapidly up the lee sides of hilly terrain in extreme bushfire contexts [46].

Controlled burning, another landscape strategy, also needs further research having received varied commentary. A recent publication, whilst questioning clearing burns deep within the forests, strongly promotes burns proximal to infrastructure: Arguing that lower fuel loads near housing reduces heat flux levels and fire spread, making fringe control more achievable [87]. Indigenous burning and land care practices have recently been acknowledged as effective through changing the density and species of forestation: opening the landscape, slowing down encroaching fires and reducing their intensity [88–90]. Only this latter approach would appear to have a broader influence upon ember attack reduction from extreme bushfire events—through potentially limiting the likelihood of such events. It is suggested here that these practices should be promoted as part of resilient and regenerative communities: as retrofitting beyond the boundaries of structure and personal land holdings.

8. Conclusions and Retrofitting Limitations

As most previous writing on retrofitting suggests [57,59,61,62], such actions have their limitations: many alluded to, or boldly stated, in the sections above. Not least is the cost. In 2017 Penman et al. [91] found that preparing occupants and homes for bushfires ranged from AUD 8500 to as high as AUD 47,000. With 2020 construction industry annual price growth being 3.6% [92], and CPI averaging 2% over the past 4 years [93], a rise of approximately 10% on these prices is reasonable: i.e., the contemporary cost for retrofitting approximates AUD 9350 to AUD 51,700. To which end of that spectrum any retrofit comes from being totally dependent upon the depth of actions taken.

Cost, however, is not the only limitation. Retrofitting is about working with what is, towards a frequently unobtainable ideal. In many cases, gaps will remain, likewise timber window frames and walls, and the tiles will remain tiles. Installing sarking behind existing cladding is a major exercise at which many will balk. Fitting out a house with appropriate spray heads, fire pumps and adequate water storage is full of complexities—and for some, it can be as simple as a question of aesthetics on a heritage home, foolish as that may sound.

This paper has sought to consolidate existing knowledge surrounding retrofitting actions for Australian homes facing bushfire threats: augmenting that knowledge through inferences drawn from otherwise uncorrelated fields. Additionally, the concept of retrofitting has been expanded beyond singular structures, conceptualizing its role in resilient communities and collective defense strategies. In offering directions for further research, the influence of the immediate landscape and other structures has been included. The purpose of retrofitting has also been broadened to consider sheltering occupant's air quality in light of recent evidence showing significant mortalities, hospitalizations and in utero risks from smoke inhalation. More significantly, the work expands our knowledge of why homes are open to ember attacks in the first instance. In so doing, a pressure differential hypothesis has been explored using the existing engineering and modeling data. This data suggests that such a hypothesis has a level of merit that should be pursued through detailed engineering modeling and practical testing. At a minimum, the existing data strongly suggests attention to retrofitting actions that may counter the influences, however minor, of these potential pressure differentials upon glazing, claddings and openings. This, because such influences, coupled with component movements brought about by extreme temperatures and wind forces, may create small openings through which embers and smoke may encroach upon a structure's interior, leading to its demise.

Funding: This research received no external funding.

Informed Consent Statement: Not applicable.

Data Availability Statement: Not applicable.

Conflicts of Interest: The author declares no conflict of interest.

References

1. Barrow, G.J. A Survey of Houses Affected in the Beaumaris Fire, January 14, 1944. *J. Counc. Sci. Ind. Res.* **1945**, *18*, 21.
2. Haynes, K.; Handmer, J.; McAneney, J.; Tibbits, A.; Coates, L. Australian bushfire fatalities 1900–2008: Exploring trends in relation to the 'Prepare, stay and defend or leave early' policy. *Environ. Sci. Policy* **2010**, *13*, 185–194. [CrossRef]
3. Haynes, K.; Tibbits, A.; Coates, L.; Handmer, J.; McAneney, J. *100 Years of Australian Civilian Bushfire Fatalities: Exploring the Trends in Relation to the 'Stay or Go Policy'*; Bushfire CRC: Melbourne, Australia, 2008.
4. Leonard, J.; Bowditch, P. *Findings of Studies of Houses Damaged by Bushfire in Australia*; CSIRO Manufacturing & Infrastructure Technology: Melbourne, Australia, 2003.
5. Kathryn, C.; Trent, P.; Owen, P. Some Wildfire Ignition Causes Pose More Risk of Destroying Houses than Others. *PLoS ONE* **2016**, *11*, e0162083.
6. Leonard, J.; Opie, K.; Blanchi, R.; Newnham, G.; Holland, M. *Wye River/Separation Creek Post-Bushfire Building Survey Findings*; CSIRO Manufacturing & Infrastructure Technology: Melbourne, Australia, 2016; p. 65.
7. Syphard, A.D.; Keeley, J.E. Factors Associated with Structure Loss in the 2013–2018 California Wildfires. *Fire* **2019**, *2*, 49. [CrossRef]
8. Teague, B.; McLeod, R.; Pascoe, S. *The 2009 Victorian Bushfires Royal Commission Final Report*; 2010, Parliament of Victoria; Victorian Bushfires Royal Commission: Melbourne, Australia, 2009.
9. Dengate, C. Why do some people decide to stay at home during bushfire emergencies? In *The Signal*; ABC Australian Broacasting Corporation: Ultimo, Australia, 2019.
10. Handmer, J.; Tibbits, A. Is staying at home the safest option during bushfires? Historical evidence for an Australian approach. *Glob. Environ. Chang. Part B Environ. Hazards* **2005**, *6*, 81–91. [CrossRef]
11. McLennan, J.; Every, D.; Bearman, C.; Wright, L. On the concept of denial of natural hazard risk and its use in relation to householder wildfire safety in Australia. *Int. J. Disaster Risk Reduct.* **2017**, *21*, 176–186. [CrossRef]
12. Venn, T.; Quiggin, J. Early evacuation is the best bushfire risk mitigation strategy for south-eastern Australia. *Aust. J. Agric. Resour. Econ.* **2017**, *61*, 40. [CrossRef]
13. CFA. Plan & Prepare: Leave Early. 2021. Available online: https://www.cfa.vic.gov.au/plan-prepare/leave-early (accessed on 28 March 2021).
14. RFS. Bushfire Safety: Staying and Defending. 2021. Available online: https://assessmyrisk.rfs.nsw.gov.au/docs/Stay_and_Defend_Factsheet.pdf (accessed on 28 March 2021).
15. Eburn, M. Legality of Forced Evacuations during NSW Bushfires. Criminal Law, Fire, Legislation and Plans 2014. Available online: https://emergencylaw.wordpress.com/2014/01/10/legality-of-forced-evacuations-during-nsw-bushfires/ (accessed on 28 March 2021).
16. Strahan, K.W.; Whittaker, J.; Handmer, J. Predicting self-evacuation in Australian bushfire. *Environ. Hazards* **2019**, *18*, 146–172. [CrossRef]

17. Tolhurst, K.G.; Chatto, K. Development, behaviour and threat of a plume-driven bushfire in west-central Victoria. In *Development, Behaviour, Threat and Meteorological Aspects of a Plume-Driven Bushfire in Westcentral Victoria: Berringa Fire 25–26 February 1995*; Research Report No. 48; Chatto, K., Ed.; Fire Management, Department of Natural Resources & Environment: Creswick, Australia, 1999.
18. Sharples, J.; Cary, G.; Fox-Hughes, P.; Mooney, S.; Evans, J.; Fletcher, M.-S.; Fromm, M.; Grierson, P.; McRae, R.; Baker, P. Natural hazards in Australia: Extreme bushfire. *Clim. Chang.* **2016**, *139*, 85–99. [CrossRef]
19. Richards, L.; Brew, N.; Smith, L. *2019–20 Australian Bushfires—Frequently Asked Questions: A Quick Guide*; Department of Parliamentary Services: Canberra, Australia, 2020.
20. Arriagada, N.B.; Palmer, A.J.; Bowman, D.M.; Morgan, G.G.; Jalaludin, B.B.; Johnston, F.H. Unprecedented smoke-related health burden associated with the 2019–20 bushfires in eastern Australia. *Med. J. Aust.* **2020**, *213*, 282–283. [CrossRef] [PubMed]
21. Handmer, J.; Van der Merwe, M.; O'Neill, S. The risk of dying in bushfires: A comparative analysis of fatalities and survivors. *Prog. Disaster Sci.* **2019**, *1*, 100015. [CrossRef]
22. Mackee, N. Harms from Bushfire Smoke: "Yesterday Was the Time to Talk about It". InSight+ 2020. Available online: https://insightplus.mja.com.au/2020/37/harms-from-bushfire-smoke-yesterday-was-the-time-to-talk-about-it/ (accessed on 15 February 2021).
23. Abdo, M.; Ward, I.; O'Dell, K.; Ford, B.; Pierce, J.R.; Fischer, E.V.; Crooks, J.L. Impact of Wildfire Smoke on Adverse Pregnancy Outcomes in Colorado, 2007–2015. *Int. J. Environ. Res. Public Health* **2019**, *16*, 3720. [CrossRef] [PubMed]
24. Whitington, P. No Smoke, No Ash: Here's How One House Survived the Fires. Housing 2020. 11 September 2020. Available online: https://thefifthestate.com.au/innovation/residential-2/no-smoke-no-ash-heres-how-one-house-survived-the-fires/ (accessed on 28 March 2021).
25. Vardoulakis, S.; Marks, G.; Abramson, M.J. Lessons Learned from the Australian Bushfires: Climate Change, Air Pollution, and Public Health. *JAMA Intern. Med.* **2020**, *180*, 635–636. [CrossRef]
26. ABCB. National Construction Code Series/Australian Building Codes Board. 2011. Available online: https://ncc.abcb.gov.au/ncc-online/NCC/2011?vol=47a1ef09-419e-42c9-88c6-580a106d9e9e (accessed on 28 March 2021).
27. Weir, I. *AS 3959:2018: Construction of Buildings in Bushfire-Prone Areas*, 4th ed.; Standards Australia AS3959:2018; Standards Australia: Sydney, Australia, 2018.
28. Roberts, M.E.; Sharples, J.J.; Rawlinson, A.A. Incorporating ember attack in bushfire risk assessment: A case study of the Ginninderry region. In Proceedings of the MODSIM2017, 22nd International Congress on Modelling and Simulation, Hobart, Tasmania, 3–8 December 2017; pp. 1145–1151.
29. Geoscience-Australia. *Canberra Bushfires Fieldwork Report—18 January 2003*; Commonwealth of Australia Symonston ACT; Geoscience-Australia: Canberra, Australia, 2003.
30. Macindoe, L.; Sargeant, A.J.; Bowditch, P.A. *Under Flame Fire Tests on Timber Decks: Report to Bushfire CRC*; CSIRO Manufacturing & Infrastructure Technology: Melbourne, Australia, 2007.
31. Wakefield, T.; He, Y.; Dowling, V.P. An experimental study of solid timber external wall performance under simulated bushfire attack. *Build. Environ.* **2009**, *44*, 2150–2157. [CrossRef]
32. Bowditch, P.A.; Sargeant, A.J.; Leonard, J.E.; Macindoe, L. *Window and Glazing Exposure to Laboratory-Simulated Bushfires: Report of Bushfire CRC*; Bushfire CRC: Melbourne, Australia, 2006.
33. Ghaderi, M.; Ghodrat, M.; Sharples, J.J. LES Simulation of Wind-Driven Wildfire Interaction with Idealized Structures in the Wildland-Urban Interface. *Atmosphere* **2021**, *12*, 21. [CrossRef]
34. Honey, S.; Rollo, J. Air pressure and wind velocity—Modelling ember attack within the urban-rural interface. In Proceedings of the State of Australian Cities National Conference, Melbourne, Australia, 29 November–2 December 2011.
35. Bergsma, W. Tathra: 18 Months on from the March 2018 Bushfires—Rebuilding a Bushfire Ravaged Coastal Town. In Proceedings of the Australian Bushfire Building Conference, Leura, Australia, 23–25 October 2019.
36. Chen, K.; McAneney, J. Quantifying bushfire penetration into urban areas in Australia. *Geophys. Res. Lett.* **2004**, *31*. [CrossRef]
37. Caird Ramsay, N.A.; McArtheur, N.A.; Dowling, V.P. How bushfires set houses alight—Lessons from Ash Wednesday. *Ecos* **1985**, *43*, 3–7.
38. Caird Ramsay, N.A.; McArtheur, N.A.; Dowling, V.P. Preliminary Results from an Examination of House Survival in the 16 February 1983 Bushfires in Australia. *Fire Mater.* **1987**, *11*, 49–51. [CrossRef]
39. Morris-Marr, L. Great Ocean Road fires: Wye River in Ruins while Lorne Has Lucky Escape. In Herald Sun. 2015: Melbourne, Online. Available online: https://www.heraldsun.com.au/news/victoria/great-ocean-road-fires-wye-river-in-ruins-while-lorne-has-lucky-escape/news-story/f1ea00f2eb2a1fff38fa53c5a9997d0f (accessed on 12 February 2021).
40. Institute, C.C. Exterior Fire Sprinklers Saved 188 Properties—Wet Homes Don't Burn. In *California Chaparral Blog*; CA Chaparral Institute: Escondido, CA, USA, 2018.
41. Standards-Australia. *AS 1684.3-2010: Residential Timber-Framed Construction—Cyclonic Areas*; Standards Australia: Sydney, Australia, 2010.
42. Standards-Australia. *AS 4055-2012: Wind Loads for Housing*; Standards Australia: Sydney, Australia, 2012.
43. McRae, R.H.D.; Sharples, J.J.; Fromm, M. Linking local wildfire dynamics to pyroCb development. *Nat. Hazards Earth Syst. Sci.* **2015**, *15*, 417–428. [CrossRef]

44. ACT-ESA. Fire Tornado Video. Available online: https://esa.act.gov.au/cbr-be-emergency-ready/bushfires/fire-tornado-video# (accessed on 28 March 2021).
45. Thistleton, J. Researchers confirm first 'fire tornado' during 2003 bushfires. In *The Sydney Morning Herald*; The Sydney Morning Herald: Sydney, Australia, 2012.
46. Sharples, J.J.; McRae, R.H.D.; Wilkes, S.R. Wind-terrain effects on the propagation of wildfires in rugged terrain: Fire channelling. *Int. J. Wildland Fire* **2012**, *21*, 282–296. [CrossRef]
47. Geoscience-Australia. Severe Wind. Community Safety 2021. Available online: https://www.ga.gov.au/scientific-topics/community-safety/severe-wind (accessed on 28 March 2021).
48. Băetu, G.; Gălățanu, T.-F.; Băetu, S.-A. Behavior of Steel Structures under Elevated Temperature. *Procedia Eng.* **2017**, *181*, 265–272. [CrossRef]
49. Wu, C.L.; Carvel, R. *The Influence of Compartment Temperature on Backdraught Dynamics*; Springer: Singapore, 2020.
50. Goodman, N.B.; Steinemann, A.; Wheeler, A.J.; Paevere, P.J.; Cheng, M.; Brown, S.K. Volatile organic compounds within indoor environments in Australia. *Build. Environ.* **2017**, *122*, 116–125. [CrossRef]
51. Fleischmann, C.M.; Pagni, P.J.; Willamson, R.B. Exploratory backdraft experiments. *Fire Technol.* **1993**, *29*, 298–316. [CrossRef]
52. Hinchcliffe, J.; Boseley, M. House Explodes as Massive Bunyip Blaze Threatens to Burn More Homes. In *The Age*; Online; The Age: Melbourne, Australia, 2019.
53. Standards-Australia. *AS 1684.2-2010: Residential Timber-Framed Construction—Non-Cyclonic Areas*; Standards Australia: Sydney, Australia, 2010.
54. Ross-Squire-Homes. Building in a Bushfire Prone Area. 2019. Available online: https://www.rsh.com.au/build-with-us/building-bushfire-prone-area (accessed on 28 March 2020).
55. Hilton, J.E.; Leonard, J.E.; Blanchi, R.; Newnham, G.J.; Opie, K.; Power, A.; Rucinski, C.; Swedosh, W. Radiant heat flux modelling for wildfires. *Math. Comput. Simul.* **2020**, *175*, 62–80. [CrossRef]
56. Standards-Australia. *AS/NZS 1170.2:2011 (R2016): Structural Design Actions—Wind Actions*; Standards Australia: Sydney, Australia, 2011.
57. Leonard, J.; Blanchi, R.; Weir, I. *Bushfire Resilient Building Guidance for Queensland Homes*; Queensland Government (State of Queensland): Brisbane, Australia, 2020.
58. NASH. *NASH Standard—Steel Framed Construction in Bushfire Area*; NASH: Hartwell, Australia, 2014.
59. Parackal, K.; Wehner, M.; Ginger, J.; Hyeuk, R.; Henderson, D.; Edwards, M. *Improving the Resilience of Existing Housing to Severe Wind Events: Annual Report 2019–2020*; Bushfire and Natural Hazards CRC: Melbourne, Australia, 2020.
60. Holmes, J.; Ginger, J.; Harper, B. What's the gust wind speed? In *Proceedings of the 16th Australasian Wind Engineering Society Workshop, Brisbane, Australia,18–19 July 2013*; Australasian Wind Engineering Society: Brisbane, Australia, 2013.
61. CFA, Building Commission. *A Guide to Retrofit Your Home for Better Protection from a Bushfire*; Country Fire Authority & Building Commission: Melbourne, Australia, 2010.
62. VBA; CFA. *A Guide to Retrofitting Certain Existing Class 9 Buildings for Better Protection from Bushfire Ember Attack*; Victorian Building Authority: Melbourne, Australia, 2019.
63. Standards-Australia. *AS 1562.1:2018 Design and Installation of Sheet Roof and Wall Cladding—Metal*; Standards Australia: Sydney, Australia, 2018.
64. Station, C.T. *Draft Guide to LHL Cyclic Testing*; James Cook University: Townsville, Australia, 2009.
65. Bahrani, B.; Hemmati, V.; Zhou, A.; Quarles, S.L. Effects of natural weathering on the fire properties of intumescent fire-retardant coatings. *Fire Mater.* **2018**, *42*, 413–423. [CrossRef]
66. Kempster, T. TBA Firefly Bushfire Presentation BMEE 24th October 2019. In Proceedings of the Australian Bushfire Building Conference, Leura, Australia, 23–25 October 2019; Blue Mountains Economic Enterprise: Leura, Australia, 2019.
67. Standards-Australia. *AS 1530 (Suite: 1530.1-4) Methods for Fire Tests on Building Materials, Components and Structures*; Standards Australia: Sydney, Australia, 2014.
68. Lyon, R.E.; Safronava, N.; Crowley, S. Thermal analysis of polymer ignition. *Fire Mater.* **2018**, *42*, 668–679. [CrossRef]
69. Summers, J.W. The melting temperature (or not melting) of poly(vinyl chloride). *J. Vinyl Addit. Technol.* **2008**, *14*, 105–109. [CrossRef]
70. Shutters, B.R. BAL-FZ Compliant Roller Shutters: German Engineered, Made in Australia. 2021. Available online: https://bushfirerollershutters.com/?gclid=CjwKCAjwjbCDBhAwEiwAiudBy3Hd_RQBKLw37DWmZd-c1CVO6CzGVZVSxKpw5OKlvANN8yrRLpx6AxoCzmIQAvD_BwE (accessed on 28 March 2021).
71. Green, A. Sprinkler Systems for the Protection of Buildings from Wildfifire. Ph.D. Thesis, Sustainable Buildings Research Centre, Faculty of Engineering and Information Sciences 2019, University of Wollongong, Wollongong, Australia, 2019; p. 229.
72. Green, A.; Cooper, P. Experimental Characterisation of Wildfire Sprinkler Sprays using High-Speed Videography. *At. Sprays* **2019**, *29*, 381–402. [CrossRef]
73. Green, A.; Kaye, N.B. On the use of sprays to intercept airborne embers during wildfires. *Fire Saf. J.* **2019**, *108*, 102842. [CrossRef]
74. Viegas, C.; Batista, R.; Albino, A.; Coelho, M.; Andrade, J.; Alves, D.; Viegas, D.X. Active Barrier Combining Fire-Resistant Fiberglass Fabric and Water Sprinkler System for Protection Against Forest Fires. *Fire Technol.* **2021**, *57*, 189–206. [CrossRef]
75. Wilkinson, C.; Eriksen, C. Fire water and everyday life: Bushfire and household defence in a changing climate. *Fire Saf. J.* **2015**, *78*, 102–110. [CrossRef]

76. Government-of-South-Australia. *Minister's Specification SA 78: Additional Requirements in Designated Bushfire Prone Areas*; Building Policy Branch Department of Planning and Local Government South Australia: Adelaide, Australia, 2011.
77. CFA (Ed.) *Guidelines for Remote Outlets on Water Tanks in the Bushfire Management Overlay*; CFA, Country Fire Authority: Burwood East, Australia, 2017.
78. RFS (Ed.) Building in Bush Fire Prone Areas: Single Dwelling Application Kit. In *NSW Rural Fire Service*; NSW Government: Granville, NSW, Australia, 2020.
79. Blanchi, R.; Leonard, J.; White, N.; Sargeant, A.J.; Bicknell, A.; Anderson, S. *Research into the Performance of Water Tanks in Bushfire*; CSIRO: St Lucia, Australia, 2007; p. 146.
80. Malo, J. Australia Bushfires: How a Gippsland Family Defended Their Home from Fire as Bush Burned Metres Away. In Domain 2020, Domain Media: Australia. Available online: https://www.domain.com.au/news/australia-bushfires-how-a-gippsland-family-defended-their-home-from-fire-while-bush-burned-metres-away-924640/ (accessed on 28 March 2021).
81. Potter, M.; Leonard, J.E. *Spray System Design for Ember Attack—Research Findings and Discussion Paper*; Bushfire and Natural Hazards CRC: East Melbourne, Australia, 2010.
82. Platypus-Fire-Pty-Ltd. Ember Protection Sprinkler System. 2021. Available online: https://platypussprinkler.com/ (accessed on 28 March 2021).
83. Rainmaker-Bushfire-Sprinklers. Simple Solutions: Rainmaker Bushfire Sprinklers Will Help to You Protect Your Home. 2021. Available online: https://rainmakerbushfiresprinklers.com.au/ (accessed on 28 March 2021).
84. Merah, N. Natural weathering effects on some properties of CPVC pipe material. *J. Mater. Process. Technol.* **2007**, *191*, 198–201. [CrossRef]
85. Nam, S. Fire tests to evaluate CPVC pipe sprinkler systems without fire resistance barriers. *Fire Saf. J.* **2005**, *40*, 595–609. [CrossRef]
86. CFA. *Landscaping for Bushfire: Garden Design and Plant Selection*; Country Fire Authority: Burwood East, Australia, 2011.
87. Mackey, B.; Lindenmayer, D.; Gould, S.; Taylor, C.; Norman, P. *Does Prescribed Burning of Native Forests Reduce the Risk to Infrastructure From Bushfires?* Bushfire Recovery Project Report No.5; Griffith University, The Australian National University: Canberra, Australia, 2021. Available online: https://www.bushfirefacts.org/ (accessed on 28 March 2021).
88. Fernandes, R.b.P.M. Fire Country: How Indigenous Fire Management Could Help Save Australia. *Int. J. Wildland Fire* **2020**, *29*, 1052–1053. [CrossRef]
89. Neale, T. Digging for Fire: Finding Control on the Australian Continent. *J. Contemp. Archaeol.* **2018**, *5*, 79–90. [CrossRef]
90. Nikolakis, W.; Roberts, E.; Hotte, N.; Ross, R.M. Goal setting and Indigenous fire management: A holistic perspective. *Int. J. Wildland Fire* **2020**, *29*, 974–982. [CrossRef]
91. Penman, T.D.; Eriksen, C.; Horsey, B.; Green, A.; Lemcke, D.; Cooper, P.; Bradstock, R.A. Retrofitting for wildfire resilience: What is the cost? *Int. J. Disaster Risk Reduct.* **2017**, *21*, 1–10. [CrossRef]
92. Dabu, B. Housing Construction Costs Rise as Demand Persists. 24 February 2021. Available online: https://www.smartpropertyinvestment.com.au/research/22329-housing-construction-costs-rise-as-demand-persists (accessed on 28 March 2021).
93. Macrotrends. Australia Inflation Rate 1960–2021. 2021. Available online: https://www.macrotrends.net/countries/AUS/australia/inflation-rate-cpi (accessed on 28 March 2021).

energies

MDPI

Article

Camera-Driven Probabilistic Algorithm for Multi-Elevator Systems

Yerzhigit Bapin [1,*], Kanat Alimanov [1] and Vasilios Zarikas [1,2]

1 School of Engineering and Digital Sciences, Nazarbayev University, Nur-Sultan 010000, Kazakhstan; kanat.alimanov@nu.edu.kz (K.A.); vasileios.zarikas@nu.edu.kz (V.Z.)

2 General Department, University of Thessaly, 38221 Volos, Greece

* Correspondence: yerzhigit.bapin@nu.edu.kz; Tel.: +7-701-444-8411

Received: 31 August 2020; Accepted: 10 October 2020; Published: 24 November 2020

Abstract: A fast and reliable vertical transportation system is an important component of modern office buildings. Optimization of elevator control strategies can be easily done using the state-of-the-art artificial intelligence (AI) algorithms. This study presents a novel method for optimal dispatching of conventional passenger elevators using the information obtained by surveillance cameras. It is assumed that a real-time video is processed by an image processing system that determines the number of passengers and items waiting for an elevator car in hallways and riding the lifts. It is supposed that these numbers are also associated with a given uncertainly probability. The efficiency of our novel elevator control algorithm is achieved not only by the probabilistic utilization of the number of people and/or items waiting but also from the demand to exhaustively serve a crowded floor, directing to it as many elevators as there are available and filling them up to the maximum allowed weight. The proposed algorithm takes into account the uncertainty that can take place due to inaccuracy of the image processing system, introducing the concept of effective number of people and items using Bayesian networks. The aim is to reduce the waiting time. According to the simulation results, the implementation of the proposed algorithm resulted in reduction of the passenger journey time. The proposed approach was tested on a 10-storey office building with five elevator cars and traffic size and intensity varying from 10 to 300 and 0.01 to 3, respectively. The results showed that, for the interfloor traffic conditions, the average travel time for scenarios with varying traffic size and intensity improved by 39.94% and 19.53%, respectively.

Keywords: smart building; smart city; Bayesian networks; elevator control algorithm; intelligent elevator system; decision theory; decision support systems

1. Introduction

As of 2018, approximately 4.2 billion people, or 55% of the world's population, live in urban areas [1]. The fact that people tend to move from rural to urban areas, along with the overall increase in global population, will result in approximately 6.7 billion people, or 68% of the global population, residing in cities by 2050 [1]. Taking this into account, we should focus on reshaping conventional approaches towards efficient energy use, reduction of waste, and improving reliability of urban infrastructure. Given this perspective, the utilization of intelligent and eco-friendly technologies is essential for sustainable growth of urban settlements.

Some cities have already taken an attractive yet challenging path to broad adoption of smart technologies. The smart city transformation is a continuous process that requires radical changes in both technology and policy frameworks. Transformation of the transportation system is one of the core components in the context of technological renovation, since timely and efficient movement of people and goods positively affects the growth and development of the cities. Innovative solutions related to public transportation and vehicle traffic management have already been implemented in many cities around the world. For instance, the city of Pittsburgh (USA) implemented the Scalable Urban Traffic Control (SURTRAC) system that analyzes data from each street intersection and adjusts the traffic lights in real time [2]. The implementation of SURTRAC helped to reduce the intersection wait time by 41%, number of stops by 31%, travel time by 26%, and vehicle emissions by 21% [3].

Total smart city transformation will require integration of intelligent systems into both the city infrastructure and the building systems, such as heating, ventilation, air conditioning, vertical transportation, and so on. The means of public transportation in buildings, specifically elevator systems, are of particular interest to the research community since implementation of smart solutions in these systems may have similar impact as in city transportation systems.

Currently, the state-of-the-art elevator control and dispatch methods are driven by two objectives to a large extent. First-the reduction of energy consumption while keeping the same level of comfort and usability. Second-the reduction of passenger wait time by improving the elevator dispatch strategy. The growth of computer power in recent years has enabled the application of sophisticated video-aided elevator control systems based on artificial intelligence (AI).

One of the earliest works describing utilization of video cameras in conventional passenger elevators was presented in a paper by [4]. The study is limited to the people-counting problem and is not intended to address elevator control optimization. Similarly, the studies of [5–8] present different methods of video camera utilization whose main purpose was not focused on improving elevator dispatch strategies. Other works propose the application of video cameras to partially optimize elevator control and dispatch. For instance, the study by [9] proposes utilization of surveillance cameras to reduce overcrowding during emergency evacuations, whereas the study by [10] proposes utilization of in-car cameras to determine the available capacity of the elevator car and reduce the number of unnecessary stops.

Although the utilization of information obtained by the video or image acquisition systems located inside the elevator cars or in hallways may significantly improve performance of conventional passenger lifts, there is still no clear vision of how we can take maximum advantage of these data in order to improve elevator technology. In the study of [11], the authors propose an elevator control system that utilizes the information from hallway surveillance cameras and adjusts the dispatching function using a generic algorithm. The study uses the information from cameras to predict the number of passengers going in upward and downward directions. However, the passenger direction prediction algorithm is rather simple and is based on an assumption that all passengers are going the same direction as the passenger who last pressed the call button. The reported results show that the information obtained from the surveillance cameras helped to reduce the average waiting time by over 8% when the average crowding was 55%. The study also reports that utilization of cameras in scenarios with very low and very high passenger load (e.g., less than 45% or higher than 75% of total capacity) resulted in insignificant wait time reduction. In a study by [12], the authors propose an elevator group control algorithm that uses information from the passenger detection and tracking system. The algorithm utilizes Haar-like feature-based passenger detection, while the passenger motion tracking is achieved through utilization of the unscented Kalman filter. The primary goal of this algorithm is to minimize consumption of electricity by the elevator system and reduce the passenger wait time. Another elevator control algorithm utilizing the information obtained by the hallway cameras is proposed in [13]. The study focuses on minimization of the passenger wait time by taking into account the number of passengers waiting for an elevator, directions of passenger movement, and availability of the elevator cars. The gathered data are analyzed by the region-based convolutional neural network and transferred to the conventional elevator control system to perform elevator dispatch.

The aforementioned studies utilize video recording systems to determine the number of passengers and predict their movement directions in order to adjust to the present passenger traffic patterns. The main disadvantage of these approaches lies in their inability to account for uncertainties associated with the passenger flow and the video/image processing systems. The states of the passengers, in these studies, are represented as deterministic values, yet it is obvious that the predictions made by the proposed systems cannot be 100% accurate. This problem can be tackled by representation of the passenger traffic using Bayesian networks where the number of passengers waiting for an elevator as well as their movement directions are represented as probabilistic variables.

One of the first studies describing the implementation of Bayesian network (BN) theory in elevator control and dispatch was a paper by Provan [14]. The paper presents a BN framework for stochastic discrete-event control applications. The author claims that the proposed hybrid variable-based BN representation of the stochastic discrete-event system is more compact and efficient compared to traditional probabilistic finite state machine. In a study by Cheng et al. [15], the authors propose a Bayesian reinforced learning algorithm for optimal scheduling of the elevator group control (EGC) systems. The proposed algorithm reduces the state space of the dispatch optimization problem by constructing a low-dimension abstract state space. The authors use a BN to conduct inference and obtain values for each of the abstract states. The final step of the proposed algorithm, involved utilization of a neural network to calculate the optimal state-action value function on the basis of the results obtained by the BN. The study assumes that the elevator loading time follows a 20th order truncated Erlang distribution.

The contribution and novelties of this study are as follows: In the present work we did not make a prediction following one statistical/machine learning method. We used real-time information with uncertainly that it was supposed to be generated by cameras and image recognition software to drive an intelligent decision about the motion of a multi-elevator system with the aim of reducing the waiting time.

Novel combination of conventional algorithm for the "closed car approach" together with probabilistic code was generated by BayesiaLab software integrated in Python.

The rest of this paper is organized as follows: Section 2 describes the proposed methodology and provides some comparison between the existing and proposed EGC methods, Section 3 provides a discussion on the results of the case study, and finally Section 4 presents the conclusion.

2. Model Description

It should be noted that all elevators discussed in this paper are conventional lifts cabins with collective control strategy and one up and down pushbuttons.

2.1. EGC with Conventional Nearest Car Algorithm

The main disadvantage of existing conventional EGC algorithms lies in their inability to account for uncertainties associated with passenger traffic. For instance, let us take a look at the conventional nearest car (NC) algorithm. The NC control policy is suitable for 2–3 elevator cars working in a building that is around 7 floors [16]. After call registration, the NC algorithm constantly searches for the best elevator car until the call is served. The search for the best option to serve the call is performed by calculating the figure of suitability (FS) according to the following rules:

(1) When the elevator is moving towards the passenger and its' direction as well as the passenger's desired direction are the same, the algorithm gives a position bias to this elevator, as if the elevator was one floor nearer to the passenger. In this case, the FS is calculated as follows:

$$\mathrm{FS} = (N + 1) - (\mathrm{d} - 1) = (N + 2) - \mathrm{d} \tag{1}$$

where $(N + 1)$ is the number of floors in a building, and d is the number of floors between the passenger and the elevator car.

(2) If the elevator is committed to moving towards the passenger whose desired direction is opposite to that of the elevator, then

$$\mathrm{FS} = (N + 1) - \mathrm{d} \tag{2}$$

Similarly, Equation (2) can be applied to calculate FS for an idle elevator car.

(3) If the elevator is committed to moving away from the passenger, the algorithm sets FS to a value that does not depend on the distance.

The algorithm allocates the call to the elevator with highest FS value; if there are several elevators with the same FS, the algorithm allocates the call to the closest car. The flowchart of conventional NC algorithm is presented in Figure 1.

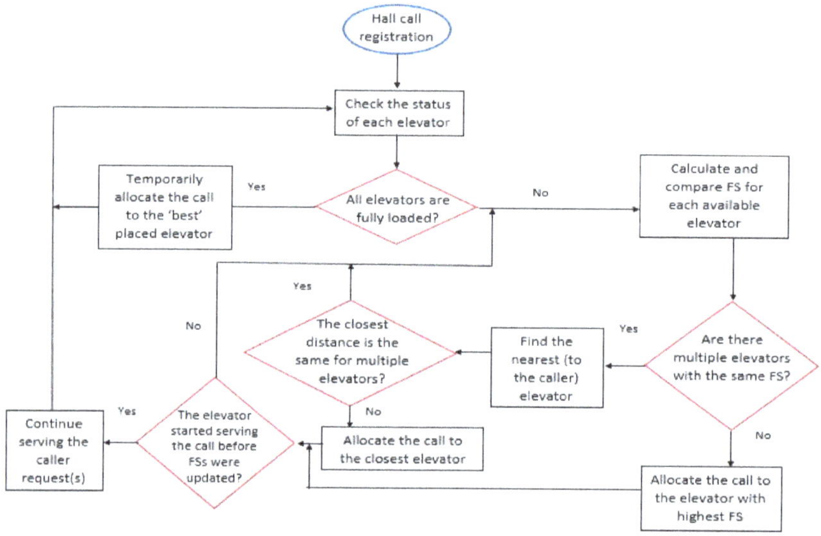

Figure 1. Flowchart of conventional nearest car (NC) algorithm.

As reported by reference [16], the main disadvantage of the conventional NC algorithm is its inability to adequately handle downpeak and uppeak traffic conditions. Thus, this algorithm can only be applied to the low-rise buildings where distribution of elevator cars is not very important.

2.2. Proposed EGC with Modified NC Algorithm

As opposed to conventional EGC with traditional NC algorithm, the proposed EGC uses surveillance cameras for the purpose of determining the number of passengers waiting in the hallways and riding the elevator cars.

The algorithm allocates the call to a single car or multiple elevator cars depending on the number of passengers waiting for an elevator and the number of available spaces inside the elevator cars. The system updates the statuses of each elevator car and every floor every 1 second. In other words, the elevator control system determines the locations and available capacity of each elevator car, as well as the number of passengers on every floor every 1 second (we need to figure out how to deal with movement direction of every passenger).

- Once the call is registered, the system determines how many passengers are waiting on the floor where the call was made.
- Next, the system checks the status of each elevator car (location, available capacity, movement direction).
- (a) When all elevator cars are fully loaded, the system checks the status of each elevator car again after 1 second. (b) When there is at least one elevator car available, the system orders elevators from highest FS to lowest; elevators with equal FS are ordered by distance to the landing floor, with the closest first.

- The system allocates the call to the elevators according to the order in 3B (Figure 2) until all passengers of the registered floor fit into the elevator cars.

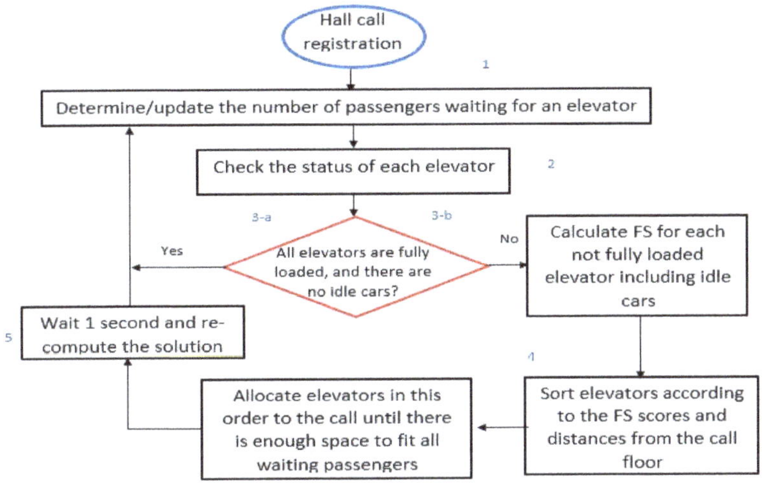

Figure 2. Flowchart of modified NC algorithm.

- The system will loop through the whole elevator allocation procedure every second looking for a better solution until all passengers get inside the elevator cars.

2.3. *Proposed EGC with BN Subroutine*

The algorithm presented in the previous section utilizes the information from surveillance cameras, yet this information is considered to be deterministic. In other words, the previous algorithm assumes that the number of people provided by the image recognition system is 100% certain. However, in reality, an image recognition system would not always provide the actual number of passengers waiting in a hallway or standing inside the elevator car. There is always a possibility for an error caused by people movements or objects that the system recognizes as humans (balloons, luggage, dogs, etc.). Thus, we have worked under the assumption that the image recognition software can provide us the number of adults, children, and small and big items with one associated probability that expresses certainty.

Therefore, in this subsection we present an algorithm that can utilize such uncertain information. The only appropriate way (that is mathematically consistent) to deal with this uncertainty is to utilize BNs. It follows is the general description of the pseudocode of the non-probabilistic part of the BN subroutine.

First of all, we define N as the effective number of "people" waiting for an elevator in one floor. If N is lower or equal than 3, we run the previous algorithm. If N is greater than 3, then N_1 idle cars go to these floors (choosing first the floor with larger N), where

$$N_1 = \frac{N}{E_C} \tag{3}$$

with E_c being the capacity of an elevator car.

If there are no idle cars, then N_2 cars are sent to this floor. These cars are selected from the elevators that move towards the landing floor.

The following pseudocode represents the remaining part of the non-probabilistic BN subroutine, where N_2 is the number of loops of the code.

If (capacity of one car—effective number of people inside the closest approaching car i) > N then send this car and go to the end line of this code

If (capacity of one car—effective number of people inside the closest approaching car j) < N then send this car and go to the next line

If (capacity of one car—effective number of people inside the second closest approaching car j+1) < N —(capacity of one car—effective number of people inside the previous closest approaching car j) then go to the next line.

...

Count loops = N_2

End.

The probabilistic part of the BN subroutine generates the effective number of "people" waiting for an elevator on the Nth floor. It is evaluated by the BN presented in Figure 3. The effective number (EN) of people inside an elevator car is also evaluated by a same BN but with a different number of states and with a different maximum number of people.

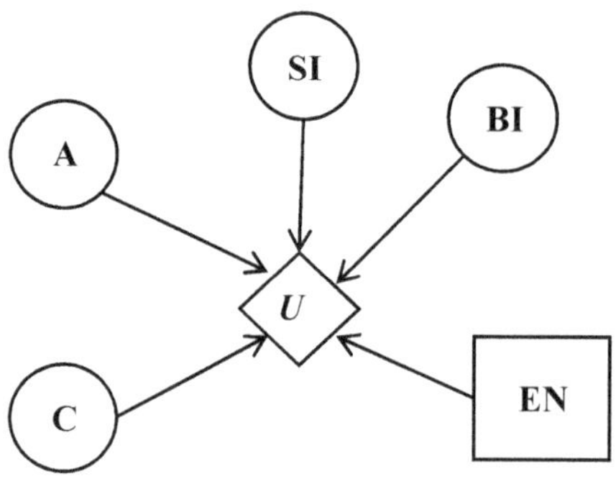

Figure 3. Bayesian network (BN) for effective number of people.

The BNs consist of the following nodes where the elevator car maximum capacity is assumed to be equal to 10 people, and the maximum capacity of the hall where people are waiting for an elevator is equal to 20. Here, with the word "people" we mean adults, children, and small or big items (passenger belongings).

- Number of Adults (A)
  - for the cameras on each floor—21 states from 0 to 20;
  - for the cameras inside an elevator car—11 states from 0 to 10.
- Number of Children (C)
  - for the cameras on each floor—21 states from 0 to 20;
  - for the cameras inside an elevator car—11 states from 0 to 10.
- Number of Small Items (SI)
  - for the cameras on each floor—21 states from 0 to 20;
  - for the cameras inside an elevator car—11 states from 0 to 10.
- Number of Big Items (BI)
  - for the cameras on each floor—21 states from 0 to 20;
  - for the cameras inside an elevator car—11 states from 0 to 10.

Children and small items are supposed to weigh 0.5 units, while adults and big items are supposed to count as a weight of 1 unit.

The hypothetical camera image recognition software provides a probability for all the states for one node. For example, the image recognition will generate the following output for one specific floor: 9 adults with a certainty of 70%, 8 adults with a certainty of 20%, and 10 adults with a certainty of 10%, with 0% for all other possible states/numbers of adults. In addition, the probability it is provided for one big item with a certainty of 65% and no big items with a 35% certainty, no children with a certainty of 100%, 2 small items with a certainty of 80%, and one small item with a certainty of 20%, etc.

The decision node suggests a decision for the effective number of people. This number is a real variable from 0 to the sum of the maximum values of the weights of each of the four categories. Thus, in our implementation, a maximum of 20 adults is allowed, or a maximum of 20 kids who carry a maximum of 20 big or small items (children and small items have a weight of 0.5 units and adults and big items have a weight of 1 unit).

Thus, it is an integer but with steps 0.5. A decision that gives 13.5 means that the largest expected utility "observes" 13 adults and 1 child waiting, or 10 adults and 6 children and 1 small item, or any other combination. Note that the effective number of people means people together with items.

The prior probabilities of the nodes are completed automatically and with equal values for all states. The utility values are estimated by the following formulae:

$$U = anX1 + knX0.5 + bnX1 + snX0.5. \tag{4}$$

where *an* is the adult state number, *kn* is the child state number, *bn* is the big item number, and *sn* is the small item number.

The expected utility $EU(A_i, H_j)$ for the decision node is given by the formula:

$$EU(A_i) = \sum_{a=1}^{4}\sum_{j=1}^{N} U(A_i, H_j^a)\, p(H_j^a) \tag{5}$$

where A_i is the mutually exclusive variable and $i = 1, \ldots, n$, represents action commands along with four variables H_a with possible states H_j, where $j = 2, \ldots, m$ represents the hypothesis influencing the decision. Another important feature of the proposed algorithm is that the action commands do not have any correlation with the determining variables H. Here, we have four determining variables: C, A, SI, and BI.

The BN is updated with new evidence from cameras every second. An example of evidence is that there are 10 adults with a probability of 69%, 3 children with a probability of 45%, and 1 item with a probability of 95%, etc.

The flowchart of the modified NC algorithm with BN is illustrated in Figure 4.

If information coming from experts or from historic data suggest that children have a priority compared to items (this can be due to the fact that adults with children enter the elevator with priority compared with adults with items due to a fair/polite attitude) then the designer of the BN can enlarge the utility of these cases. A combination of 2 children and 1 adult effectively weighs more than 2 small items and 1 adult or 1 big item and one adult. Thus, for the case with 1 adult and 1 child, the resulting number of effective passengers will be bigger compared to 1 adult and 1 small item. This strategy was not implemented in this study, rather, it is mentioned at this point in order to exhibit the role of utility.

In this section, first, the conventional NC algorithm; second, the modified NC algorithm assuming an image recognition software that provides a deterministic number of people (does not distinguish among kids and adults and if they carry any items); and, finally, a third probabilistic NC algorithm with BNs generating an effective number of waiting "passengers/units" are compared.

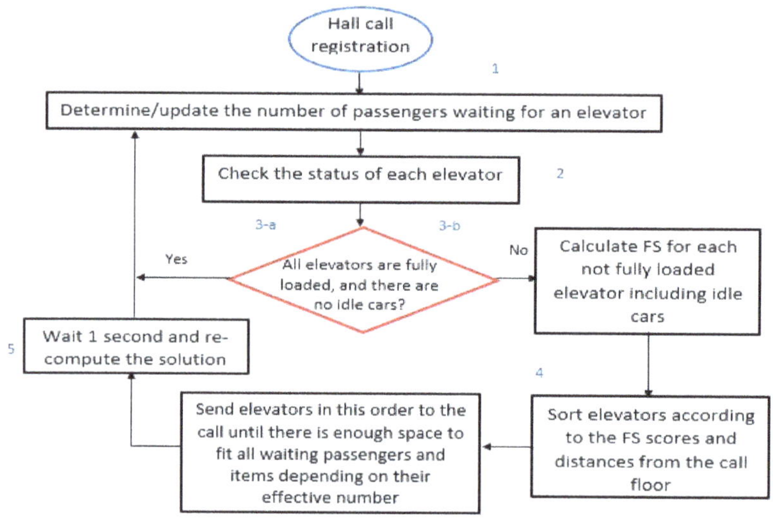

Figure 4. Flowchart of modified NC algorithm with BN.

First, the camera associates probabilities to 4 different categories of passengers (adults, children, or small big items). Second, the effective number (a chosen decision state) is the one that has the maximum expected utility that counts the probability value of each state and in addition the utility priority weight that is set by the designer of the BN.

A general comparison of EGC algorithms based on NC, modified NC, and probabilistic NC with BNs shows that the later EGC algorithm has higher performance compared to the other two. The comparison would require conducting the quantitative evaluation of each EGC algorithm and comparing their performance regarding waiting time on similar set of scenarios.

3. Results

3.1. Elevator Performance Metrics and Traffic Patterns

This section presents a case study performed on a 10-storey office building with five elevators. The main goal of this case study was to quantify and compare the following elevator performance metrics for the EGC algorithms presented in Section 2:

Passenger average travel time (ATT) is the time, measured in seconds, spent by a passenger travelling in an elevator car, starting from the moment of boarding the lift until the moment of stepping on the destination floor.

Passenger average journey time (AJT) is the time, measured in seconds, spent by a passenger starting from the moment of pressing the call button, or joining other people who already pressed the call button, until the time when an elevator opens the doors at the destination floor.

Passenger average waiting time (AWT) is the time, measured in seconds, spent by a passenger starting from the moment of pressing the call button, or joining other people who already pressed the call button, until the time when an elevator opens the doors at the boarding floor.

The calculation of the elevator performance metrics was conducted for three sets of scenarios, each representing the following passenger traffic patterns:

Uppeak traffic is the traffic pattern that occurs when people mostly move in an upward direction. In office buildings, this pattern occurs in the morning when people come to work and, to a lesser extent, when people return at the end of their lunch break.

Downpeak traffic is the traffic pattern that occurs when people mostly move in a downward direction. As opposed to the uppeak pattern, the downpeak occurs at the end of the business day, when people leave the office and, to a lesser extent, at the beginning of the lunch break.

Random interfloor traffic is the traffic pattern that exists when there is no distinctive pattern of passenger movement. A random interfloor traffic pattern usually occurs due to normal circulation of people in the building during a business day.

It was also assumed that 30% of the people are children and 50% of adults carry baggage, half of which are small bags, with the other half being large items.

The conventional EGC algorithms (NC and modified NC) were simulated in Python (Version 3.7, manufacturer: Python Software Foundation, Wilmington, Delaware, United States) [17], whereas the proposed EGC algorithm was implemented in Python and Bayesialab (Version: 9.1 PE-L 00000, manufacturer: Bayesia S.A.S., Changé, France) [18].

The python simulation was written using a discrete event simulation framework SimPy. A number of scenarios were randomly generated with various passenger spawn rates, total passenger numbers, and traffic patterns. The base system was implemented by adding event triggers to events such as passenger arrivals, button presses, and elevator car arrivals. Individual algorithms were then written and added as callbacks to the event triggers. This allowed for seamless interchangeability between multiple elevator control algorithms.

3.2. *Dependence on Traffic Size and Intensity*

Testing of EGC algorithms was conducted on scenarios with different traffic sizes and intensity. The traffic size or the total number of passengers varied from 100 to 300. The traffic intensity or the average interval of passenger arrival varied from 0.01 to 2 seconds. Table 1 presents the results of the simulations.

Table 1. Simulation results for the uppeak traffic pattern.

Traffic Size	Interval	NC [1]			MNC [2]			MNCBN [3]		
		ATT	AJT	AWT	ATT	AJT	AWT	ATT	AJT	AWT
100	0.01	8.58	35.20	26.62	8.66	29.36	20.70	8.68	29.30	20.62
200	0.01	8.86	100.94	92.08	9.03	59.99	50.96	9.03	59.87	50.84
300	0.01	7.79	171.52	163.73	8.05	94.98	86.93	8.05	94.98	86.93
100	0.1	8.54	35.83	27.30	8.59	28.72	20.13	8.53	28.47	19.94
200	0.1	7.87	100.02	92.15	8.02	50.49	42.47	8.04	50.20	42.16
300	0.1	8.24	167.53	159.29	8.50	86.88	78.38	8.45	86.33	77.88
200	1	8.14	24.18	16.05	8.20	21.19	12.99	8.20	18.84	10.64
100	2	8.45	15.89	7.44	8.51	16.21	7.71	8.42	15.93	7.52
200	2	8.28	18.27	9.99	8.30	18.40	10.11	8.29	17.24	8.95
300	2	8.10	19.56	11.47	8.13	18.28	10.16	8.07	17.81	9.74

[1] Nearest car elevator group control (EGC) algorithm. [2] Modified nearest car EGC algorithm. [3] Modified nearest car EGC algorithm with BN.

Table 1 presents the simulation results obtained for the uppeak traffic pattern. As can be seen from the table, NC showed the best performance in terms of ATT in cases with high traffic intensity. Modified nearest car EGC algorithm (MNC) and modified nearest car EGC algorithm with BN (MNCBN), on the other hand, show better performance in terms of AJT and AWT in scenarios with low passenger arrival interval. All algorithms tended to show somewhat similar performance when the traffic intensity was very low.

Table 2 presents the results obtained for the downpeak traffic pattern. The results of the NC algorithm under downpeak conditions were the worst in terms of every elevator performance index. The MNC and MNCBN showed similar results, especially in terms of AJT and AWT. Similarly, as in the previous case, all algorithms tended to show similar results with increasing passenger arrival interval.

Table 2. Simulation results for the downpeak traffic pattern.

Traffic Size	Interval	NC [1]			MNC [2]			MNCBN [3]		
		ATT	AJT	AWT	ATT	AJT	AWT	ATT	AJT	AWT
100	0.01	10.61	51.00	40.39	9.01	39.80	30.78	8.54	35.72	27.18
200	0.01	9.61	78.49	68.89	8.47	54.38	45.91	8.37	55.58	47.21
300	0.01	9.35	107.15	97.80	7.90	86.09	78.19	7.92	81.68	73.76
100	0.1	9.64	36.00	26.36	8.89	34.72	25.84	7.65	32.24	24.60
200	0.1	9.99	68.28	58.29	8.56	59.44	50.88	8.70	61.18	52.48
300	0.1	9.82	92.94	83.12	8.14	75.63	67.49	8.17	75.45	67.27
200	1	13.31	18.35	5.04	9.97	17.69	7.72	9.27	16.56	7.29
100	2	11.03	15.84	4.81	8.36	14.67	6.31	8.08	16.58	8.50
200	2	11.40	16.06	4.66	8.14	14.24	6.10	8.21	15.67	7.46
300	2	11.20	16.34	5.15	8.27	14.77	6.50	8.46	16.74	8.28

[1] Nearest car EGC algorithm. [2] Modified nearest car EGC algorithm. [3] Modified nearest car EGC algorithm with BN.

Table 3 presents the simulation results obtained for the random interfloor traffic pattern. Conventional NC showed the worst results, except for the case traffic size equal to 200 people. In most of the scenarios, MNC and MNCBN showed similar scenarios for ATT and AJT. However, MNCBN outperformed NC and MNC in terms of AWT in scenarios with 100 and 300 people.

Table 3. Simulation results for the random interfloor traffic pattern.

Traffic Size	Interval	NC [1]			MNC [2]			MNCBN [3]		
		ATT	AJT	AWT	ATT	AJT	AWT	ATT	AJT	AWT
100	0.01	9.24	20.94	11.70	6.07	15.73	9.67	6.32	16.51	10.20
200	0.01	9.14	29.34	20.20	5.82	23.75	17.94	5.70	22.35	16.65
300	0.01	8.79	37.49	28.70	6.21	32.87	26.66	5.93	31.11	25.18
100	0.1	8.99	16.72	7.73	6.00	14.84	8.85	5.60	14.27	8.67
200	0.1	8.93	20.75	11.82	5.75	16.92	11.18	5.26	15.92	10.67
300	0.1	8.91	26.32	17.41	5.38	19.60	14.22	5.54	20.35	14.80
200	1	7.59	10.08	2.49	6.06	9.19	3.13	5.88	10.49	4.60
100	2	7.31	9.85	2.54	6.03	9.16	3.13	6.47	10.13	3.66
200	2	7.37	9.67	2.30	6.13	9.13	3.00	6.22	10.50	4.29
300	2	7.79	10.44	2.65	6.49	9.77	3.28	6.58	10.97	4.39

[1] Nearest car EGC algorithm. [2] Modified nearest car EGC algorithm. [3] Modified nearest car EGC algorithm with BN.

As it was mentioned before, in office buildings, the uppeak and downpeak traffic patterns usually occur twice a day each, with total duration of 2–3 hours/day. The random interfloor traffic pattern prevails throughout the day, and we thus focused more on analyzing the performance of EGC algorithms in this traffic condition.

As can be seen from Figure 5, MNCBN showed the best results in terms of average travel time for traffic size with total number of people varying from 10 to 300 and average passenger arrival time equal to 0.1. On the other hand, MNC and NC showed somewhat similar results with slight overperformance of MNC.

Dependence of the average journey time and average waiting time on the traffic size are presented in Figures 6 and 7, respectively. Similarly, as in the previous case, both AJT and AWT of the proposed MNCBN were lower than those of the other two algorithms. However, in this case, the curves of the MNC algorithm were much steeper than the curves of NC and MNCBN. Both MNCBN and NC resulted in somewhat similar values with slight overperformance of the MNCBN algorithm.

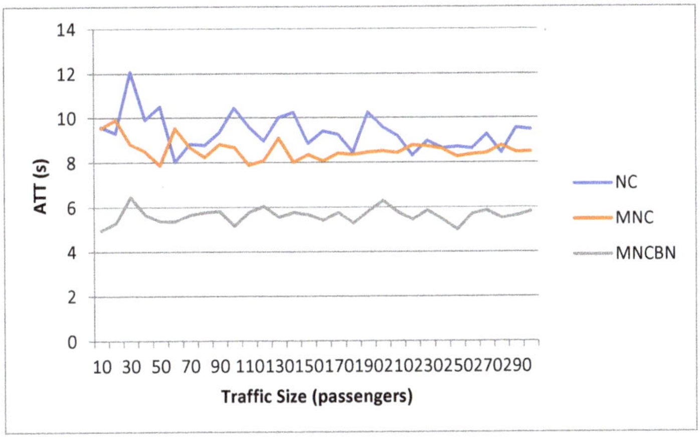

Figure 5. Dependence of average travel time (ATT) on change in traffic size.

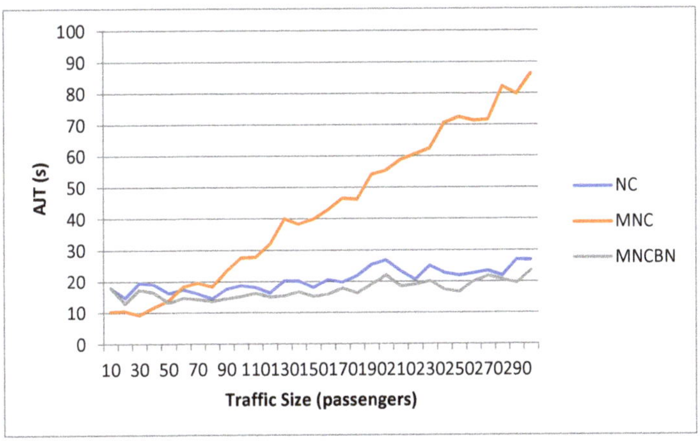

Figure 6. Dependence of average journey time (AJT) on change in traffic size.

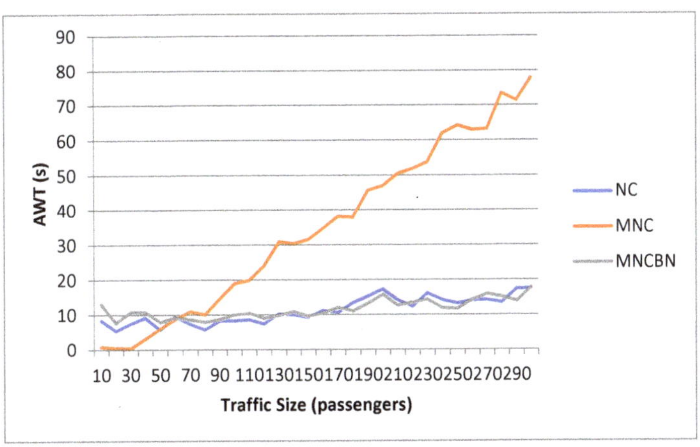

Figure 7. Dependence of average waiting time (AWT) on change in traffic size.

Figure 8 illustrates the dependence of the average waiting time on the traffic intensity. As can be seen from the figure, both MNC and MNCBN showed somewhat similar performances, whereas the ATT of the conventional NC was higher. However, it is noticeable that the curves tend to approach each other with increasing traffic intensity, which means that both MNC and MNCBN tended to outperform in cases with highly intense interfloor traffic.

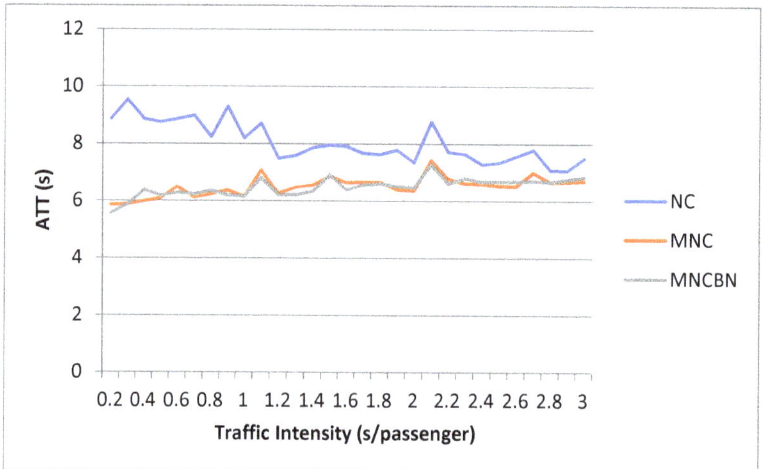

Figure 8. Dependence of ATT on change in traffic intensity.

Dependence of the average journey time and average waiting time on the traffic intensity is presented in Figures 9 and 10, respectively. Notice that these results were obtained under the same conditions as described previously. Interestingly, the average journey and waiting times of the proposed MNCBN tended to increase with increasing traffic intensity, whereas the AJT and AWT of NC and MNC tended to decrease, meaning that the performance of the proposed MNCBN algorithm worsened with increasing interfloor traffic intensity.

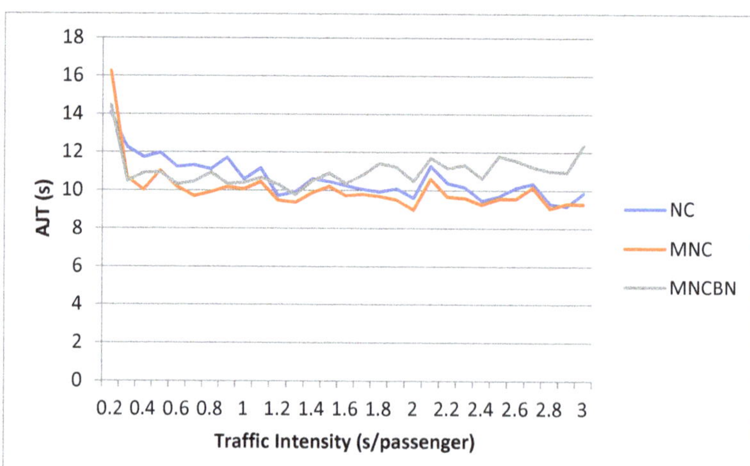

Figure 9. Dependence of AJT on change in traffic intensity.

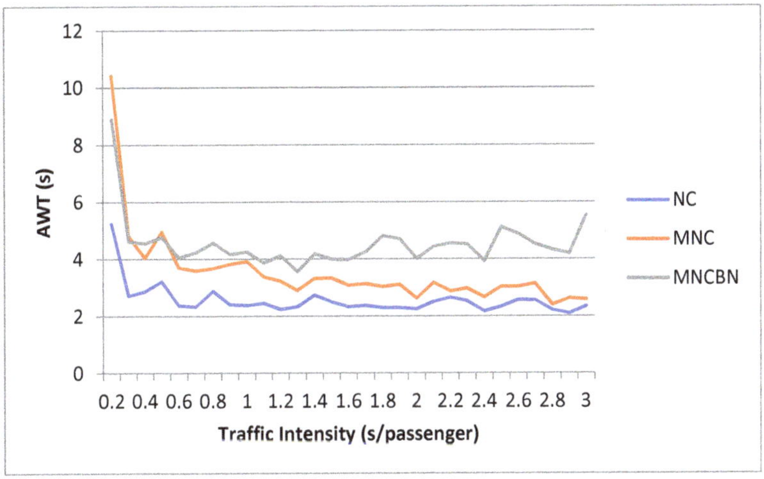

Figure 10. Dependence of AWT on change in traffic intensity.

On the basis of the results presented in Figure 10, the performance of NC, in terms of the average waiting time, was better during high interfloor traffic intensity.

3.3. Dependence on the Building Height

Building height is an important factor that plays a key role in an EGC algorithm performance. For instance, according to [16], the conventional NC algorithm is suitable for mid-rise buildings (7–10 floors). In this section, the performance of MNC and MNCBN EGC algorithms is evaluated for buildings with 11–30 floors and 5 elevators and is compared to the performance of the conventional NC EGC algorithm. Similarly, as in the previous section, the case in which the total number of 300 people arrived with an interval of 0.1 seconds was taken into consideration. Figure 11 illustrates the dependence of ATT on the building height.

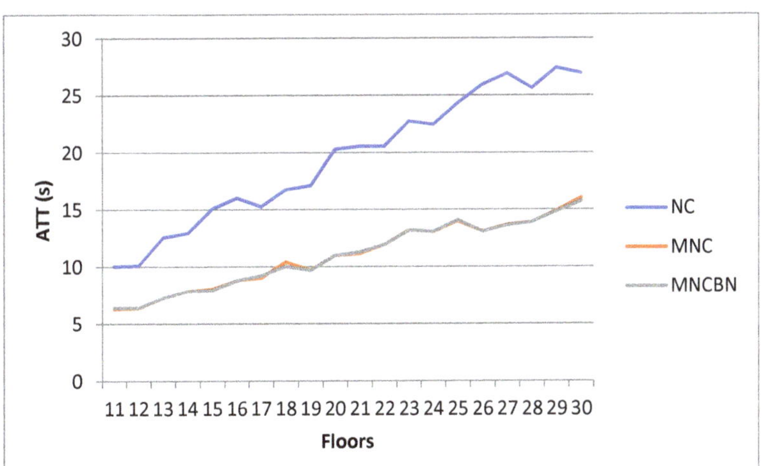

Figure 11. Dependence of ATT on the building height for the random interfloor traffic pattern.

ATT of all three algorithms increased with increasing building height. However, the curve of the conventional NC has a higher gradient as compared to those of MNC and MNCBN. The same can be observed in AJT and AWT, which are presented in Figures 12 and 13, respectively.

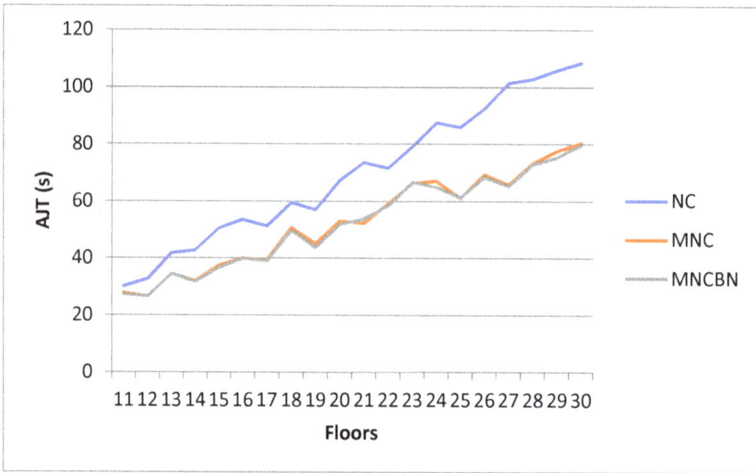

Figure 12. Dependence of AJT on the building height for the random interfloor traffic pattern.

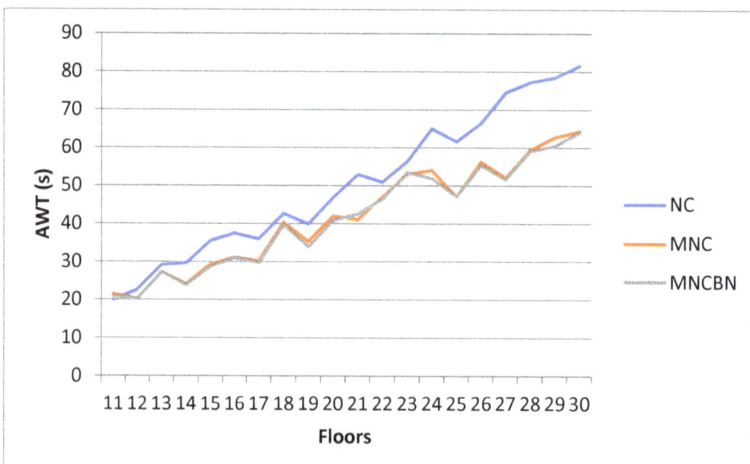

Figure 13. Dependence of AWT on the building height for the random interfloor traffic pattern.

Similarly, as in the case of ATT, we found that AJT and AWT increased with an increasing number of building floors, yet in this case, the MNC and MNCBN curves are steeper. This fact leads to a reasonable conclusion that the conventional NC algorithm is not suitable for a high-rise building. In fact, a conventional approach for improving the performance of an NC algorithm in high-rise buildings is to implement building clustering.

3.4. Comparison of Results

A statistical analysis was performed to compare ATT, AJT, and AWT among the three different algorithms. For the interfloor traffic pattern test case (values of Table 3), we display the comparisons in Table 4.

Table 4. Hypothesis test summary.

#	Null Hypothesis	Test	Significance [1]	Decision
1	Distribution of ATT_NC, ATT_MNC, and ATT_MNCBN are the same	Related samples Friedman's two-way analysis of variance by ranks	0.001	Reject the null hypothesis
2	Distribution of AJT_NC, AJT_MNC, and AJT_MNCBN are the same		0.027	
3	Distribution of AWT_NC, AWT_MNC, and AWT_MNCBN are the same		0.905	Retain the null hypothesis

[1] The significance level is 0.05.

The Friedman's test revealed that there were statistically significant differences among the three algorithms for ATT and AJT, but not for AWT.

4. Conclusions

This paper presented a probabilistic EGC algorithm based on the nearest car elevator dispatch strategy. The proposed algorithm uses information obtained from hypothetical surveillance cameras located in the hallways and in elevator cars. The elevator control system uses the information on the number of people waiting in hallways or riding the elevator cars together with their associated probabilities. This information is updated every second and is utilized for optimal elevator dispatch.

The optimization of elevator dispatching is conducted via combination of calculations of the figure of suitability and determining the nearest elevator car. The proposed algorithm dispatches the elevator cars such that the maximum number of people is collected from the crowded floors. Moreover, the algorithm takes into account the size of each person and whether or not they carry any baggage. BNs generate a value that is called the effective number of passengers, which is further used for optimization of elevator dispatch. The goal of the proposed algorithm is to minimize the average travel, journey, and waiting time.

The case study implemented on a 10-storey office building containing five elevator cars showed that the proposed algorithm performs the best in scenarios with small traffic size (less than 200 people). The proposed algorithm mainly underperforms in uppeak traffic conditions with high traffic intensity and a large number of people. The best performance of the proposed algorithm was determined for scenarios with random interfloor conditions. According to the results, the average travel time for scenarios with varying traffic size and intensity improved by 39.94% and 19.53%, respectively.

Comparison of results obtained for the interfloor traffic pattern using the related samples Friedman's two-way analysis of variance by ranks showed that ATT and AJT are different among all the algorithms, whereas AWT is not different.

The future work will be focused on the integration of a camera-driven probabilistic BN-based algorithm into the fixed and dynamic sectoring elevator control strategies.

Author Contributions: Conceptualization, V.Z. and Y.B.; methodology, Y.B. and V.Z.; software, K.A.; writing—original draft preparation, Y.B.; writing—review and editing, V.Z. and Y.B.; supervision, V.Z. All authors have read and agreed to the published version of the manuscript.

Funding: This research received no external funding.

Acknowledgments: The authors are thankful to the administration of Nazarbayev University for the support provided.

Conflicts of Interest: The authors declare no conflict of interest.

References

1. United Nations Organization. *World Urbanization Prospects: The 2018 Revision (Key Facts)*; United Nations: New York, NY, USA, 2019.
2. Smith, S.F.; Barlow, G.J.; Xie, X.-F.; Rubinstein, Z.B. SURTRAC: Scalable Urban Traffic Control. In Proceedings of the TRB 2013 Annual Meeting, Washington, DC, USA, 13–17 January 2013.
3. Rapid Flow Technologies LLC. RAPID FLOW BLOG. Rapid Flow Technologies LLC, 30 August 2018. Available online: https://www.rapidflowtech.com/blog/surtrac-deployment-at-urban-grid-networks-in-pittsburgh-neighborhoods (accessed on 5 August 2020).
4. Bapin, Y.; Zarikas, V. Smart building's elevator with intelligent control algorithm based on Bayesian networks. *Int. J. Adv. Comput. Sci. Appl.* **2019**, *10*, 16–24. [CrossRef]
5. Amrin, A.; Zarikas, V.; Spitas, C. Reliability analysis and functional design using Bayesian networks generated automatically by an "Idea Algebra" framework. *Reliab. Eng. Syst. Saf.* **2018**, *180*, 211–225. [CrossRef]
6. Zarikas, V.; Papageorgiou, E.; Pernebayeva, D.; Tursynbek, N. Medical decision support tool from a fuzzy-rules driven Bayesian network. In Proceedings of the 10th International Conference on Agents and Artificial Intelligence, Funchal, Portugal, 16–18 January 2018.
7. Yu, D.C.; Nguyen, T.; Haddawy, P. Bayesian Network Model for Reliability Assessment of Power Systems. *IEEE Trans. Power Syst.* **1999**, *14*, 426–432. [CrossRef]
8. Yongli, Z.; Limin, H.; Liguo, Z.; Yan, W. Bayesian Network Based Time-sequence Simulation for Power System Reliability Assessment. In Proceedings of the Seventh Mexican International Conference on Artificial Intelligence, Atizapan de Zaragoza, Mexico, 27–31 October 2008.
9. Ebrahimi, A.; Daemi, T. A Novel Method for Constructing the Bayesian Network for Detailed Reliability Assessment of Power Systems. In Proceedings of the International Conference on Electric Power and Energy Conversion Systems, Sharjah, UAE, 10–12 November 2009.
10. Shofield, A.; Stonham, T.; Mehta, P. Automated people counting to aid lift control. *Autom. Constr.* **1997**, *6*, 437–445. [CrossRef]
11. Shao, H.; Li, L.; Xiao, P.; Leung, M.K. ELEVIEF: An Active Elevator Video Surveillance System. In Proceedings of the Workshop on Human Motion, Austin, TX, USA, 7–8 December 2000.
12. Lee, Y.; Song, T.; Kim, H.; Han, D.K.; Ko, H. Hostile Intent and Behaviour Detection in Elevators. In Proceedings of the 4th International Conference on Imaging for Crime Detection and Prevention 2011 (ICDP 2011), London, UK, 3–4 November 2011.
13. Zou, J.; Zhao, Q. Occupancy Detection in Elevator Car by Fusing Analysis of Dual Videos. In Proceedings of the 13th IEEE Conference on Automation Science and Engineering (CASE), Xi'an, China, 20–23 August 2017.
14. Sun, Z.; Xu, B.; Wu, D.; Lu, M.; Cong, J. A real-time video surveillance and state detection approach for elevator cabs. In Proceedings of the International Conference on Control, Automation and Information Sciences (ICCAIS), Chengdu, China, 23–26 October 2019.
15. Ding, N.; Luh, P.B.; Chen, T.; Zhang, H. Optimization of Elevator Evacuation Considering Potential Over-Crowding. In Proceedings of the 11th World Congress on Intelligent Control and Automation, Shenyang, China, 29 June–4 July 2014.
16. Barney, G.; Al-Sharif, L. *Elevator Traffic Handbook Theory and Practice*; Routledge: New York, NY, USA, 2016.
17. Python 3.7–Python Releases for Windows. Available online: https://www.python.org/downloads/windows/ (accessed on 16 August 2020).
18. Bayesialab 8–Bayesian Networks for Research, Analytics, and Reasoning. Available online: https://www.bayesia.com/introduction?__hstc=221168007.d6327a006492fd03c47179bdedf277c4.1549810675300.1549810675300.1565871411196.2&__hssc=221168007.17.1565871411196&__hsfp=1049575405 (accessed on 24 August 2020).

Publisher's Note: MDPI stays neutral with regard to jurisdictional claims in published maps and institutional affiliations.

MDPI
St. Alban-Anlage 66
4052 Basel
Switzerland
Tel. +41 61 683 77 34
Fax +41 61 302 89 18
www.mdpi.com

Energies Editorial Office
E-mail: energies@mdpi.com
www.mdpi.com/journal/energies

www.ingramcontent.com/pod-product-compliance
Lightning Source LLC
Chambersburg PA
CBHW040840250926
51850CB00007BA/1367
* 9 7 8 3 0 3 6 5 5 2 6 9 9 *